Studies in Computational Intelligence

Volume 769

Series editor

Janusz Kacprzyk, Polish Academy of Sciences, Warsaw, Poland
e-mail: kacprzyk@ibspan.waw.pl

The series "Studies in Computational Intelligence" (SCI) publishes new developments and advances in the various areas of computational intelligence—quickly and with a high quality. The intent is to cover the theory, applications, and design methods of computational intelligence, as embedded in the fields of engineering, computer science, physics and life sciences, as well as the methodologies behind them. The series contains monographs, lecture notes and edited volumes in computational intelligence spanning the areas of neural networks, connectionist systems, genetic algorithms, evolutionary computation, artificial intelligence, cellular automata, self-organizing systems, soft computing, fuzzy systems, and hybrid intelligent systems. Of particular value to both the contributors and the readership are the short publication timeframe and the world-wide distribution, which enable both wide and rapid dissemination of research output.

More information about this series at http://www.springer.com/series/7092

Andrzej Sieminski · Adrianna Kozierkiewicz
Manuel Nunez · Quang Thuy Ha
Editors

Modern Approaches
for Intelligent Information
and Database Systems

 Springer

Editors
Andrzej Sieminski
Department of Information Systems
Wrocław University of Science
and Technology
Wrocław
Poland

Manuel Nunez
Department of Information Systems
and Computing
Complutense University of Madrid
Madrid
Spain

Adrianna Kozierkiewicz
Department of Information Systems
Wrocław University of Science
and Technology
Wrocław
Poland

Quang Thuy Ha
Faculty of Information Technology
Vietnam National University
Hanoi
Vietnam

ISSN 1860-949X   ISSN 1860-9503 (electronic)
Studies in Computational Intelligence
ISBN 978-3-030-09398-3   ISBN 978-3-319-76081-0   (eBook)
https://doi.org/10.1007/978-3-319-76081-0

Printed on acid-free paper

This Springer imprint is published by the registered company Springer International Publishing AG
part of Springer Nature
The registered company address is: Gewerbestrasse 11, 6330 Cham, Switzerland

Preface

Intelligent information and database systems are a very vibrant research area for over thirty years now. Over the years, the researchers have proposed more and more complex theoretical models. These models provide a theoretical background for numerous applications. The applications, on the one hand, have a profound influence on almost all areas of human activity and on the other hand, enable us to validate the underlying theoretical concepts.

In the recent years, we witness an enormous growth of available data ranging from textual repositories of the Internet to the overwhelming flow of data generated by the IoT. The data can be analyzed in a variety of ways, and some of the already achieved goals like reliable, speaker-independent speech transcription ten years ago belonged to the realm of science fiction. This all was possible due to the remarkable progress on both intelligent information and database systems. The resulting systems are complex and perform data-intensive and resource-consuming tasks. To cope with the flood of data, we need to acquire a profound understanding of old issues, to rethink previous paradigms, and to develop new concepts and approaches. The aim of the book is to provide readers with a carefully selected collection of research reports to facilitate the comprehension of the state of the art of such systems, thus promoting new research.

The area of intelligent information and database systems is very wide. This book presents the theory and practice of the ongoing research in its most active sections. Nowadays, we witness the integration of artificial intelligence and classic database technologies. In recent years, due to the advances in technology amounts of multimedia, social media, and IoT data are available. All this makes it possible to develop a novel class of innovative information systems. Their main goal is to offer the end users quasi-intelligent operation. They combine advanced learning techniques, knowledge engineering, NLP, decision support systems, IoT, computer vision, and tools and techniques for intelligent information systems to name some of used techniques.

The chapters in this book cover research work on these diverse topics. They are presented and discussed both from the practical and theoretical points of view and are extended versions of the poster presentations of the 10th Asian Conference on

Intelligent Information and Database Systems—ACIIDS 2018 which was held in Dong Hoi City, Vietnam, from March 19th until 21st, 2018.

The volume consists of 45 chapters that are divided into seven parts:

Part I "Knowledge Engineering and Semantic Web" includes five chapters that focus on uncertainty elicitation of experts using belief function, storing hypergraph-based data models in non-hypergraph data storage, using a three-stage consensus-based method for collective knowledge determination, modeling of fuzzy ontology by utilizing fuzzy set and fuzzy description logic, and recommending group experts for question and answering sites.

Part II "Natural Language Processing and Text Mining" consists of seven chapters that deal with predicting the popularity of presidential candidates using a fuzzy logic approach, representing DNA sequences by discrete wavelet transformation known from text similarity recognition, predicting the type of a DBpedia entity, tweet integration, or event detection, predicting the length of written responses to open-ended questions, combining inner approach and context-based approach to extract features of medical record data.

In Part III "Machine Learning and Data Mining" which encompasses nine chapters, we have collected research on: robust scale-invariant normalization and similarity measurement for time series data; attributes of game AI using fuzzy logic, building a detection model for water quality, a deep learning approach to case-based reasoning to the evaluation and diagnosis of cervical carcinoma, fast and memory-efficient mining of periodic frequent patterns, development of seawater temperature announcement system for red tide estimation, a fuzzy approach for the diagnosis of depression, a coupling support vector machines with the feature learning of deep convolutional neural networks for classifying microarray gene expression data, and finally on a weighted approach for class association rules.

Part IV "Decision Support Systems" contains seven chapters. They focus on supporting product development, supporting investments decision making on the basis of system dynamics, improvement of the community bus operation management system, and predicting consumer choices based on product brand. The very important topic of the e-commerce is discussed in the context of dynamic configuration of same-day delivery, the current trends in online shopping in the Czech Republic, and achieving lean and agile supply chain.

Part V "Computer Vision Techniques and Applications" comprises seven chapters and concentrates upon the identification of persons by his/her actions or more conventionally by face in the surveillance applications. They also discuss some industrial applications such as video stream magnification for touchless object vibration measurement or CNN-based classification for small specific datasets. The computer vision techniques are used also in the medical research for the breast cancer detection.

In Part VI "Sensor Networks and Internet of Things and Tools" part, we have collected five research reports. They discuss a multi-metric routing protocol of mobile ad hoc networks, integrating data access to heterogeneous data stores for IoT cloud, path estimation from smartphone sensors, localization of patients in urgent admission department and the design of universal hardware node board for smart home and the IoT.

Part VII "Techniques for Intelligent Information Systems." It encompasses five chapters. Their authors propose and analyze new methods and techniques for securing our data in public cloud, forecast load using leveraging database technology, analyze privilege control system with data mining techniques, use agent programming languages and logics in agent-based simulation, and propose a tool for computing the leakage of multi-threaded programs.

We sincerely do hope that this volume should be a valuable source of reference data and provide ample inspiration for your future research work. It should be also useful for students interested in computer science and in particular in artificial intelligence, big data, multimedia processing, and advanced databases.

We would like to express our sincere thanks to Prof. Janusz Kacprzyk, the Editor of this series, and Dr. Thomas Ditzinger from Springer for their interest and support for our project. Our thanks are due to all reviewers, who helped us to guarantee the highest quality of the chapters included in the book. Finally, we cordially thank all the authors for their valuable contributions to the content of this volume.

Wrocław, Poland   Andrzej Sieminski
Wrocław, Poland   Adrianna Kozierkiewicz
Madrid, Spain   Manuel Nunez
Hanoi, Vietnam   Quang Thuy Ha
April 2018

Contents

Part I
Knowledge Engineering and Semantic Web

A Three-Stage Consensus-Based Method for Collective Knowledge Determination

Dai Tho Dang, Van Du Nguyen, Ngoc Thanh Nguyen
and Dosam Hwang

Abstract Nowadays, the problem of referring knowledge from a large number of autonomous units for solving some problems in the real world has become more and more popular. The need for new techniques to process knowledge in collectives has become urgent because of the rapidly increasing in size of collectives. Many methods for determining the knowledge of collectives have been proposed; however, the traditional data processing methods are inadequate to deal with big collectives. In the present study, we propose a three-stage consensus-based method to determine the knowledge of a big collective. In particular, in the first stage, the sequence partitioning method is applied to partition a big collective into chunks having the same size. Then, the k-means algorithm is used for clustering each chunk into smaller clusters. The knowledge of each chunk is determined based on the knowledge of these clusters. Finally, the knowledge of the big collective is determined based on a set of the knowledge of the chunks. Simulation results have revealed the effectiveness of the proposed method in terms of the running time as well as the quality of the final collective knowledge of a big collective.

Keywords Collective knowledge · Consensus method · Three-stage
consensus-based

D. T. Dang · D. Hwang (✉)
Department of Computer Engineering, Yeungnam University,
Gyeongsan, Republic of Korea
e-mail: dshwang@yu.ac.kr

D. T. Dang
e-mail: daithodang@ynu.ac.kr

V. Du Nguyen · N. T. Nguyen
Department of Information Systems, Faculty of Computer Science and Management,
Wrocław University of Science and Technology, Wrocław, Poland
e-mail: van.du.nguyen@pwr.edu.pl

N. T. Nguyen
e-mail: ngoc-thanh.nguyen@pwr.edu.pl

© Springer International Publishing AG, part of Springer Nature 2018
A. Sieminski et al. (eds.), *Modern Approaches for Intelligent Information
and Database Systems*, Studies in Computational Intelligence 769,
https://doi.org/10.1007/978-3-319-76081-0_1

1 Introduction

Nowadays, along with the rapid development of information technology, there is a dramatic growth of the data produced especially on the Internet and the Web [1]. It is common that to solve some problems one may refer knowledge from a large number of members. For such an approach, one can deal with the so-called big collectives that involve a huge number of members. These members are often autonomous units such as agent systems or even human. Moreover, traditional data processing tasks such as data analysis, data classification, data calculation are inadequate to deal with it [2].

In this work, a set of opinions (or solutions) from a number of autonomous members on the same problem in the real world can be treated as a set of knowledge states in a collective. They can be inconsistent with each other because of incompleteness and uncertainty [3, 4]. On the basis of these knowledge states, there exists a need to integrate them into a consistent one (called collective knowledge) that can be considered as their representative.

Previous studies have proposed many approaches for determining the collective knowledge such as single-stage, two-stage, and even multi-stage approaches. In [5, 6], many consensus-based algorithms have been developed for such a task. Of course, these algorithms are based on the so-called single-stage approach, which aims at determining a knowledge state that can be considered as the representative of the collective as a whole. In case of big collectives, however, collectives are often very big. Thus the conventional single-stage approach may not be sufficient to deal with the task of collective knowledge determination. Instead, this task can be divided into smaller ones and performed through many stages and even on many nodes in a distributed system. In particular, the two-stage method for collective knowledge determination has been proposed in [7, 8]. In these works, the k-means algorithm is used to cluster a big collective into smaller ones; then the representatives of these collectives are determined by using consensus method. Notice that, these representatives later are treated as the input for the second stage of collective knowledge determination. Similarly, with the multi-stage approach, the process of clustering and determining the knowledge of smaller collectives will be repeated until the number of stages is reached [9]. However, it can be noted that the previous methods only perform efficiently on small collectives. In the present study, therefore, we will propose a new consensus-based method for determining the knowledge of a big collective. The proposed method will be not only efficient in computational time but also the quality of final collective knowledge. First, a big collective is partitioned into smaller chunks having the same size that can be later processed in parallel [1, 2, 10]. Next, each chunk is clustered into clusters by using the k-means algorithm and consensus choice method will be utilized to determine the knowledge of these clusters [5, 11]. Then these knowledge states will be considered as elements of the collective formed in the last stage and the knowledge of the big collective will be determined based on this collective.

The rest of this paper is constructed as follows. In Sect. 2, an overview of related work is mentioned. Section 3 briefly presents some basic concepts. The proposed method is presented in Sect. 4. Experimental results and their evaluation are explained in Sect. 5. Finally, conclusions and future works are presented in Sect. 6.

2 Related Work

For big collectives, collective knowledge is often determined by using the multi-stage approaches [4, 7]. In fact, it is a quite new approach and it has not been widely investigated in the literature [9]. In [5, 6], many single-stage consensus-based algorithms are proposed for determining the knowledge of a collective with different representations of knowledge structures. These algorithms have recently been used as a base for most of the multi-stage collective knowledge determination methods [7].

In [8] the authors have proposed a two-stage consensus-based method for determining the knowledge of a big collective by using the k-means algorithm. The experimental evaluation confirmed that this method had a positive impact on reducing the difference between collective knowledge determined by using the two-stage and the single-stage methods. Nguyen et al. [7] improved the two-stage method by taking into account the problem of susceptibility to consensus. The experimental results proved that this improvement was practicable in minimizing the difference between the single-stage method and the two-stage method in the process of collective knowledge determination. It can be seen that in the two-stage approach, the k-means is often used for clustering a collective into smaller ones. However, applying the k-means to classify the whole collective into groups is not efficient for a big collective [1, 2]. In other words, the two-stage method is incapable of dealing with a big collective.

In [9] the authors have presented a multi-level approach for determining the knowledge of a collective by using consensus theory. This approach considered to classify a collective into groups that could be processed independently. Two main important priorities of this method were determining the number of steps for processing and dividing the big collective into smaller groups. In this work, the authors only calculated the number of groups relied upon Fleiss's Kappa that is based on the whole collective members. It is clear that the calculation of the Fleiss's Kappa is incapable in case of big collectives. Besides, the experiments were considered only a two-level approach with a collective containing just nine members.

In those previous studies, the data analysis is relatively complicated and limited to small sizes of collectives. Therefore, the present study proposes a consensus-based method for determining the knowledge of a big collective by combining the sequence partitioning, the k-means algorithm. The goal of the proposed method is to obtain better running time and quality of collective knowledge determined.

3   Basic Notions

3.1   Collective of Knowledge States

Let U denote a finite set of objects representing all potential knowledge states for a given real world. Let $\prod_k (U)$ be a set of all k-element subsets of set U for $k \in \aleph$ ($\aleph$ is the set of natural numbers), and let

$$\prod (U) = \cup_{k \in \aleph} \prod_k (U).$$

where $\prod (U)$ is the set of all non-empty subsets with repetitions of set U. Each element of $\prod (U)$ is called a collective that involves a set of knowledge states on the same real world [11]. As the structure of set U, we assume a distance function:

$$d: U \times U \to [0, 1]$$

which is

- Nonnegative: $\forall x, y \in U: d(x, y) \geq 0$,
- Reflexive: $\forall x, y \in U: d(x, y) = 0$ iff $x = y$,
- Symmetrical: $\forall x, y \in U: d(x, y) = d(y, x)$

where [0,1] is the closed interval of real numbers between 0 and 1. Pair (U, d) is called a distance space. The knowledge of a collective is determined by knowledge functions [11].

   In this study, each knowledge state is represented by a binary vector, which describes a member's opinion about a subject in the real world [6, 12]. The formula below expresses the collective:

$$X = \{x_1, x_2, \ldots, x_m\}$$

where x_i is a binary vector, m is the number of members in the collective (the size of the collective), and every member x_i is the same length.

3.2   Collective Knowledge Determination

In general, consensus choice has usually been considered in the context of a general concurrency in which participants have not agreed on some matters [13]. In [6], the authors developed many consensus-based algorithms for such a collective knowledge determination process for different knowledge representations. According to [6, 11], criteria "1-*Optimality*" and "2- *Optimality*" (or O_1 and O_2 for short) are the most important and popular ones for determining the knowledge of a collective. In this paper, it is assumed that the knowledge of a collective is considered as the consensus of members' knowledge states [11].

For a given collective X, let x^* be its collective knowledge. By satisfying criterion O_1, it demands that the sum of distances from the collective knowledge to the collective members to be minimal.

$$d(x^*, X) = \min_{y \in U} d(y, X)$$

Similarly, by satisfying criterion O_2, it requires the sum of squares of distances from the collective knowledge to the collective members to be minimal.

$$d^2(x^*, X) = \min_{y \in U} d^2(y, X)$$

3.3 Quality of Collective Knowledge

The quality of collective knowledge is a main aspect that expresses how good a collective knowledge is [14]. In this paper, its measure is based on the distances from collective knowledge to the collective members [6, 11, 15]. Its definition is as follows:

$$\hat{d}(x^*, X) = 1 - \frac{d(x^*, X)}{card(X)}$$

where $card(X)$ is the number of members in collective X and x^* is the knowledge of collective X.

According to above measure, the quality of a collective knowledge reflects the average distance from the collective knowledge to the elements of the collective. A collective whose knowledge states are more closer to each other will have a better quality of collective knowledge.

4 Proposed Method

In this section, a new consensus-based method is proposed to determine the knowledge of a big collective. It is a combination of the sequence partitioning, the k-means algorithm, and the single-stage method (see Fig. 1).

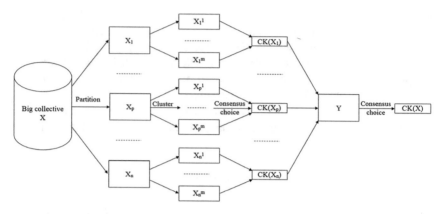

Fig. 1 Proposed method

Firstly, the sequence partitioning is applied to divide a big collective into smaller chunks with the same size $(card(X_1) \approx card(X_2) \approx \ldots \approx card(X_n))$ that can be parallelly processed in later stages. Tasks dealing with data such as analyzing, classifying and calculating data are only performed on chunks.

In the next stage, the k-means algorithm, which is the simplest and most commonly used data clustering method [16], will be used to cluster chunks into smaller clusters such that members within a cluster have a large similarity and members in different clusters have a great dissimilarity [17, 18]. The single-stage method is used for determining the knowledge of clusters, which later is treated as the inputs for the final collective knowledge determination. For binary vector representation, the single-stage method determines the knowledge of each cluster by calculating the number of occurrences of the 0 and 1 in the members' knowledge states. This method can be found elsewhere (see [9]).

To determine the knowledge of a chunk, each cluster of this chunk is assigned a weight value depending on the cluster's number of members. The weight approach has a positive impact on decreasing the difference between the two-stage and the single-stage consensus-based methods in determining the knowledge of a collective in comparison to a non-weight approach [8]. After completing this stage, the set of knowledge states of all chunks is determined as $Y = \{CK(X_1), CK(X_2), \ldots, CK(X_n)\}$. The final step is to determine the knowledge of the big collective $CK(X)$. $CK(X)$ is determined by the single-stage method based on Y.

The algorithm for determining the knowledge of a big collective is presented below:

Input: X - big collective
Output: $CK(X)$ – knowledge of X
Begin
    $Y = \emptyset$ - a set of knowledge states
    1. Partition a big collective X into n chunks $X_1, X_2,.., X_n$ with the same size:
        $X = X_1 \cup X_2 \cup ... \cup X_n$
    2. For each chunk $X_i \in X$ perform sequential tasks:
        - Cluster chunk X_i into m clusters $X_i^1, X_i^2, ..., X_i^m$:
            $X_i = X_i^1 \cup X_i^2 \cup ... \cup X_i^m$
        - Determine the knowledge of clusters $X_i^1, X_i^2, ..., X_i^m$
        - Determine the knowledge of chunk X_i is $CK(X_i)$
        - $Y = Y \cup CK(X_i)$
    3. Determine the knowledge of X ($CK(X)$) based on Y
End.

The collective knowledge satisfying criterion O_1 or criterion O_2 has maximal quality [9, 10]. In this study, criterion O_1 will be used to determine the knowledge of a big collective.

5 Experiments and Evaluation

5.1 Settings

In this section, we conduct simulation experiments to evaluate the effectiveness of the proposed method with respect to the quality of collective knowledge and the running time. The settings are as follows: we use binary vectors for such a unified representation of knowledge structure. The length of each vector is 10 and collective sizes are 600.000, 800.000, 1.000.000, 1.200.000, 1.400.000, 1.600.000, 1.800.000, 2.000.000, 2.200.000 and 2.400.000. Moreover, we run 10-repetition for each setting.

In the first stage, the number of knowledge states in each chunk is 200.000 (corresponding to 4 MB). Each chunk is independently analyzed on each node and the nodes are processed in parallel. Initially, all chunks of the collective are clustered into 3 clusters for determining the knowledge of each chunk and then the determination of the knowledge of the collective based on the knowledge of the chunks from the nodes. Besides, we also run with other numbers of clusters in each chunk such as 5, 7, and 9.

5.2   *Experimental Results and Their Evaluation*

One of the challenges of collective knowledge determination through a multi-stage approach is how to minimize the difference between the collective knowledge determined using single-stage and multi-stage approaches. Therefore, in this paper, in order to evaluate the effectiveness of the proposed method, we compare the obtained results with those obtained by using single-stage approach (see Table 1). In particular, for a given big collective X, we determine the collective knowledge, which satisfies criterion O_1. Let x_1^* be the knowledge of collective X determined by the single-stage method, and $d(x_1^*, X)$ be the total of distances from x_1^* to members of collective X. Similarly, let x_2^* be the knowledge of collective X determined by the three-stage method, and $d(x_2^*, X)$ be the total of distances from x_2^* to members of collective X. In case of $d(x_1^*, X) = 0$, X only contains one element x with a number of its occurrences. Therefore $d(x_2^*, X)$ must be zero and $x_1^* = x_2^* = x$. In case of $d(x_1^*, X) > 0$, the difference between the three-stage methods to the single-stage method is expressed as follows:

$$Dis = \frac{|d(x_2^*, X) - d(x_1^*, X)|}{d(x_1^*, X)} \times 100\%$$

As can be seen in Table 1, the differences between the qualities of collective knowledge determined by single-stage and three-stage approaches are much less than 0.1%. According to the Shapiro-Wilk test, the obtained results do not come from a normal distribution. Therefore, the Wilcoxon signed-rank test is chosen for such a task. For simplicity, the quality of collective knowledge determined by using the single-stage method is denoted Data_1 and the quality of collective knowledge determined by using the three-stage method is denoted Data_2. According to the Wilcoxon signed-rank test, in this case, the *p-value* is greater than 0.05

Table 1 Experimental results

Collective (members)	Number of chunks	Number of clusters in each chunk			
		3	5	7	9
600.000	03	0.100	0.063	0.021	0.023
800.000	04	0.065	0.045	0.000	0.001
1.000.000	05	0.073	0.026	0.025	0.023
1.200.000	06	0.030	0.029	0.016	0.015
1.400.000	07	0.064	0.037	0.036	0.021
1.600.000	08	0.075	0.015	0.008	0.015
1.800.000	09	0.050	0.017	0.051	0.028
2.000.000	10	0.037	0.013	0.014	0.025
2.200.000	11	0.037	0.015	0.043	0.059
2.400.000	12	0.038	0.047	0.031	0.032

Fig. 2 Running time of 3
clusters

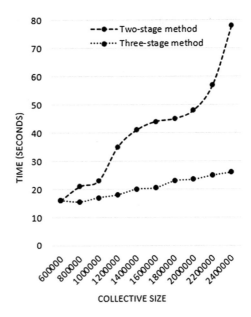

(the significance level). It means that the difference between the pairs of Data_1 and Data_2 is not statistically significant. From this finding, it can be concluded that there is no difference between collective knowledge determined by using the single-stage and three-stage methods.

Beyond the quality of collective knowledge, the running time is also an important issue in collective knowledge determination. For this aim, we compare the running time of the proposed method with that of the two-stage method [8] (see Figs. 2, 3, 4, and 5).

As can be seen in Figs. 2, 3, 4, and 5, the increase in running time is caused by an increase in the collective size. The running time of the two-stage method is much higher than that of the three-stage method. In case of the three-stage method, the running time gradually increases with the collective size because the collective was divided into chunks with the small size and the chunks were processed in parallel. Furthermore, in order to compare effectiveness of the three-stage method and the two-stage method, we computed total running time of all collectives (the numbers of clusters used are 3, 5, 7, and 9). The results show that the running time of three-stage method is faster than that of the two-stage method by 72.1%. From these findings, we can conclude that the proposed method is an effective approach for determining the knowledge of a big collective.

Fig. 3 Running time of 5 clusters

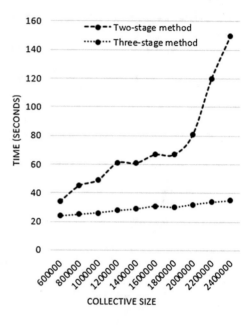

Fig. 4 Running time of 7 clusters

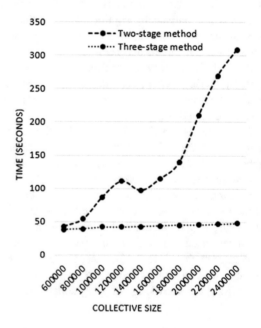

Fig. 5 Running time of 9 clusters

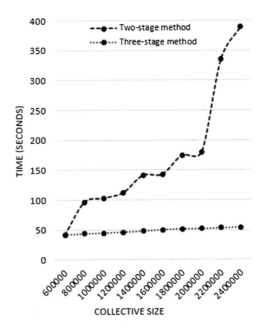

6 Conclusions and Future Works

In this paper, a new consensus-based method for determining the knowledge of a big collective was presented. The proposed method is the combination of the sequence partitioning, the k-means algorithm, and the consensus method. The simulation results have revealed the effectiveness of the proposed method in determining the knowledge of a big collective. In particular, regarding the quality of collective knowledge, we did not found a significant difference between the collective knowledge determined by using the single-stage method and that by using the three-stage method. In addition, the three-stage method shows much shorter running time in comparison to the two-stage method. Future works should examine this study with other knowledge structures such as relational structures, interval numbers.

Acknowledgements This research was supported by Basic Science Research Program through the National Research Foundation of Korea (NRF) funded by the Ministry of Science, ICT & Future Planning (2017R1A2B4009410).

References

1. Subbu, K.P., Vasilakos, A.V.: Big data for context aware computing—perspectives and challenges. Big Data Res. **10**, 33–43 (2017)
2. Ramesh, C.D.: Handbook of Research on Economic, Financial, and Industrial Impacts on Infrastructure Development. IGI Global, USA (2017)
3. Nguyen, V.D., Nguyen, N.T.: A method for temporal knowledge integration using indeterminate valid time. J. Intell. Fuzzy Syst. **27**(2), 667–677 (2014)
4. Maleszka, M., Nguyen, N.T.: Integration computing and collective intelligence. Expert Syst. Appl. **42**(1), 332–340 (2015)
5. Nguyen, N.T.: Methods for Consensus Choice and their Applications in Conflict Resolving in Distributed Systems, Wroclaw University of Technology Press (2002)
6. Nguyen, N.T.: Advanced Methods for Inconsistent Knowledge Management. Springer, London (2008)
7. Nguyen, V.D., Nguyen, N.T., Hwang, D.: An improvement of the two-stage consensus-based approach for determining the knowledge of a collective. Proc. of ICCCI **2016**, 108–118 (2016)
8. Nguyen, V.D., Nguyen, N.T.: A two-stage consensus-based approach for determining collective knowledge. Proc. of ICCSAMA **2015**, 301–310 (2015)
9. Kozierkiewicz, H.-A., Pietranik, M.: Assessing the quality of a consensus determined using a multi-level approach. In: Proceedings of INISTA 2017, pp. 131–136. IEEE (2017)
10. Ramakrishnan, R., Johannes, G.: Database Management Systems. McGraw-Hill, UK (2002)
11. Nguyen, N.T.: Inconsistency of Collective of Knowledge and Collective Intelligence. Cybernet. Syst. Int. J. **39**(6), 542–562 (2008)
12. Nguyen, N.T.: Processing inconsistency of knowledge in determining knowledge of collective. Cybernet. Syst. **40**(8), 670–688 (2009)
13. Day, W.H.E.: The consensus methods as tools for data analysis. In: Bock, H.H. (eds.) Classification and Related Methods of Data Analysis, Proceedings of IFCS 1987, pp. 317–324, North-Holland (1987)
14. Gebala, M., Nguyen, V.D., Nguyen, N.T.: An analysis of influence of consistency degree on quality of collective knowledge using binary vector structure. In: Proceedings of ICCCI 2014, pp. 3–13. Springer (2015)
15. Nguyen, N.T.: Using Distance Functions to Solve Representation Choice Problems. Fundamenta Informat. **48**(4), 295–314 (2001)
16. Oded, M., Lior, R.: Data Mining and Knowledge Discovery Handbook. Springer, US (2005)
17. Mac Queen, J.E.: Some methods for classification and analysis of multivariate observations. Proc. Fifth Berkley Sympos. Math. **1**(14), 281–297 (1967)
18. Wang, J., Su, X.: An improved K-means clustering algorithm. In: Proceedings of ICCSN 2011, pp. 44–46, IEEE (2011)

Fuzzy Ontology Modeling by Utilizing Fuzzy Set and Fuzzy Description Logic

Xuan Hung Quach and Thi Lan Giao Hoang

Abstract In this paper, we propose a fuzzy ontology model by using fuzzy set and description fuzzy logic in order to support fuzzy inference model and fuzzy ontology integration. As in previous studies, elements of a fuzzy ontology are fuzzed by using a class member function based on the fuzzy set theory. The significant contribution of this work is to supplement the fuzzification of concepts with respect to the corresponding ontology. Besides, we clarify the set Z by using fuzzy description logic representing for constrains among ontology elements which have never been mentioned before. In this paper, we also execute the proposed fuzzy ontology implemented in OWL2 language. We employ Protégé to present the fuzzy ontology design in weather domain for demonstration.

Keywords Fuzzy ontology · Fuzzy logic · Fuzzy description logic
Fuzzy OWL2

1 Introduction

Ontology based traditional description logic is not enough to describe fuzzy information and may not fully characterize and process uncertain knowledge in various application domains. In other words, the traditional ontology is unable to describe a measure which a sentence aims to deliver. As a result, this leads to the appearance of more and more research on fuzzy ontology field.

The original version of this chapter was revised: Reference has been removed. The erratum to this chapter is available at https://doi.org/10.1007/978-3-319-76081-0_46

X. H. Quach (✉) · T. L. G. Hoang
Faculty of Information Technology, Hue University of Sciences, Hue, Vietnam
e-mail: tiasang70@yahoo.com

T. L. G. Hoang
e-mail: hlgiao@hueuni.edu.vn

© Springer International Publishing AG, part of Springer Nature 2018   15
A. Sieminski et al. (eds.), *Modern Approaches for Intelligent Information and Database Systems*, Studies in Computational Intelligence 769,
https://doi.org/10.1007/978-3-319-76081-0_2

In 2006, Straccia [1] based on fuzzy set theory of Zadeh and the platform of description logic [2], has represented fuzzy description logic to supply handling uncertain knowledge on the Semantic Web. Since then, the study and development of fuzzy description logic as well as the fuzzy description logic integration into the ontology as a basis for knowledge representation and argumentation has been pushed forward. Lee et al. [3] proposed that the method to generate fuzzy ontology in order to synthesize daily news in China as well as query the events or news based on the fuzzy ontology [4]. They proposed a fuzzy ontology framework (Foga) for uncertain information [5]. This framework is developed on the idea of fuzzy set theory and formal concept analysis (FCA). Abulaish and Lipika [6] suggested a fuzzy ontology concepts which are represented as a fuzzy relationship and are encoded by the degree of attributes using fuzzy membership functions. To express and reason a fuzzy ontology, [7] mentioned a new fuzzy extension of description logic called fuzzy description logic with comparable expressions (FCDLs).

However, the existing fuzzy ontology model presents fuzzy concepts derived from conventional fuzzification, but does not focus on the semantic relationships between the fuzzy concepts. Thus, it causes troubles to apply fuzzy ontology on fuzzy knowledge processing and fuzzy ontology integration.

Currently, there is a great number of papers discussing how to apply fuzzy description logic into OWL to build a fuzzy ontology as the construction method [8]. The main idea is to define fuzzy ontology and construct fuzzy OWL language through fuzzy syntax and fuzzy semantics of SHOIN (D).

In this paper, we introduce an integration approach of fuzzy description logic and fuzzy set into OWL2 to construct a new fuzzy ontology form. The main contribution is to remodel fuzzy ontology based on fuzzy description logic, fuzzy set theory and yield a few new definitions. Most previous studies [8–13] have not mentioned the fuzziness concept. The fuzzy measurement expresses the significant or correlative level to an ontology. This usually occurs when an ontology is used for forecasting or fuzzy inference purposes. Therefore, we supplement the fuzzy characteristic to each concept. Besides, most of the previous studies, including ours [9–12] have not clearly defined set Z. We will employ fuzzy description logic to define the axioms or constrains in the set Z. Moreover, taking advantage of the previous studies [14], we model the defined fuzzy ontology using OWL2.

2 Fuzzy Ontology Definition

In this section, the definitions of fuzzy ontology which rely on the ontology development for semantic web and knowledge system expressed through OWL2 language using description logic.

We implement the integration of fuzzy logic and fuzzy description logic into ontology to build a fuzzy ontology model in which components of fuzzy ontology were fuzzy using a membership function. This fuzzy ontology definition is an extension of the earlier proposed ontology [9] (We extend the specific fuzzy concept, to describe the conceptual cases mentioned. Similarly, we have proposed more specific fuzzy attribute, instead of weighted attribute):

Definition 1 (*Fuzzy ontology*) Let (A, V) be some representation of real world, where A is a finite set of attributes and V is the value domain of A. Fuzzy ontology is determined by a set of four parameters (C, R, I, Z), in which:

- C is a finite set of fuzzy concept. $C = \{(c_i, f^C(c_i)) \mid i = 1 \ldots n\}$, where $f^C(c_i)$ is the value of membership function f^C of the concept $c_i \in C$, $f^C(c_i) \in [0, 1]$.
- R is a set of fuzzy relations between concepts $R = \{R_1, R_2, \ldots, R_m\}$, $R_i \subseteq C \times C \times [0, 1]$, i = 1, .., m. A relation is a set having a pair of concepts and fuzzy value representing the correlation between them. The relation R_i between two concepts in ontology is represented by a unique fuzzy value, it means that:

  - If (c, c', v) $\in R_i$ and (c, c', v') $\in R_i$ then v = v'.
  - I is a set of fuzzy entities, $I = \{I_1, I_2, \ldots, I_k\}$, $I_i = \{(i, f^{I_i}(i)) \mid i \in c_i\}$, in which $f^{I_i}(i)$ is the value of membership function f^{I_i} of case $i \in c_i, f^{I_i}(i) \in [0, 1]$.

- Z is a set of axioms, the integrity constraints or relationships between concepts and entities in ontology that can not be represented by the relationships in R or the conditions (necessary and sufficient) to determine the concept C.

Definition 2 (*Specific fuzzy concept*) The specific fuzzy concept cf $\subset$ C is defined as a set of five parameters: $\left(cf, V_{cf}, V'_{cf}, L_{cf}, f_{cf}\right)$.

where cf is a unique identifier of the concept. $V_{cf} \subset V$ is a specific set, called value domain of a concept. $V'_{cf} \subset [0, 1]$ demonstrate the fuzzy values of the specific set V_{cf}. f_{cf} is a fuzzy membership function, determined by: $f_{cf}: V_{cf} \to V'_{cf}$, with $\forall v \in V_{cf}, f_{cf}(v) \in V'_{cf}$. $L_{cf} \subset V$ is the language value to determined attribute value level in V_{cf}.

An example of a specific fuzzy concept is: Nil, Trace, Light, Moderate, and Heavy to describe the fuzzy concept rainfall.

Definition 3 (*Instance*) An instance of a concept c is described by the attribute of set A^c, with the values in set V^c pair(id, v), in which id is the identifier of instance and v is the value the mapping instance $v: A^c \to V^c, v(a) \in V^c, \forall a \in A^c$.

Definition 4 (*The specific fuzzy attribute*) The specific fuzzy attribute d is defined as a pair: $d = (c_d, c_r)$, or set of five parameters: $d = (c_d, V_d, V'_d, L_d, f_d)$. Where, d is the unique identifier of the attribute. $c_d \in C$ is a fuzzy concept called attribute domain d, c_r is the specific fuzzy concept. $V_d \subset V$ is a specific set called value ranged domain d. $V'_d \subset [0, 1]$ which present the fuzzy values of the specific set V_d. f_d is the fuzzy membership function, determined by: $f_d: V_d \to V'_d$, with $\forall v \in V_d, f_d(v) \in V'_d$. $L_d \subset V$ is the languages value to be determined attribute value level in V_d.

Example: the concept Rain{hasRainfall, hasDepth} has two specific fuzzy attributes hasRainfall and hasDepth, where hasRainfall has the value range (0, 100], hasDepth has the value range Nil, Trace, Light, Moderate and Heavy.

Definition 5 (*Fuzzy concept*) A concept of fuzzy ontology is defined as a set of four parameters: $(c_i, A^{c_i}, V^{c_i}, f^{c_i})$, where c_i is a set of objects or instances, $A^{c_i} \subseteq A$ is a set of concept description, $V^{c_i} \subseteq V$ is the value domain of attribute $V^{c_i} = \bigcup_{a \in A} V_a(V_a)$ is value domain of attribute a and f^{c_i} is a fuzzy membership function: $f^{c_i}: A^{c_i} \to [0, 1]$ represents the important/determined degree of an attribute of a concept c_i. The fuzzy structure of c_i is defined by the set $(A^{c_i}, V^{c_i}, f^{c_i})$.

3   A Fuzzy Ontology Modeling Method Using OWL2

In this section we present a fuzzy ontology modeling method using OWL2. Our method uses syntax inherited from OWL/XML and the methodology of fuzzy ontologies representation which were presented in [14]. OWL2 provides annotations for axioms and entities (OWL DL only supplies annotation of ontology for entities).

In [14] their methodology to represent fuzzy OWL uses OWL2 ontology and extends its elements with annotation properties representing the features of the fuzzy ontology that OWL2 can't directly encode such as: fuzzy modifier, fuzzy data types, fuzzy concepts, fuzzy roles, fuzzy axioms. In this paper, we will alter their fuzzy concepts.

3.1   Annotation

We use annotation property fuzzyLabel instead of default annotation property in OWL2. Each annotation will be separated by a starting tag <fuzzyOwl2> and end tag </fuzzyOwl2>, with attribute fuzzyType to determine fuzzy labeled factors.

3.2   Fuzzy Modifiers

We usually can compute the value μA (x) based on the following fuzzy modifiers: Trapezoidal function, Triangular function; L-function, R-function. The fuzzy modifiers parameters are a, b, c. In this case, the value of fuzzyType is 'modifier', and a tag named 'Modifier' having attribute 'type' value could be linear or triangular, and attributes a, b, c, depending on the type of modifier.

Table 1 Fuzzy datatype

D1	left (k_1, k_2, a, b)
D2	right (k_1, k_2, a, b)
D3	triangular (k_1, k_2, a, b, c)
D4	trapezoidal (k1, k2, a, b, c, d)
D5	ɯ(d)

3.3 Fuzzy Datatype

According to ontology definition, (A, V) is a real world, where A is a finite set of attributes, V is the value domain of A. Thus, in addition to the usual datatype as in a crisp ontologies, a fuzzy ontology V includes fuzzy data types as shown in Table 1.

A fuzzy modifier 'ɰ' is defined as a function fɰ: [0, 1] → [0, 1], which members are linear, or triangular function:

ɰ → linear (c)
    Triangular (a, b, c)

- The fuzzy atomic data types (specific):

The specific fuzzy data type is often used to declare specific fuzzy attributes. The fuzzy data type has following parameters k1, k2, a, b, c, d. The four parameters (k1, k2, a, b) are common for all fuzzy data types, c appears in (D4) and (D5); d is only used in (D5).

The fuzzy transformed data type: In a specific fuzzy property or concept, there often exists the complement (adjective or adverb) used to describe the fuzzy degree of a properties or concept.

In this case there are only two parameters: type of transformation and transformed data type. Type of transformation modifies the other fuzzy data types.

3.4 Fuzzy Concept C

We produce a new concept C and put in an annotation property describing the type of initialized methods and the values of their parameters. At this time, the value of fuzzyType is the concept, and there is a corresponding tag 'concept' with an attribute 'type', and another property depending on the concept initialization. The general rule does not allow recursive case, means that, C is not defined in the instances of C, so that C is not a valid value for the property.

- The specific fuzzy concepts:

A specific fuzzy concept is a set of five parameters: $\left(cf, V_{cf}, \left[V'_{cf} \right], \left[L_{cf} \right], f_{cf} \right)$

where the parameters $\left[V'_{cf} \right]$ and $[L_{cf}]$ could exist or not.

In case L_{cf} exists, the values of type is modified. There are two added attributes is 'modifier' (fuzzy modifier), and 'base' (name of the changed fuzzy concept). The value range of annotation: a concept OWL2.
The syntax of the annotation:

<fuzzyOwl2 fuzzyType ="concept">
<MODIFIED_CONCEPT >
</fuzzyOwl2 >
Which:
<MODIFIED_CONCEPT>:=<Concept   type   ="modified"   modifier
="<STRING>"base="<STRING >"/>

- modifier is defined as a fuzzy modifier
- base has a different name in comparison to the annotated concept.

In case $\left[V_{cf}' \right]$ exists, the value of the attribute type is weighted. There are two added attributes value (the value is a real number in range (0,1]), and 'base' (the name of the weighted fuzzy concept).
The value range of annotation: a concept of OWL2.
The syntax of the annotation:

<fuzzyOwl2 fuzzyType ="concept">
<WEIGHTED_CONCEPT >
</fuzzyOwl2 >
Which
<WEIGHTED_CONCEPT   >   :=   <Concept   type
="weighted"value="<DOUBLE>"base="<STRING >"/>
Where

- value in the range (0, 1].
- base has a different name with the annotated concept.

  - Fuzzy concept (Weighted sum): A fuzzy concept in ontology is defined as a set of four parameters: $(c_i, A^{c_i}, V^{c_i}, f^{c_i})$, where A^{c_i} is the attribute set of concept c_i. Each property goes with a fuzzy value. As a result, the value of attribute type is weightedSum.

The value range of annotation: a concept OWL2.
The syntax of the annotation:

```
<fuzzyOwl2 fuzzyType ="concept">
<Concept type ="weightedSum ">
     <WEIGHTED_CONCEPT >
</Concept ></fuzzyOwl2 >
Where
```

<WEIGHTED_CONCEPT>:=   <Concept   type   ="weighted"   val-
ue="<DOUBLE>"base="<STRING >"/>

where k is the number of specific fuzzy concepts.

3.5 The Fuzzy Role R

We create a role R and add a descriptive annotation attribute as it is in the initialized methods and parameter values. At this time, the value of 'fuzzyType' is 'role'. There is a 'role' tag corresponding to an attribute 'type', and other properties depend on the initialized method of 'role'. The general rule is not to allow the recursive case. Currently this paper only supports fuzzy transformation.

In this case, the value of attribute 'type' is 'modified'. There are two added properties 'modifier' (fuzzy modifier), and 'base' (the name of fuzzy role changed).

3.6 Fuzzy Axiom Z

In this case, the value of the fuzzyType is axiom, and there is a tag 'Degree' (optional) with an attribute 'value'. If we skip this tag, it will be assigned the value of 1.

The value range of annotation: an axiom of OWL2 including the following categories: concept assertion, role assertion.

4 Fuzzy Ontology Construction with Protégé

We present fuzzy weather ontology construction process and apply fuzzy description logic into OWL. We first create the structure of the weather ontology using the Protégé 5.0 software (Sect. 4.1). We then based on the fuzzy description logic to add the description using the annotation properties in section Create description logic for fuzzy weather ontology.

4.1 Fuzzy Weather Ontology Construction

As defined on the fuzzy ontology already proposed, the fuzzy weather ontology structure (FOW) is depicted in Fig. 1.

Set of concepts C: Include set of fuzzy concepts like Wind, Rain, Storm … or set of specific fuzzy concepts HeavyRain, LightWind … These sets describe rainfall or wind intensity by a dependent function (Fig. 1) with the syntax of fuzzy logic. We portray HeavyRain function as follow:

HeavyRain = Rain ⊔ ∃ hasPrecipitationRate.rs5-7.5

The Storm concept set by weather factors that include at least one of the following cases:

S1: Rain ⊔ ∃ hasPrecipitationRate.rs5-7.5
S2: Wind ⊔ ∃ hasWindSpeed.rs10-30
S3: Wind ⊔ ∃ hasWindSpeed.les10 ⊔ Rain ⊔ ∃ hasPrecipitationRate.very(rs5-7.5)

The different weather conditions determine the capability (degree) of forming "Hurricane" variously. For example, in some areas, winds cause more heavy rain and vice versa. As a result, each weather condition case will have a certain influence on the concept of "Hurricane" diversely. We describe this section based on the attribute annotation of the Storm concept (Fig. 1, Set C).

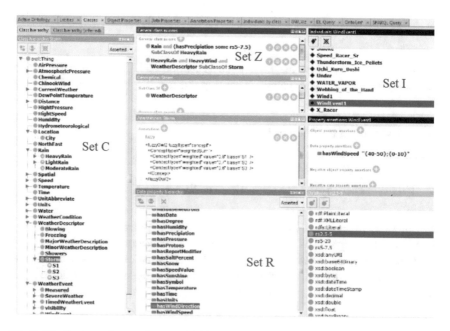

Fig. 1 The fuzzy weather ontology model; Set C: Set of concepts; Set I: Set of individuals; Set R: Set of roles. Set Z: Set of axioms

Table 2 Attribute description of concepts

Concept	Sub-concept	Attribute	Value range/Unit
Cloud cover		hascloudcover	m
	Partly cloudy		1, 2, 3, 4
	Mostly cloudy		5, 6, 7
	Overcast		8,9
Humidity		hasHumidity	–
	Dry		[0.3; 0.4)
	Average humidity		[0.4; 0.7]
	Moist		(0.7; 0.8]
Rain		hasPrecipitationRate	Mm/h
	Light rain		[2.5, 5]
	Moderate rain		(5, 5, 7.5]
	Heavy rain		>7.5
Wind		HasWindSpeed	m/s
	Light wind		(5, 20]
	Medium wind		(20, 40]
	Heavy wind		>40
		HasWindDirection	°
	North wind		[0; 45) [315; 360)
	East wind		[45; 135)
	South wind		[135; 225)
	West wind		[225; 315)

The set of fuzzy relation R is constructed as (Fig. 1, Set R). For each concept, there is a set of fuzzy separate attributes. Partial attribute of concepts in the fuzzy weather ontology is described in Table 2: The concept Wind {HasWindSpeed, HasWind Direction} has some particular fuzzy attributes as HasWindDirection, value range [0, 36); HasWindSpeed, value range {Light, Medium and Heavy} (Fig. 2).

We define data (datatype) such as the level of rainfall, winds as in Fig. 3

The set of instance I is constructed as (Fig. 1, Set I). Each instance is characterized by determined attributes. For example, the instance WindEvent1 is an instance of the concept WindEvent and has a set of dependent values on the depending function.

The set of axioms Z includes definitions of the concepts and attributes as well as the relationship between them (Fig. 1, Set Z), for example:

$$\text{HeavyRain} = \text{Rain} \sqcup \exists \, \text{hasPrecipitationRate.very} \, (\text{rs5} - 7.5)$$

$$\text{VeryHeavyRain} = \text{Rain} \sqcup \exists \, \text{hasPrecipitationRate.very} \, (\text{rs5} - 7.5)$$

$$\text{With very}(x) = \textbf{linear}(3).$$

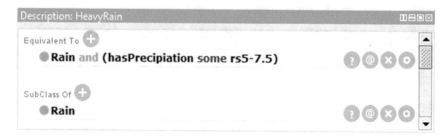

Fig. 2 The description of HeavyRain concept

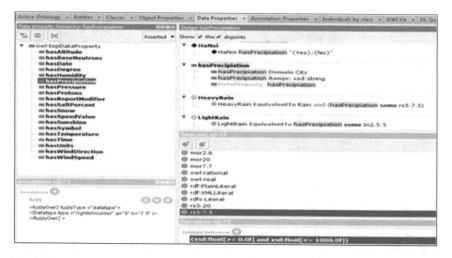

Fig. 3 Fuzzy description logic construction for datatype

4.2 Create Description Logic for Fuzzy Weather Ontology

After initializing the fuzzy weather ontology, we continue to create fuzzy description for the ontology by using annotation attributes in annotation property. Instead of manually producing annotations for each attribute or concept on ontology, we use Fuzzy OWL plug-in. Plugging allow users to select annotation and corresponding description logic function, as shown in Fig. 3.

As a result, the generated fuzzy weather ontology comprises 121 classes and 148 individuals. In fuzzy description logic, we also produce 20 fuzzy annotations, 6 of them are fuzzy classes and the rest are fuzzy data type.

5 Conclusion

In this paper, we propose a new definition of fuzzy ontology, in which all the components are fuzzed by utilizing a membership function. The fuzzy ontology is proposed by the integration of fuzzy logic and fuzzy description logic into ontology to effectively address the reasoning methods. In the fuzzy ontology definition, in addition to concept fuzziness, relationship, individuality, we are concerned with the degree of importance/determination of the properties. We present the syntax to model the fuzzy description language in OWL2 version. We present the syntax and interpretation on OWL2 based on the extension of the syntax and semantics of fuzzy SHOIN (D). The main contribution is the proposal of adding fuzziness to the fuzzy concept on ontology. The clarification set Z (set of axioms) is used in fuzzy description logic for axiom expression on fuzzy ontology. The inheritance of the fuzzy methods on ontology components using fuzzy set.

This paper presents the experimental methods to develop fuzzy weather ontology. First, we build the core of ontology relying on the fuzzy ontology structure proposed in Sect. 3 by using an ontology editor that supports OWL2 (Protégé). Finally, we employ the attribute annotation to create the fuzziness for ontology.

References

1. Straccia, U.: A fuzzy description logic for the semantic web. In: Sanchez, E., (ed.) Fuzzy Logic and the Semantic Web. Capturing Intelligence, pp. 73–90. Elsevier (2006)
2. Zadeh, L.A.: Fuzzy sets. Inform. Control **8**, 338–353 (1965)
3. Lee, C.S., Jian, Z.W., Huang, L.K.: A fuzzy ontology and its application to news summarization. IEEE Trans. Syst. Man Cybernet. (Part B) **35**(5), 859–880 (2005)
4. Lee, C.S., Chen, Y.J., Jian, Z.W.: Ontology based fuzzy event extraction agent for Chinese e-news summarization. Expert. Syst. Appl. **25**(3), 431–447 (2003)
5. Tho, Q.T., Hui, S.C., Fong, A.C.M., Cao, T.H.: Automatic fuzzy ontology generation for semantic web. IEEE Trans. Knowl. Data Eng. **18**(6), 842–856 (2006)
6. Abulaish, M., Lipika, D.: Interoperability among distributed overlapping ontologies—a fuzzy ontology framework. In: Proceedings of the 2006 IEEE/WIC/ACM International Conference on Web Intelligence, Hong Kong, pp. 397–403 (2006)
7. Kang, D., Xu, J., Baowen, L., Yanhui, L.: Description logics for fuzzy ontologies on semantic web. J. Southeast Univ. (English Ed.) **22**(3), 343–347 (2006)
8. Silvia, C., Davide, C.: Fuzzy ontology, fuzzy description logics and fuzzy-OWL. In: Applications of Fuzzy Sets Theory, of the series Lecture Notes in Computer Science, vol. 4578, pp. 118–126
9. Truong, H.B., Quach, X.H.: An overview of fuzzy ontology integration methods based on consensus theory. ICCSAMA 217–227 (2014)
10. Truong, H.B., Duong, T.H., Nguyen, N.T.: A hybrid method for fuzzy ontology integration. Cybernet. Syst. **44**(2–3), 133–154 (2013)
11. Truong, H.B, Nguyen, N.T.: A multi-attribute and multi-valued model for fuzzy ontology integration on instance level. ACIIDS (1) 187–197 (2012)
12. Truong, H.B, Nguyen, N.T.: A framework of an effective fuzzy ontology alignment technique. SMC 931–935 (2011)

13. Jun, Z., Yan, C., Wang, Q., Miao, L.: Fuzzy ontology models using intuitionistic fuzzy set for knowledge sharing on the semantic web. In: 12th International Conference on Computer Supported Cooperative Work in Design, 2008. CSCWD 2008. pp. 465–469 (2008)
14. Fernando, B., Straccia, U.: Representing fuzzy ontologies in OWL 2. In: Proceedings of the 2010 International Conference on Fuzzy Systems (FUZZ-IEEE-10), pp. 1–6 (2010)

An Approach for Recommending Group Experts on Question and Answering Sites

Dinh Tuyen Hoang, Ngoc Thanh Nguyen, Huyen Trang Phan
and Dosam Hwang

Abstract Question-and-answer (Q&A) sites can be understood as information systems where users generate and answer questions. Also, they can determine the top answers using the number of positive and negative votes from crowd knowledge and experts. Knowledge sharing sites have been rapidly growing in recent years. It is difficult for a user to find experts who can write great answers to their questions. Recent approaches have focused on recommendations from a single expert. However, a question may contain several topics. Thus, finding the experts group to answer the questions is the best solution. In this paper, we propose a new expert group-recommendation method for Q&A systems. First, the users' profiles are built to determine experts and non-experts. Second, a topic modeling method is used to identify the topic of the question and matches it to corresponding experts. Third, a social graph is generated to find expert groups. In order to increase knowledge and avoid following the crowd, we require that the members of expert groups not only match the skill requirements to answer the questions but also be diverse. Diversity is an essential factor to promote the development of Q&A sites. Experimenting on Quora dataset shows that the method achieves promising results.

Keywords Recommending-expert · Question answering · Quora-expert-finding

D. T. Hoang · H. T. Phan · D. Hwang (✉)
Department of Computer Engineering, Yeungnam University, Gyeongsan, Korea
e-mail: dosamhwang@gmail.com

D. T. Hoang
e-mail: hoangdinhtuyen@gmail.com

H. T. Phan
e-mail: huyentrangtin@gmail.com

N. T. Nguyen
Faculty of Computer Science and Management, Wroclaw University
of Science and Technology, Wroclaw, Poland
e-mail: Ngoc-Thanh.Nguyen@pwr.edu.pl

© Springer International Publishing AG, part of Springer Nature 2018   27
A. Sieminski et al. (eds.), *Modern Approaches for Intelligent Information
and Database Systems*, Studies in Computational Intelligence 769,
https://doi.org/10.1007/978-3-319-76081-0_3

1 Introduction

Social question-and-answer (Q&A) sites, in which users can share their knowledge by posting questions, getting answers, and following experts, have been growing more and more quickly in recent years. Benefitting from this, many social Q&A platforms developed, such as Quora, Yahoo Answers, Zhihu, Google Answers, and Stack Overflow. However, some of them, such as Yahoo Answers, have stalled and begun to shrink or shut down (e.g., Google Answers). There are three main reasons for this. (i) Most users are not enthusiastic about answering questions, or are not experts. (ii) There are users willing to respond to the questions, but they are not aware of new questions of interest to them. (iii) Once websites grow, a significant number of low-value questions flood the system. It is hard for users to find useful or interesting content. However, unlike other Q&A systems, Quora is proliferating. One of the main differences between Quora and other Q&A sites is the way users can connect. While users in other Q&A sites exist in a global search space, Quora enables users to follow each other to make a social network. Relationships on Quora are, in fact, not necessarily bi-directional. A user can choose to follow other users' posts, becoming a follower of experts, but the act of following might not be reciprocated. Additionally, users can follow the topics they care about, and get updates on the questions and answers on this topic. In particular, Quora[1] is a rapidly growing service that improves on a regular Q&A system with social connections among users. As of April 2017, Quora claimed 190 million unique visitors monthly, up from 100 million a year earlier.[2] Due to a large dataset, it is challenging for a user to find experts who can write great answers to questions. Recommender systems are at the core of this problem [13]. Generally, recommender systems aim at offering a list of recommendations to someone by using content-based or collaborative-based filtering. Collaborative-based filtering develops a model to predict items a user is most likely interested in. Content-based filtering recommends items based on features and user historical activities. Some recommender systems have been proposed for finding experts on Q&A sites [6, 10, 13, 14]. However, most of them try to find a list of experts who are most likely to write great answers to specific questions. In some cases, the question contains several topics that are not in the person's area of expertise. Existing methods tend to recommend the most suitable expert for the question. This leads to a list of highly recommended experts with the same expertise, but they will not cover other subtopics in a question. For example, given a question: "How do I create a deep neural network for Alzheimer's detection?" and asked for 2 topics, the results might return the following: 70% Topic "Deep neural network", 30% Topic "Medicine and Healthcare". In this case, most of current method will return a list of experts who have expertise is "Deep neural network". This leads to the system ignoring experts who know very well about Alzheimer's. Fortunately, Surowiecki [11] said, "Under the right circumstances, groups are remarkably intelligent, and are often smarter than the smartest people in them." Other research on

---

[1]https://www.quora.com/.

[2]https://www.quora.com/How-many-people-use-Quora-7/answer/Adam-DAngelo.

inconsistent knowledge management [9] showed that "The intelligence of the crowd is better than the smartest." This indicates that the combination of expert groups gives the best solutions, compared with a single expert [7]. Also, in order to decrease many potential social impacts among group members and to avoid just following the crowd, we demand that the members of the group not only match the skill requirements for answering the question but also be diverse. Diversity provides an important aspect of the process of decision-making by an expert group. Lack of diversity may appear in potential conflicts of interest, and also brings about a situation where experts begin ignoring their expertise and following the crowd, because of social pressure to go along or the information waterfall effect. In this study, we propose a new expert group recommendation method for Q&A systems. The method consists of three steps. First, users' profiles are built to determine experts and non-experts. Second, a topic modeling method is used to identify the topics of the question and match the questions to corresponding experts. Third, a social graph is generated to find expert groups. Quora dataset was used to evaluate the accuracy of the proposed method. The rest of this paper is constructed as follows. In the next section, we briefly investigate related work in expert finding on Q&A systems. In Sect. 3, the details of the proposed method are presented. The experimental results and evaluations are shown in Sect. 4. Lastly, conclusions and future work are in Sect. 5.

2   Related Works

In the last few years, Q&A sites have successfully obtained increasingly large knowledge repositories, each directed by a lot of questions and answers, not only from the user community but also from experts. Despite their success, however, with a vast and rich dataset, it is difficult for a user to find experts who can write great answers. Recommender systems are the key to solve this problem. The existing work for finding an expert on a Q&A system can be divided into two groups: the users' activities approach, and the topics modeling approach [4]. The users' activities approach for finding experts is based on investigating the ask-answer relationship between users in the rating matrix. A personalized learning-to-rank algorithm is used to find the experts for the target user. The topics modeling approach for finding experts is based on topic modeling methods for the content of the questions. Also, some approaches have been worked out for recommending experts to the most important questions. For example, GunWoo et al. [6] presented the InfluenceRank algorithm to recommend credible users to answer the target question on a Q&A site. The InfluenceRank algorithm is based on the users' communications on the Q&A sites. Activity and trust were determined as primary factors to measure the impact between users. Another approach using a deep neural network for Q&A communities was proposed by Azzam et al. [1]. They built a question routing system named QR-DSSM in which textual features are used to predict the semantic similarity between the users' profiles and the posted questions. In the case of Quora, some approaches have been worked out for recommending several recommendation problems. Yang and Amatriain [13] described several recommendation problems on Quora, such as recommending home feed, digest answers, and related questions. They also presented their approaches solving these recommendation problems with different machine learning models, such as personalized learning-to-rank and ask-to-answer (A2A). Patil and Lee [10] built a system that follows three steps to detect experts on Quora. First, users' profiles from Quora are collected to investigate the attributes of experts and non-experts. Second, the querying user's behaviors are analyzed to qualify the answer features. Then, statistical models are developed based on the proposed features to detect experts on a particular topic and on general topics. Besides, authors also considered extracting features from users who have a Twitter account. They achieved 97% accuracy on a Quora dataset. Zhao et al. [14] proposed a method to find experts for a new user by taking into account following relations. The relationships between users and topical interests were explored to develop a user-to-user graph. Their experiments on Quora dataset showed that the method performs correctly when addressing the cold-start problem. In summary, on existing methods, finding an expert to answer a new question involves two steps. First, the users' profile is built by extracting the past question-answer and user activities. The Q&A community votes on the quality of the question-and-answer activities through up-votes and down-votes. Second, the performance of users answering new questions is predicted by means of a users' profile. The users with the highest predicted performance are selected for answering questions.

Most of the existing methods tend to find a list of experts who are most suitable to write excellent answers on a particular question. However, in some cases, questions contain different subtopics that are not domains of the expert. Current methods tend to recommend the most proper expert for the question. This leads to a list of highly recommended experts with the same expertise, but they will not cover other subtopics in a question. This is also known as the *long tail problem* in recommendations. To solve the problem, we propose a group experts recommendation method that can combine to give the best solution. We also consider the diversity of group members to increase knowledge and to avoid following the crowd.

3   Group Experts Recommendation Method

3.1   Problem Formulation

This section discusses how to build a recommender system to find group experts to answer a given question. Let $E = \{e_1, e_2, \ldots, e_n\}$ be a set of candidates consisting of n experts and let $T = \{t_1, t_2, \ldots, t_m\}$ be a set of m arbitrary skills. Each expert e_i has a connection to a set of skills T_i, so that $T_i \subseteq T$. Expert e_i has expertise t_j if and only if $t_j \in T_i$; otherwise, expert e_i does not have expertise t_j. Let T_Q be a set of topics that is determined from given question Q, $(T_Q \subseteq T)$ by using topic modeling method. If $t_j \in T_Q$, we say skill t_j is required for question Q. Let $C(E_Q, T_Q)$ be the cover of a group

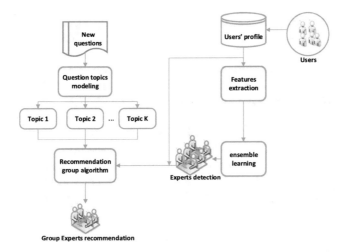

Fig. 1 The workflow of the experts group recommendation method

of experts E_Q, with respect to T_Q topics. That is, $C(E_Q, T_Q) = T \cap (\bigcup_{e_i \in E_Q} T_i)$. For each topic t_j, we find a set of experts, $E(t_j)$, who are most interactive and relevant to that topic: $E(t_j) = \{e_i | e_i \in E \wedge t_j \in T_i\}$. Let $S(e_i, e_j)$ be the value of social influence between experts e_i and e_j. Since given a question Q has T_Q topics, we need to find a group of experts $E_Q \subseteq E$, where $C(E_Q, T_Q) = T_Q$, and social influence $S(E_Q)$ is minimized.

In summary, our proposed method consists of the three following steps:

(i) The expert detection methodology is applied to determine that a user is an expert or a non-expert.
(ii) The topic modeling technique is used to classify given question Q into K topics.
(iii) The group experts recommender system is built by taking into account the topics of the given question and the experts' skills and activities.

Figure 1 shows the workflow of the experts' group recommendation method. The details of the proposed method are presented in the following subsections.

3.2 Expert Detection

In this subsection, an expert detection methodology is proposed by taking into account users' profiles and behaviors to determine that if each user is an expert or a non-expert. A Quora user profile consists of user activity-associated information, such as the number of followers, the number of users followed, the number of answers, and the number of topics that the user has followed, etc. Unlike other social networks, such as Facebook and LinkedIn, where a relationship has to be confirmed by both parties before being set up, a Quora user can decide to follow other users' posts, becoming their follower, but the action of following might not be reciprocated. Therefore, the number of expert followers is usually greater than the number of non-expert followers [10]. Also, the number of answers and edits can reflect how active a user is on Quora. Usually, experts have a greater number of answers and edits than non-experts, showing that experts are users who are more active than non-experts. Thus, we also consider the number of answers and edits as features to distinguish between an expert and a non-expert. Another thing we need to care about is the quality of the answers. Many features can reflect the quality of the answers, such as up-votes, down-votes, an answer's length, references, and the answers ratio. However, we only select three important features to determine the quality of the answer: up-votes, down-votes, and answer's length. After determining the features, ensemble learning methods are used to determine whether a user is an expert or a non-expert.

3.3 Question Topic Modeling

We assume that a question can contain a mixture of topics, and each topic is a distribution of words. Since questions typically include multiple topics, the presumption is suitable for a Quora dataset. There are several methods of question modeling, such as Latent Dirichlet Allocation (LDA) [2], Word2Vec [8]. The goal is to find the topics in a given question. Let K be the number of topics in given question Q. We need to find the experts' groups where their expertises can cover K topics. By means of Quora datasets where questions usually have several topics and the requirements of the problem, we choose the LDA algorithm for question topic modeling. We model an expert's profile as a mixture of topics; meanwhile, a question is represented as a vector of N_w words. Each w_i is selected from a vocabulary of size V. Let μ be a matrix of expert-profile probabilities for K topics picked independently from a symmetric Dirichlet α prior. Let β be a matrix of topic probabilities for all the words in the vocabulary picked from a symmetric Dirichlet η prior. Let Z be the topic assigned to word W from the μ distribution, in which W is drawn from the topic distribution corresponding to Z. The process of creating the user's profile includes three steps: (i) select a topic $K \in \{1, \ldots, K\}$ from the μ distribution. (ii) Pick word w from distribution β. (iii) Repeat the process S_w times, in which S_w is the total number of words in a user's profile. Gibbs sampling [5] is used to estimate the number of topics in a given question.

3.4 Group Experts Recommendation

Group recommender systems provide suggestions in contexts where users operate in groups. The purpose of this subsection is to provide a group of experts to answer a new question. Let $G(V, E_v, W)$ be a social graph among experts, where V is a set of vertices, which considers as a set of experts; E_v is a set of edges that represents connections among experts; and W is the weight of the graph, which represents the interaction strength values among experts. Here, we define $S(e_i, e_j)$ to measure the interaction strength between experts e_i and e_j as follows:

$$S(e_i, e_j) = \frac{Upvote(e_i, e_j)}{PossibleUpvote(e_i)} \tag{1}$$

where $Upvote(e_i, e_j)$ is the number of up-votes, where expert e_j votes for expert e_i's posts. $PossibleUpvote(e_i)$ is the largest number of up-votes that a user can cast for expert e_i's posts. The value of $S(e_i, e_j)$ is lower for a correspondingly lower mutual influence between the experts, thereby avoiding the tendency to follow the crowd, but freely expressing their opinions. The value of $S(e_i, e_j)$ ranges from 0 to 1. We set a threshold δ for $S(e_i, e_j)$ when running experiments to select the best output. There are some expert groups that meet the diversity criterion; thus, we compute the activity level of each group by taking into account the number of answers, the number of followings, the number of followers, and the number of edits. After finding a list of groups of experts, we rank the active value of each group to get the recommending list.

For example, given question Q, which covers the following topics: $T_Q = \{Machine\ Learning,\ Deep\ Learning,\ Data\ Science,\ Python\}$, also assume there are seven expert candidates, $E = \{e_1, e_2, e_3, e_4, e_5, e_6, e_7\}$, with the following expertise:

$T_{e_1} = \{Machine\ Learning,\ Deep\ Learning,\ Data\ Science,\ Python\}$
$T_{e_2} = \{Deep\ Learning,\ Artificial\ Intelligence,\ Machine\ Learning\}$
$T_{e_3} = \{Data\ Mining,\ Python\}$
$T_{e_4} = \{Deep\ Learning,\ Data\ Science,\ Machine\ Learning\}$
$T_{e_5} = \{Python,\ Artificial\ Intelligence\}$
$T_{e_6} = \{Deep\ Learning,\ Data\ Science,\ Data\ Mining\}$
$T_{e_7} = \{Python,\ Information\ Retrieval\}$

In order to cover the topics of question Q, we can choose the following sets: $E_1 = \{e_1\}$, $E_2 = \{e_4, e_3\}$, $E_3 = \{e_4, e_5\}$, $E_4 = \{e_6, e_5\}$, $E_5 = \{e_6, e_3\}$, $E_6 = \{e_6, e_7\}$ or $E_7 = \{e_4, e_7\}$. Compare them with the diversity criterion; we remove E_1, and six solutions remain: $E_2, E_3, E_4, E_5, E_6, E_7$. Then, we consider the diversity of each group. Assume that after compute the interaction strength between experts, we have the social network graph shown in Fig. 2. The structure of the graph in Fig. 2 shows

Fig. 2 Network of connections between experts in $\{e_1, e_2, e_3, e_4, e_5, e_6, e_7\}$

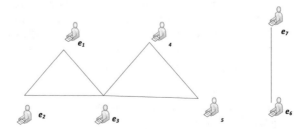

that there is no connection between e_5 and e_6, e_3 and e_6; and e_4 and e_7. Thus, they are free to write their answers to the question without being affected by each other. By ranking the activity levels from high to low in expert's groups E_4, E_5, and E_7, we obtain the experts' groups recommendation list to answer the question.

4 Experiments

4.1 Dataset

There is no Quora official application programming interface (API), so we developed a crawler to collect a Quora dataset. We collected the question dataset based on a dataset released by Quora[3] that consists of over 400,000 questions. From each question, we collected users profiles, which included user activity-associated information, such as a list of answers, the number of answers, the number of followings, the number of followers, etc. We removed users who have no answers or questions. Also, questions that have no answers, or only one answer, were deleted. Then, we manually labeled the dataset to get the ground truth. In total, we collected over 85,000 users including 10,000 experts and over 75,000 non-experts, and 800,000 answers. The dataset was divided into two parts: the training dataset and the testing dataset.

4.2 Evaluations and Results

First, we evaluate the experts detection method. The prediction accuracy method is used to evaluate the performance of the proposed method. We measure the accuracy and area under the curve (AUC). As shown in confusion matrix, Table 1, *True positive* (TP) is the number of exactly classified experts; *False positive* (FP) is the number of misclassified experts; *False negative* (FN) is the number of misclassified non-experts and *True negative* (TN) is the number of exactly classified non-experts. The accuracy is $(TP + TN)/(TP + FP + FN + TN)$. We converted users' profile to feature values and used the Python module Scikit-learn[4] for implementation and evaluation. We achieved promising results that is 95.2% accuracy.

Second, we evaluate the group recommender system. There is no standard method to assess group recommendation approaches. For evaluation, we required the true group ratings for all experts on the topics of the question. We use Normalized Discounted Cumulative Gain (*nDCG*) [12], that is used to evaluate the ranked relevance of the recommended items. The formulations of *nDCG* and Discounted Cumulative Gain (*DCG*) [3] are defined as follows:

---

[3]https://data.quora.com/First-Quora-Dataset-Release-Question-Pairs.
[4]http://scikit-learn.org.

Table 1 Confusion matrix

Actual	Predicted	
	Expert	Non-expert
Expert	True positive	False positive
Non-expert	False negative	True negative

Table 2 Experimental result

Group size	Mean $nDCG$	Precision@10
K	0.20	0.12
$2 \times K$	0.22	0.11
$3 \times K$	0.15	0.04
$4 \times K$	0.05	0.02

$$nDCG_{k,u} = \frac{DCG_{k,u}}{Ideal\_DCG_{k,u}} \tag{2}$$

$$DCG_{k,u} = \sum_{i=1}^{K} \frac{2^{rel_i} - 1}{log_2(1 + i)} \tag{3}$$

where rel_i represents the binary relevant of the recommended results at the Kth ranking, which is set to a value of 1 if the expert is a suitable recommendation for u and 0 otherwise. $Ideal\_DCG_{k,u}$ is the ideal obtained value on a given list. To calculate the $nDCG$ value of a given group, we computed the $nDCG$ value of each group's members, and obtained the average $nDCG$ values of all groups' members. The greater the value of $nDCG$, the more relevant the group recommendations list. In this case, many experts will be recommended to write answers for many questions, but typically, most experts are not willing to answer all questions. An expert only writes answers to a few questions. Therefore, in the offline scenario, we only measured the questions that the expert answered.

The results of the proposed method are shown in Table 2. The accuracy of the results is quite good at a group size of K or $2 \times K$ and decreases as the scale of the group increases. This is understandable, because normally, the question only contains a few topics. For each topic, if an expert has answered well, then other experts do not reply any more, but just up-vote. Besides, the expert only answered a few questions from the many questions created by the users. There are no previous works on Q&A that had taken into account group recommendations and also diversity. Furthermore, since there is no standard answer set, it is difficult to evaluate the quality of the combination of expert answers to obtain the most comprehensive solution. However, our primary goal is to create new knowledge. We do not focus on sorting out existing knowledge. Therefore, diversity is essential. It promotes the development of Q&A sites.

5   Conclusion and Future Work

This work proposes a new approach for recommending an expert group to answer new questions. First, users' profiles are built to determine experts and non-experts. Second, a topic labeling method is employed to identify the topic of a question and match it to corresponding experts. Third, a social graph is generated to find expert groups. In addition to increasing knowledge, and to avoid just following the crowd, we require that the members of expert groups not only match the skill requirements to answer the questions but also be diverse. Diversity plays a significant role in the process of decision-making by an expert group. The experimental results show that the proposed method gives higher accuracy, in comparison to other methods. In future work, we plan to measure the satisfaction of users with the answers by using a sentiment analysis technique. Also, the questions' topic labeling can be improved by using a deep neural network.

Acknowledgements This research was supported by Basic Science Research Program through the National Research Foundation of Korea (NRF) funded by the Ministry of Science, ICT & Future Planning (2017R1A2B4009410).

References

1. Azzam, A., Tazi, N., Hossny, A.: A question routing technique using deep neural network for communities of question answering. In: International Conference on Database Systems for Advanced Applications, pp. 35–49. Springer (2017)
2. Blei, D.M., Ng, A.Y., Jordan, M.I.: Latent Dirichlet allocation. J. Mach. Learn. Res. 3(Jan), 993–1022 (2003)
3. Burges, C., Shaked, T., Renshaw, E., Lazier, A., Deeds, M., Hamilton, N., Hullender, G.: Learning to rank using gradient descent. In: Proceedings of the 22nd International Conference on Machine Learning, pp. 89–96. ACM (2005)
4. Cain, J.O.: Using topic modeling to enhance access to library digital collections. J. Web Librariansh. **10**(3), 210–225 (2016)
5. Griffiths, T.L., Steyvers, M.: Finding scientific topics. Proc. Nat. Acad. Sci. **101**(suppl 1), 5228–5235 (2004)
6. GunWoo, P., SoungWoung, Y., SooJin, L., SangHoon, L.: Credible user identification using social network analysis in a Q&A site. In: Proceedings of ICOMP, pp. 1–7 (2011)
7. Hoang, D.T., Tran, V.C., Nguyen, T.T., Nguyen, N.T., Hwang, D.: A consensus-based method to enhance a recommendation system for research collaboration. In: Asian Conference on Intelligent Information and Database Systems, pp. 170–180. Springer (2017)
8. Mikolov, T., Chen, K., Corrado, G., Dean, J.: Efficient estimation of word representations in vector space. arXiv:1301.3781 (2013)
9. Nguyen, N.T.: Advanced Methods for Inconsistent Knowledge Management. Advanced Information and Knowledge Processing. Springer (2008)
10. Patil, S., Lee, K.: Detecting experts on Quora: by their activity, quality of answers, linguistic characteristics and temporal behaviors. Soc. Netw. Anal. Min. **6**(1), 5 (2016)
11. Surowiecki, J.: The Wisdom of Crowds. Anchor (2005)
12. Wang, Y., Wang, L., Li, Y., He, D., Chen, W., Liu, T.Y.: A theoretical analysis of NDCG ranking measures. In: Proceedings of the 26th Annual Conference on Learning Theory (COLT 2013) (2013)

13. Yang, L., Amatriain, X.: Recommending the world's knowledge: application of recommender systems at quora. In: Proceedings of the 10th ACM Conference on Recommender Systems, pp. 389–389. ACM (2016)
14. Zhao, Z., Wei, F., Zhou, M., Ng, W.: Cold-start expert finding in community question answering via graph regularization. In: International Conference on Database Systems for Advanced Applications, pp. 21–38. Springer (2015)

A Method for Uncertainty Elicitation of Experts Using Belief Function

Tuan Nha Hoang, Tien Tuan Dao and Marie-Christine Ho Ba Tho

Abstract The reliability of the biomedical data plays an essential role in the translation of the computational models and simulations of the human body systems into clinical decision support. Numerical models are commonly linked to the hypotheses on the data range of values due to the lack of in vivo data for some biomaterial variables. However, the reliability of these data is still not fully understood due to a lack of a systematic evaluation approach. The objective of this present study was to assess the reliability of biomedical data using expert judgment and belief theory. A systematic evaluation framework was developed using belief theory to perform the expert elicitation process. Seven parameters related to the muscle morphology and mechanics and motion analysis were selected. Twenty data sources related to these parameters were acquired using a systematic review process on the reliable search engines. A questionnaire was established including four main questions and four complementary questions related to the confidence levels. Eleven experts participated into the evaluation process via Google Form. A transformation process was developed to convert qualitative expert judgments to the numeric representations of the mass functions in the framework of belief theory. Two combination rules (Demspter and Dubois-Prade) were used to fuse the responses of multiple experts. At the end, data reliability was assessed using the pignistic probability to select the sources that correspond to some on-demand levels of confidence.

Keywords Biomechanical data reliability · Expert opinion · Expert elicitation
Belief theory

T. N. Hoang (✉)
Quang Binh University, Quang Binh, Vietnam
e-mail: hoang.tuan.nha@gmail.com

T. T. Dao · M.-C. Ho Ba Tho
Sorbonne University, Université de Technologie de Compiègne,
60319 Compiègne, CS, France

© Springer International Publishing AG, part of Springer Nature 2018
A. Sieminski et al. (eds.), *Modern Approaches for Intelligent Information
and Database Systems*, Studies in Computational Intelligence 769,
https://doi.org/10.1007/978-3-319-76081-0_4

1 Introduction

Biomechanical data are commonly used to describe the anatomical, mechanical and functional behaviors of biological tissues and systems. Most biomechanical data may be classified into four groups (physiological, morphological, mechanical and motion properties). Biomechanical data is crucially important in the development of numerical models in the framework of in silico medicine. Mechanistic models need biomechanical data as input data to perform multiphysics simulations [8, 10]. Machine learning models needs biomechanical data to establish statistical relationships between subject's states and decision-making indicators [4, 10, 11, 12]. Thus, the reliability of these data plays an important role in the development of good and useful models and their relevant outcomes. However, there is a lack of evaluation approach to assess the reliability of biomechanical data.

The assessment of the data reliability may be performed using internal or external approaches. The internal approach uses intrinsic information of a dataset such as its range to evaluate its reliability level according to a baseline range [1, 14]. This approach depends strongly on the available data to establish a reliable baseline range, which is not easily achieved in practice. The external approach applies the expert judgment on the reliability level of each dataset [3, 4, 9, 20]. This approach has showed its potential applications in risk assessment [2, 7, 16, 19] and biomedicine [3]. This approach needs commonly an expert elicitation process to fuse judgments from multiple experts [4]. The use of external approach relates commonly to the partial knowledge and uncertain information derived from expert opinions and this needs a powerful mathematical framework to process this specific kind of information and knowledge. The theory of belief functions, known as Dempster-Shafer theory or evidence theory is a formal framework to model and reason with uncertain information and knowledge. This theory was introduced by Dempster and Shafer [15] and then developed and generalized by Smets [17, 18]. This theory is one of the most general formalism dealing with uncertainty among other frameworks like probabilities or set, and the most used one in engineering applications. However, this powerful theory is never used for assessing the data reliability in biomechanical application. Consequently, the objective of this present work was to develop a systematic evaluation method based on the theory of belief functions and expert opinions for assessing biomechanical data reliability.

2 Materials and Methods

2.1 Theory of Belief Functions

Basic concepts and elements of belief function theory are briefly summarized in the following subsections.

Frame of discernment Θ. is a finite set of mutually exclusive elements in a domain. A hypothesis or proposition is a subset $A \subseteq \Theta$ of the frame of discernment, i.e., it is an element of the power set $P(\Theta)$.

Mass function. (also called basic belief assignment or basic probability assignment —*bba*) is in many respects the most fundamental belief representation and all other representations can be easily obtained from a mass function. Formally, a mass function m is a mapping m: $P(\Theta) \rightarrow [0, 1]$ assigning a mass value to each hypothesis $A \subseteq \Theta$ of the frame of discernment Θ such that

$$\sum_{A \subseteq \Theta} m(A) = 1 \tag{1}$$

The value $m(A)$ is the amount of belief strictly committed to hypothesis A.

Decision making. There are several methods to make decision on choosing a singleton in. The first one relates to the highest plausibility [5]. In this case, an interval of belief ($[Bel(A), Pl(A)]$) is defined as follows:

$$Bel(A) = \sum_{A, B \in 2^{\Omega}, B \subseteq A} m(B) \tag{2}$$

$$Pl(A) = \sum_{A, B \in 2^{\Omega}, B \cap A \neq \emptyset} m(B) \tag{3}$$

where $Bel(A)$ is a belief function and $Pl(A)$ is the plausibility function. A and B are focal elements. Another approach deals with the use of a pignistic transformation function ($BetP(A)$) proposed by Smets [17] as follows:

$$BetP(A) = \sum_{A, B \in 2^{\Omega}, B \subseteq A} \frac{|A \cap B|}{|B|} m(B) \tag{4}$$

Combination of evidence. Several rules have been proposed to combine evidences from multiple sources. One of the most used rules is the Dempster's rule expressed mathematically as follows:

$$K = \sum_{B \cap C = \emptyset} m_1(B) m_2(C) \tag{5}$$

$$m_{1,2}(A) = (m_1 \oplus m_2)(A) = \frac{1}{1 - K} \sum_{B \cap C = A \neq \emptyset} m_1(B) m_2(C) \tag{6}$$

$$m_{1,2}(\emptyset) = 0 \tag{7}$$

where K is defined as the degree of conflict between two mass functions m_1 and m_2. A, B, C are focal elements. $m_{1,2}$ is the fused mass function when $K < 1$. $\oplus$ denotes the Dempster's combination operator.

Another rule of combination is the Dubois and Prade rule [6]. The constitutive equations of this rule are expressed as follows:

$$m_{DP}(A) = \sum_{\substack{B, C \in 2^\Theta \\ B \cap C = A \\ B \cap C \neq \varnothing}} m_1(B)m_2(C) + \sum_{\substack{B, C \in 2^\Theta \\ B \cup C = A \\ B \cap C = \varnothing}} m_1(B)m_2(C) \tag{8}$$

$$m_{DP}(\varnothing) = 0 \tag{9}$$

$$\forall A \in 2^\Theta \backslash \varnothing \tag{10}$$

where m_{DP} is the fused mass function between two mass functions m_1 and m_2 using Dubois and Prade rule. A, B, C are focal elements.

2.2   Expert Elicitation Using the Theory of Belief Functions

A generic expert elicitation workflow based on expert opinions and the theory of belief functions was proposed in Fig. 1. The workflow includes three first steps to establish an expert opinion database from a data collection and a questionnaire. The last three steps aim to perform the expert elicitation to evaluate the reliability of each dataset.

Expert opinion database. A systematic review process was performed using reliable scientific search engines (PubMed and ScienceDirect). Seven parameters related to the muscle morphology (medial gastrocnemius physiological cross-sectional area and volume), mechanics (lateral gastrocnemius shear modulus, soleus shear modulus) and motion (stance time, cadence and maximal dorsiflexion angle) properties were selected. The total of 20 data sources (scientific papers) was used. Note that each parameter has from 2 to 7 sources.

A questionnaire was established for each data source. The questionnaire includes four main questions related to the measuring technique, experimental protocol, number of samples, and the range of values obtained. Four complementary questions related to the auto-evaluation of confidence level of each expert for each response to each main question. For each main question (Table 1), five possible responses were proposed (very high confidence, high confidence, moderate confidence, low confidence, and very low confidence) and each expert selected one response. For each complementary question (*What is the confidence level of your expertise?*), five possible responses were also proposed (very high, high, moderate, low, and very low) and each expert selected one response.

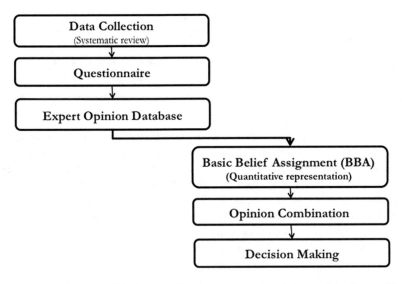

Fig. 1 The generic expert elicitation workflow based on expert opinions and the theory of belief functions

Table 1 Four main questions for each data source and their respective contents

Categories	Questions
Measuring technique	*Is the technique is appropriate to measure this property?*
Experimental protocol	*Is the protocol appropriate to measure this property?*
Number of samples	*Is the number of subjects correct for data and/or statistical analyses?*
Range of values	*Are you confident with the range of values reported?*

To establish the expert opinion database, an international panel of experts, with different domain expertise like medical imaging, motion analysis, and mechanical characterization, remotely fulfilled the questionnaire in Google Form. Each expert response was anonymized before entering into the elicitation process. Twenty experts were contacted and only eleven experts participated into the opinion making.

Expert elicitation. The theory of belief functions [15] is an effective mathematical tool to represent epistemic uncertainty due to lack of information. It is chosen to model the answers of the experts in our project. A rule is defined to convert the responses of the Likert scale into quantitative form as follows:

Scale	Value
Very high	5
High	4
Moderate	3
Low	2
Very low	1

Using this rule, the mass function of expert i for question j of each source S is defined. $m_{ij}(conf)$ represents the degree of confidence, $m_{ij}(non\_conf)$ represents the degree of distrust, $m(\Omega)$ represents the level of ignorance due to lack of information. Q_{S-ij} is the response of expert i to question j concerning source S. C_{S-ij} is the supplementary question to determine the expert's level of expertise i. C_{S-ij} is considered to be the degree of certainty of the expert i, therefore, $m_{S-ij}(conf)$ (degree of confidence of expert i in question j concerning source S) is calculated by: degree certainty multiplies the degree of response of question j.

$$m_{S-ij}(conf) = Q_{S-ij} \times 0.2 \times C_{S-ij} \times 0.2 \tag{11}$$

$m_{S-ij}(non\_conf)$ (degree of defiance of expert i to question j concerning source S) is calculated by:

$$m_{S-ij}(non\_conf) = (5 - Q_{S-ij}) \times 0.2 \times C_{S-ij} \times 0.2 \tag{12}$$

$m_{S-ij}(\Omega)$ (degree of ignorance of expert i to question j concerning source S) is calculated by: 1 minus the degree certitude.

$$m_{S-ij}(\Omega) = (5 - C_{S-ij}) \times 0.2 \tag{13}$$

For each source S, there are 4 questions (j = 1, 2, 3, 4). The conservative strategy based on the worst case has been applied [13], so the mass function of the expert i for S is that which has the value of m(conf) the smallest, m_i is defined by this equation:

$$m_i = \min_{conf} m_{ij} \tag{14}$$

Depending on the number of experts, a combination of the mass functions of all responding experts is performed to establish a final mass for each source.

$$m = \oplus m_i \tag{15}$$

The combination (called the common mass) is calculated from the two mass functions m_1 and m_2 using the Dempster rule or that of Dubois and Prade. The Dempster rule allows to normalize the masses after the combination. Then, a selection threshold is set to select reliable data sources. This threshold is calculated using the mass function and the pignistic probability [18]. The transformation of the mass functions into pignistic probability was carried out by the following formula:

$$BetP(conf) = m(conf) + \frac{m(\Omega)}{2}$$ (16)

3 Results

The questionnaires were submitted to the 20 internal and external experts of the BMBI laboratory (UTC—Université de technologie de Compiègne). To model these results, colors are used to represent response choices. A summary of all responses, the color codes and their respective responses are presented in Fig. 2.

The Fig. 3 illustrates the mass functions of eleven experts for the first questionnaire (technical question) of the sources #3 and #4. It is found that the confidence degree m(conf) is always higher than the degree of confidence m(non_conf). The value m(conf) for the eleventh expert is one, while that of the others varies between 0.3 and 0.65. So this can be considered a noise because one is very far from the range of 0.3–0.65.

An iteration is used to combine the mass functions of the experts using the combination rules of Dempster and Dubois-Prade. The Fig. 4 shows the evolution process of the values m(conf), m(non_conf) and m(Ω) using the Dempster-Shafer combination rule. The confidence degree m(conf) is close to 1. This means that this source is certainly reliable. The confidence level m(conf) positive (i.e. >0.5) of each expert helps to increase the final value of m(conf) and reduce the values of m(Ω) and m(non_conf). The pink curve shows the level of conflict between the experts, the final conflict is 0.0. The peak of conflict is 0.58.

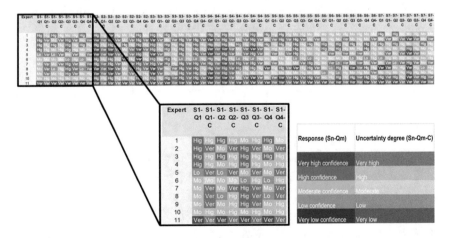

Fig. 2 Summary of answers in the form of color codes

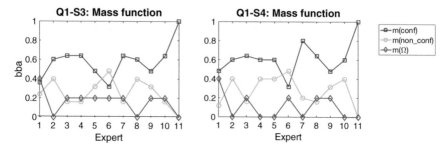

Fig. 3 Basic belief assignments (*bba*) related to data sources #S3 and #S4

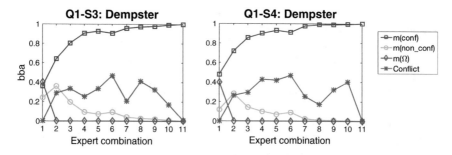

Fig. 4 Fused basic belief assignments (*bba*) related to data sources #S3 and #S4 using Dempster rule

The Fig. 5 shows the process of evolution of the values m(conf), m(non_conf) and m(Ω) by applying the Dubois and Prade combination rule. The confidences degree m(conf) final of the source #3 and #4 are 0.88 and 0.8 respectively, which means that these sources are certainly reliable, if we choose the selection threshold < 0.8. It is found that the expert number 11 contributes much to the increase of m(conf), because his degree of confidence m(conf) is 1.

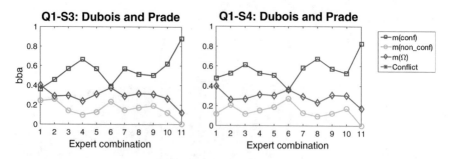

Fig. 5 Fused basic belief assignments (*bba*) related to data sources #S3 and #S4 using Dubois-Prade rule

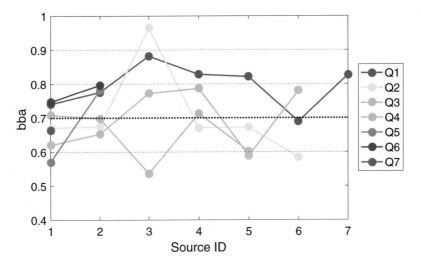

Fig. 6 Pignistic probability of the m(conf) using Dubois-Prade rule

Obtained results showed that the use of Demspter's rule allowed the fused reliability of each parameter to be converged at the highest level when several experts agreed and selected the highest level for their confidence (Fig. 4). The Dubois-Prade rule showed a more prudent behavior in the fused results (Fig. 5).

Based on the fused basic belief assignments, pignistic probabilities were computed (Fig. 6). Then, appropriate thresholds of reliability will be defined to include or exclude the data into other biomechanical models like multiphysics simulation, machine learning, etc.

4   Discussion

For the combination by the Dempster rule, the degree of reliability of all the sources converges to 1, this means that we have m(confidence) = 1, m(non_confidence) = 0, m(Ω) = 0. This is explained by the fact that most experts have chosen the answer "confidence" (i.e. ≥ moderate confidence). The relatively high number of responses (9–11 per source) also made the degree of ignorance close to zero. The combination by the Dempster rule allows us to solve and measure conflicts between experts. On the other hand, that of Dubois and Prade maintains them in the process of combination. For the combination by Dubois and Prade rule, the degree of reliability of all sources is greater than 0.69. If the selection threshold is selected at 0.55 (a reasonable threshold), all sources are considered reliable.

5 Conclusions

Obtained results showed that the use of Demspter rule allowed the fused reliability of each parameter to be converged at the highest level when several experts agreed and selected the highest level for their confidence. The Dubois-Prade rule showed a more prudent behavior in the fused results. In fact, these analyses allowed defining appropriate probabilistic thresholds to include or exclude the data into a decision-making model. This new method not only helps us to integrate expert opinions but also allows experts to express their answer with an uncertain level. It's useful in cases that we don't have enough information to confirm our answer. Luckily, this frequently happens in reality.

Acknowledgements The authors would like to thank all anonymous experts participating into the evaluation process.

Funding This work was carried out and funded in the framework of the Labex MS2T. It was supported by the French Government, through the program "Investments for the future" managed by the National Agency for Research (Reference ANR-11-IDEX-0004-02).

References

1. Aboal, J.R., Boquete, M.T., Carballeira, A., Casanova, A., Fernández, J.A.: Quantification of the overall measurement uncertainty associated with the passive moss biomonitoring technique: sample collection and processing. Environ. Pollut. **224**, 235–242 (2017)
2. Boone, I., Van der Stede, Y., Bollaerts, K., Messens, W., Mintiens, K.: Expert judgement in a risk assessment model for Salmonella spp. in pork: the performance of different weighting schemes. Prev. Vet. Med. **92**(3), 224–234 (2009)
3. Charles Osuagwu, C., Okafor, E.C.: Framework for eliciting knowledge for a medical laboratory diagnostic expert system. Expert Syst. Appl. **37**(7), 5009–5016 (2010)
4. Chatterjee, S., Bhattacharyya, M.: Judgment analysis of crowdsourced opinions using biclustering. Inf. Sci. **375**(1), 138–154 (2017)
5. Cobb, J.B.R., Shenoy, P.P.: On the plausibility transformation method for translating belief function models to probability models. Int. J. Approx. Reason. **41**(3), 314–330 (2006)
6. Dubois D., Prade, H.: Representation and combination of uncertainty with belief functions and possibility measures. Comput. Intell. (1988)
7. Hanea, D.M., Jagtman, H.M., van Alphen, L.L.M.M., Ale, B.J.M.: Quantitative and qualitative analysis of the expert and non-expert opinion in fire risk in buildings. Reliab. Eng. Syst. Saf. **95**(7), 729–741 (2010)
8. Jörg, E., Julia, H., Valentin, Q., Markus, T., Björn, R.: Biomechanical model based evaluation of Total Hip Arthroplasty therapy outcome. J. Orthop. **14**(4), 582–588 (2017)
9. Lev, V.U.: A method for processing the unreliable expert judgments about parameters of probability distributions. Eur. J. Oper. Res. **175**(1), 385–398 (2006)
10. Nicholas, T., Danielle, P., Nikhil, V.D., Robert, P.L.: Biomechanical analysis of gait waveform data: exploring differences between shod and barefoot running in habitually shod runners. Gait Posture **58**, 274–279 (2017)
11. Nicolas, R., Didier, P., Julie, C., Johanna, R., Raphael, Z.: Categorization of gait patterns in adults with cerebral palsy: a clustering approach. Gait Posture **39**(1), 235–240 (2014)

12. Pauk, J., Minta-Bielecka, K.: Gait patterns classification based on cluster and bicluster analysis. Biocybern. Biomed. Eng. **36**(2), 391–396 (2016)
13. Rustem, B., Robin Becker, G., Wolfgang, M.: Robust min–max portfolio strategies for rival forecast and risk scenarios. J. Econ. Dyn. Control **24**(11), 1591–1621 (2000)
14. Samuel, T.R., Alejandro, S.: Error correction in multi-fidelity molecular dynamics simulations using functional uncertainty quantification. J. Comput. Phys. **334**(1), 207–220 (2017)
15. Shafer, G.: A Mathematical Theory of Evidence. Princeton University Press (1976)
16. Skinner, D.J.C., Rocks, S.A., Pollard, S.J.T.: Where do uncertainties reside within environmental risk assessments? Expert opinion on uncertainty distributions for pesticide risks to surface water organisms. Sci. Total Environ. **572**, 23–33
17. Smets, P.: Data fusion in the transferable belief model. In: Proceedings of 3rd International Conference on Information Fusion, Paris, France, pp. 21–33 (2000)
18. Smets, P.: Belief functions: the disjunctive rule of combination and the generalized Bayesian theorem. Int. J. Approx. Reason. **9**(1), 1–35 (1993)
19. Wang, P., Ma, Z., Tian, Y.: Application of expert judgment method in the aircraft wiring risk assessment. Proc. Eng. **17**, 440–445 (2011)
20. Yun, Z., Norman, F., Martin, N.: Bayesian network approach to multinomial parameter learning using data and expert judgments. Int. J. Approx. Reason. **55**(5), 1252–1268 (2014)

Storing Hypergraph-Based Data Models in Non-hypergraph Data Storage

András Béleczki, Bálint Molnár and Bence Sarkadi-Nagy

Abstract The type of the data and especially the relationship among the data entities has changed in the last couple of years: beside the well-established relational data-model a new approach—the graph representation of the data and their connections—is becoming more and more common in the last couple of years. The graphs can describe large and complex networks—like social networks—but also capable of storing rich information about complex data. This was mostly of relational data-model trait before. This also can be achieved with the use of the knowledge representation tool called "hypergraphs". To check the power of this tool in practice we propose two tools for testing: the HypergraphDB which is focusing on the concrete hypergraph theory. The other solution will be SAP HANA in-memory database system which has a "Graph Core" engine besides the relational datamodel.

Keywords Data modeling · Hypergraph theory · Graph representation

1 Introduction

Designing and modeling complex Information Systems were never an easy task to complete. It is originated from many problems, but particularly the wide and heterogeneous domain the IS has to be designed for is the source in most case. Also, new data-type structures are gaining more and more ground in the organizations' life. The reason for this is automation of business processes and organization functions.

A. Béleczki · B. Molnár (✉) · B. Sarkadi-Nagy
Eötvös Loránd University, Budapest, Hungary
e-mail: molnarba@inf.elte.hu

A. Béleczki
e-mail: bearaai@inf.elte.hu

B. Sarkadi-Nagy
e-mail: bence.sarkadi@outlook.com

© Springer International Publishing AG, part of Springer Nature 2018   51
A. Sieminski et al. (eds.), *Modern Approaches for Intelligent Information
and Database Systems*, Studies in Computational Intelligence 769,
https://doi.org/10.1007/978-3-319-76081-0_5

Previously the interactive forms, the web pages, various kinds of semi-structured or structured documents were filled by "human hands". Since these tasks were rarely synchronized even inside the same departments the output of these documents often resulted in redundant data. This means a smaller set of new information and the simple knowledge is represented in way more complex structures than it should be.

An obvious solution would be some kind of automation of this process: if we have the opportunity to store the variables of these forms and documents in a heterogeneous environment there is a possibility the already specified variables can be reused. Despite this looking like a simple objective, creating an Information System with these goals can result in information loss.

The other approach is not a classic automatization process. The base concept is to find similarities between sets or sub-sets of already bounded variables [1] and map these values to each other. Thus the knowledge originated from only one set can be extended with the help of the other set. To accomplish the goals, the above-mentioned heterogeneous environment is also necessary. In previous papers [2–4] we mentioned the knowledge representational power of the hypergraphs. This paper focuses on the implementation of the hypergraphs in various system even if it does not support storing the structure of hypergraphs natively. To bridge this gap between the mathematical tool and the databases we propose to represent and store the hypergraphs as bipartite graphs. This opens up the opportunity to use basic graph-databases too besides the outdated HypergraphDB which is a purely hypergraph oriented database.

In Sect. 2 we take a brief literature review to understand the mindset behind the main idea, Sect. 3 describes the mathematical tools needed and Sect. 4 presents our solution. Section 5 will conclude the results of our systems and have an outlook about possible future progress.

2   Literature Review

The core conception of knowledge representation and data modeling is based on hypergraph theory. Bretto [5] gives a very detailed description of the whole hypergraph theory including the mathematical background, usabilities and also some of the representation.

Using a flexible tool to model business processes, datatypes, information, ontologies and much more could have a huge impact in the designing and developing phase of Information Systems, especially in the Enterprise Architecture, where the business processes and documents play important roles at the interface and at interaction levels. There are many frameworks already which help to understand the complexity of the Information Systems and its models. We examined three of them, namely Blokdijk's perception of Information Systems, the Zachman ontology and TOGAF [6–8].

The advantages of using hypergraphs and a unified designing framework are discussed in some of our previous papers [1–4].

3   Mathematical Background

Hypergraphs. There are several conceptual formalizations that are mentioned in other papers [2, 9] which can be described by a set of relationships from individual models (like UML-based class-diagram, work-flows, etc.). Since these models are representing different facets of perception of IS, and they represent a complex system through a set of complex, heterogeneous relationships. This set of relationships can be described by directed hypergraphs; the directed hypergraph applies the same basic notions as the generalized hypergraphs with the extension of direction. In this set we can separate the elements into two subsets:

- Hierarchical.
- Network-like relationships.

The hypergraphs because of their versatility can be used to represent complex data models, views and also their relationships [6, 10]. A detailed description of the definitions and opportunities about the generalized hypergraphs and their usage as Architecture Describing Hypergraphs can be found in [5, 9, 11, 12].

Graph representation. There are a lot of theorems about representing hypergraphs in many different particular ways. This case is also fitting hypergraphs. But in our case we need a specific approach to these representations: we need a specific way to depict a hypergraph in a way that it can be represented as a standard graph. The necessity behind this mindset is that this way we have a bigger chance to store a hypergraph in a common graph-database without any information loss. The most accepted data model is when the hypergraph represented as a bipartite graph. In this case the hyperedges are normal, labelled nodes. When a hyperedge is containing a node or another hyperedge, then these nodes are connected together. Figure 1 shows, how the same hypergraph can be represented in two different ways.

Fig. 1   The hypergraph on the right side is represented as a bipartite graph on the left side

4 Solution and Implementation

HypergraphDB is an extensible, portable, distributed open-source data-storage mechanism. It is a graph-database designed specifically for artificial intelligence and semantic web projects, however because of its general mindset, it is a perfect tool to represent heterogeneous relationships between different types too. There are key facts that are convincing enough to use the HypergraphDB as tool to store our model [13].

Since there aren't any first-party user interface for the HypergraphDB, the first step to start using it was to design and develop a middleware software which can create complete hypergraphs by creating the appropriate nodes and edges based on various input. These inputs can be mostly XML-based descriptors—like OWL—but can be also some custom, user-defined XML schema. From past labor works we created a tool—written in C++ using the Qt Framework—which is capable of designing Workflow Models based on Petri-nets. This tool generates a custom XML file consisting of the *places, transitions, arcs, flow relations, presets* and *offsets* of transitions and all other required data.

To test the capabilities of the database, we created a hypergraph based on a custom business process. This process was first designed in the above-mentioned tool, then the XML output was passed to the middleware. After the middleware finished the processing of the XML file it created the necessary nodes and the edges. Also utilizing the HypergraphDB efficiency the nodes and hyperedges can be labeled with custom JAVA classes, therefore the potential of the object-oriented class hierarchy is exploitable.

All of the above-mentioned things sound pretty promising at first sight but we ran into major drawbacks of the software later on. The first problems with HypergraphDB were revealed after we had a complete data model. We could not evaluate any results, we could not find similarities between documents and processes with the given default toolset. To solve these issues the first thing came in our minds was to fork the HypergraphDB repositories and implement the necessary missing features. However, the repositories were very outdated and we were barely motivated to develop something for this project. Also from its nature, the JAVA code was not robust enough so it would not be the proper choice to implement enormous graph datasets. So we had to find another tool to store hypergraphs.

SAP HANA and its Graph Core. The main points of the new solution were: the capability to handle big and complex relational and non-relational datasets, free-to-use license availability, and open-source if it is possible. Since the last point's requirement only met at HypergraphDB we continued to search for a solution outside the open-source domain.

One of today's most evolving and developing database platform is SAP's HANA platform. Beside its database capabilities, it also has a ton of services and libraries. It also has the opportunity to store graph-like datatypes with its Graph Core engine.

This allows the user to create hybrid data models, so the benefits of the two different worlds can be used simultaneously: the descriptive nature of relational data models and the relations from the graph data models [14].

4.1 Storing Hypergraph Model in Common Graph-Database

Abstract Data Structure Level. As of now, we know of four different ADS level representation of hypergraphs. Two of these are based on adjacency lists. Even though some algorithms run somewhat faster on these representations than on the other two (although not by any order of magnitudes), the list-based nature of them would make it difficult to implement them on relational databases and the resulting implementations would waste a lot—if not all—of the performance gain.

The other two ADSs are the incidence matrix and the so-called bipartite incidence structure. The incidence matrix allocates a matrix in which each row corresponds to a vertex, each column to a hyperarc and a nonzero $M[i,j]$ element of the matrix means that the jth hyperarc contains the ith vertex. This ADL has similar up- and downsides than adjacency matrix does with graphs, namely it uses a constant amount of memory for every hypergraph (even sparse ones), although it does not need to allocate any new memory segments during its runtime and will not have any redundancy.

The bipartite incidence structure (BPIS) uses a bipartite graph $G(A, B)$ where vertices in A are the vertices of the hypergraph, vertices in B representing the hyperarcs and there is an (a, b) arc $(a \in A, b \in B)$ when and only when b contains a. As one can see this is analogous to adjacency list when talking about graphs, that is it is well-suited for sparse graphs and has high redundancy on dense ones.

The latter two ADLs are perfectly reasonable choices however there are a few rationales supporting BPIS. First of all, we cannot forget that we are using SAP HANA to exploit its graph utilities. The implementation of the BPIS would result in such tables which are more easily interfaceable with those utilities and thus minimizing the unnecessary conversions.

Another reason is that real-world examples of data models are usually sparse graphs. And lastly, we considered future researches regarding the transition of the most used graph algorithms to this hypergraph representation for which an easily conceptualized model is preferable.

According to these advantages, we chose bipartite incidence structure as the ADL level representation.

Implementation Level. To represent the bipartite incidence structure we use two tables for each stored hypergraphs. The first table stores the vertices of the ADS. It has three columns; *globalId, type, userId* where *globalId* is a unique id in the table (i.e. among the vertices), *type* is a binary field for describing whether the given vertex in $G(A, B)$ was a vertex or a hyperarc originally and *userId* is a user-provided id by which he or she will be able to identify the elements of any result tables of the framework. It is important to emphasize that a vertex and a hyperarc can have the same *userId* although that does not necessarily describe any relations between them.

The second table stores the values associated with arcs in the representational graph, for which we used the columns *arcId*, *source*, *target* and *cost*. Here *arcId* is the unique identifier of each representational arc, source is the *globalId* of the starting vertex of the arc (this makes it a foreign key), target is similarly the endpoint of the arc while cost is the value of a cost function which we calculate from a user-provided one in a way described in the next paragraph.

To create a new hypergraph one has to provide all the (a, b) pairs in the hypergraph where a is a vertex and b is a hyperarc, or b is null (in which case a can appear in the provided table only once) and also another table storing a cost function c on the hyperarcs. During the building of the representational graph, we use a different cost function $c_{G(A,B)}$ on the representational graph where all the in-arcs of the hyperarcs have a value of 0 and all the out-arcs have the value of the original cost function at the hyperarc. This derived cost function has the nice property of having the same sum on every valid path (i.e. starting and ending in A) as the original cost function would have. This implementation will result in a framework which is highly general and is not limited to data modeling.

The Interface of the Framework. We have already mentioned that our design is built upon the idea that the user of the framework can identify any vertex or hyperarc appearing in any output of the framework. To associate any additional data or label (in our case data modelling attributes) one has to store and manage those separately from the framework (to make these tasks easier the framework provides views which we will cover later).

The provided functionality consists procedures for creating, modifying and deleting hypergraphs and views as replacements for better functions and to track changes in the represented hypergraphs by triggers.

There are interface level guarantees for the consistency of the representation, that is any direct operation on the tables may corrupt the overall structure of the representation, however none of the procedures can do that (in these case exceptions are raised).

Procedures

Procedure name	Description
create_hypergraph	Creates a new directed hypergraph representation based on the input sets
add_items	Adds new hyperarcs and/or vertices to the hypergraph and/or vertices to hyperarcs
modify_costs	Modifies the cost function of the graph where it differs from the provided one
delete_hyperarcs	Deletes the specified hyperarcs
delete_vertices	Deletes the specified vertices
delete_graph	Deletes the whole graph (i.e. all associated tables, views and entries)

Most of these are very straightforward, however there are a few things that need clarifications. The ubiquitous input, name is a unique—and hopefully meaningful—identifier for each hypergraphs. Any tables associated with a given hypergraph is stored in a distinct workspace identified by the same name. During its lifetime one can always refer to a hypergraph by this identifier and can check its existence by querying the meta table, where we store an entry for each of them.

The input tables in regard to *add_items* and *create_dihypergraph* have several constraints between them. The *vertexId* and *hyperarcId* are the user provided identifiers for these structural elements. A (*vertexId, hyperarcId*) pair in the sets table means that the referenced vertex is in either the head or the tail set of the implied hyperarc. To decide between these sets we use the binary *setType* column. When a vertex is in none of the hyperarcs then it is paired to the null value and in this case it can only appear only once in the table (otherwise it can show up multiple times as it can be part of multiple subsets). The *hyperarcCost* table comprises the cost function for which it has to contain all the hyperarcs existing in the sets table and none other (except for the null value).

One might wonder why there is—in contrary to the delete family of operations—a standalone *add_items* procedure instead of separating it into *add_vertices* and *add_hyperarcs*. To keep consistency during the operations we have to make sure that there is no point in time when any tail or head set of any hyperarc is empty (otherwise the underlying hypergraph would be invalid by definition). To ensure this property the interface has to enforce the user to provide a starting set for each new edge (during both *create_hypergraph* and *add_items*) and similarly has to prohibit any attempt (i.e. *delete_nodes* call) to delete all of the elements of any head or tail set. For reference we provide high level pseudocodes for the more complex procedure below.

input: - name
       - TABLE *sets(vertexId, hyperarcId, direction)*
       - TABLE *hyperarcCost(hyperarcId, cost)*
Result: Creates and populates representational tables and registers it in the *meta* table
validate inputs;
create workspace *name*;
create table *representationVertices (globalId, userId, type)*;
create table *representationArcs (id, source, target, cost)*;
add name to *meta* table;
 add_items(name, sets, hyperarcCost);
create view *vertices (vertexIds* in representationVertices where *type* = 'vertex');
create view *hyperarcs (vertexIds* in representationVertices where *type* = 'hyperarc');
create view *tailSets ((vertexId, hyperarcId)* where vertex is in the tail set of the hyperarc);
create view *headSets ((vertexId, hyperarcId)* where vertex is in the head set of the hyperarc);
Algorithm 1: *create_hypergraph*

Views are used to serve as the primary means to get information about the hypergraph and to be used as base for triggers which maintain the user-managed label tables. They are coupled with the underlying tables so they always contain

up-to-date information and do not copy their data. The following are created together
with each hypergraph representation:

vertices	Containing all vertexIds in the hypergraph
hyperarcs	Containing all hyperarcIds in the hypergraph
tailSets	Containing all (v, a) pairs where v is in the tail set of a
headSets	Containing all (v, a) pairs where v is in the head set of a

input: - name
      - TABLE *newSets(vertexId, hyperarcId, setType)*
      - TABLE *newHyperarcCost(hyperarcId, cost)*
Result: Adds new hyperarcs and/or vertices to the hypergraph and/or vertices to hyperarcs
validate inputs;
for *vertexId in newSets* **do**
    if *vertexId not in vertices* **then**
        insert (new *globalId, vertexId,* 'vertex') into *representationVertices*;
    end
end
for *hyperarcId in newHyperarcCost* **do**
    if *hyperarcId not in hyperarcs* **then**
        insert (new *globalId, hyperarcId,* 'hyperarc') into *representationVertices*;
    end
end
for *row in newSets* **do**
    if *row.hyperarcId is not null* **then**
        if *row.setType is 'head'* **then**
            insert (new *id, globalId* of row.*hyperarcId, globalId* of row.*vertexId,* cost of
            row.*hyperarcId*) into *representationArcs*;
        else
            insert (new *id, globalId* of row.*vertexId, globalId* of row.*hyperarcId,* 0) into
            representationArcs;
        end
    end
end

Algorithm 2: add_items

5   Conclusion and Future Work

With the ability to use the knowledge from our previously developed framework in
an up-to-date, versatile system creates a large room to develop the solution. Using
this new hybrid model let us exploit the richness of a hypergraph model alongside
the built-in libraries and high performance of SAP HANA.

However, the core of our new framework is not nearly completed. The graph algorithms which are shipped with the database default are simply not fully capable to handle hypergraph models. So in the near future, our first steps will be extension and implementation of the set of these methods (BFS, DFS, Dijkstra, strong components, nearest neighbors, etc.).

When the low-level implementations of the data models are finished, we can focus more on the usage of the hypergraphs again: for document modeling, process modeling, ontologies, and semantic web.

Acknowledgements This work was supported by European Commission [grant number EFOP-3.6.3-VEKOP-16].

References

1. Molnár, B., Benczúr, A., Béleczki, A.: A model for analysis and design of information systems based on a document centric approach. In: Intelligent Information and Database Systems (IIDS), pp. 290–299. Springer, Berlin (2016)
2. Molnár, B.: Applications of hypergraphs in informatics: a survey and opportunities for research. Ann. Univ. Sci. Budapest. Sect. Comput. **42**, 261–282 (2014)
3. Molnár, B., Tarcsi, A.: Architecture and system design issues of contemporary web-based information systems. In: Proceedings of the 5th International Conference on Software, Knowledge Information, Industrial Management and Applications (SKIMA 2011), pp. 8–11. Benevento, Italy, Sept 2011
4. Molnár, B., Benczúr, A.: Facet of modeling web information systems from a document-centric view. Int. J. Web Portals (IJWP), **5**(4), 57–70 (2013). (IGI Global)
5. Bretto, A.: Hypergraph Theory: an introduction. Springer (2013)
6. Zachman, J.A.: A framework for information systems architecture. IBM Syst. J. **26**(3), 276–292 (1987)
7. Blokdijk, A., Blokdijk, P · Planning and Design of Information Systems. Academic Press, London (1987)
8. Ausiello, G., Franciosa, P. G., & Frigioni, D. 2001. Directed hypergraphs: Problems, algorithmic results, and a novel decremental approach, in: *Theoretical Computer Science* pp. 312–328, Springer Berlin Heidelberg
9. Open Group, TOGAF: The Open Group Architecture Framework, TOGAF®Version 9 (2010). http://www.opengroup.org/togaf/
10. Gallo, G., Longo, G., Pallottino, S., Nguyen, S.: Directed hypergraphs and applications. Discret. appl. math. **42**(2), 177–201 (1993)
11. Iordanov, B.: Hypergraphdb: a generalized graph database. In: Web-Age Information Management, pp. 25–36. Springer, Berlin, Heidelberg (2010)
12. Molnár B., Benczúr A., Béleczki A.: Formal approach to modelling of modern information systems. Int. J. Inf. Syst. Proj. Manag. (2016) (to be published)
13. Kobrix Software.: HypergraphDB—A Graph Database. [ONLINE] http://hypergraphdb.org (2010). Accessed 27 May 2016
14. Rudolf, M., Paradies, M., Bornhövd, C., Lehner, W.: The graph story of the sap hana database. In: BTW, pp. 403–420 (2013)

Part II
Natural Language Processing and Text Mining

A Fuzzy Logic Approach to Predict the Popularity of a Presidential Candidate

Pritom Mazumder, Navid Anjum Chowdhury, Moh. Anwar-Ul-Azim Bhuiya, Shabbir Haque Akash and Rashedur M. Rahman

Abstract We are noticing a new era of social networks where in a blink of eye millions of tweets about any topic can be emerged. Especially, when an event like national election comes for a nation, the messages in social media especially twitter rises at its peak. The amount of data twitter has during that time is enormous and those tweets were never been used to analyze anyone's popularity. Our work is focused on predicting a presidential candidate's live popularity through sentiment analysis. We design the system to predict the popularity by a single day. To do this several features from tweets of a particular day have been passed through a dimensionality reduction algorithm, e.g., PCA (Principal Component Analysis). Consequently, the PCA components have been exercised into a fuzzy system. In particular, we used ANFIS (Adaptive Neuro Fuzzy Inference System) to predict a presidential candidate's popularity on a single day.

Keywords Fuzzy logic · Popularity prediction · Presidential candidate
Sentiment analysis · Feature scaling · Principal Component Analysis (PCA)
Adaptive Neuro Fuzzy Inference System (ANFIS)

P. Mazumder · N. A. Chowdhury · Moh. Anwar-Ul-Azim Bhuiya · S. H. Akash
R. M. Rahman (✉)
Department of Electrical and Computer Engineering, North South University, Plot-15,
Block-B, Bashundhara Residential Area, Dhaka, Bangladesh
e-mail: rashedur.rahman@northsouth.edu; rashedurrahman@yahoo.com

P. Mazumder
e-mail: pritom169@outlook.com

N. A. Chowdhury
e-mail: navid.rashik@northsouth.edu

Moh. Anwar-Ul-Azim Bhuiya
e-mail: anwar.bhuiyan@northsouth.edu

S. H. Akash
e-mail: shabbier.akash@northsouth.edu

© Springer International Publishing AG, part of Springer Nature 2018
A. Sieminski et al. (eds.), *Modern Approaches for Intelligent Information
and Database Systems*, Studies in Computational Intelligence 769,
https://doi.org/10.1007/978-3-319-76081-0_6

63

1 Introduction

Since the dawn of technological boom, the electronic devices has reached to millions of hands and a good percentage of time those devices are used in social media applications, like Facebook, Tweeter etc. Social media becomes an emerging public emotion container where billions of people can express their emotion or opinion by a single tap and share it with the world. The usage of social media is radically increased when a special national occasion or event happens and people show their excitement about that event. National election is undoubtedly one of the most anticipated events for a nation when people express their opinion on social media. A candidate who is fighting for the presidential post will value the public opinion about him on social media. It follows that, any presidential candidate will be concerned about his position in people's mind and would love to know how popular he is among people. Our work is to generate a presidential candidate's popularity on any particular day based on some attributes. This will help the candidate to change his campaign policies on the basis of his popularity on any single day.

As good number social media applications exist and huge number of people is interacting on a social media platform, sentiment analysis on every social media would be incredibly hard. The dissimilation and the fluctuations between different social media platforms would make any system unstable and difficult to predict anything. To give a solution, we use sentiment analysis only on twitter and develop a fuzzy system which gives results on the basis of millions of tweets. In our approach, we count the number of tweets, retweets, likes, comments and many other factors to generate the attributes. After selecting the attributes, we have reduced the dimensions so that the fuzzy system becomes more efficient. To reduce the dimensions, we applied PCA (Principle Component Analysis). We then passed the PCA components into the Adaptive Neuro Fuzzy Inference System (ANFIS) to generate the output which is the popularity of a presidential candidate on a single day.

2 Related Works

The authors in [1] have worked on predicting popularity of entities (politicians) using LDA SVA and other logistic regression optimization tool on Twitter based on the news cycle. The authors in [2] have used the multi-variable time-series approach to predict poll trends of 2016 US presidential campaign using time-series model, LTS, and averaging user positive and negative tweets. The authors in [3] have performed sentiment analysis on twitter data from 2016 Spanish General Election using Tweetinvi API. The authors in [4] did sentiment analysis with SentiWord and a natural language processor SNLP to predict popularity of the IPhone6 and its features. In the paper [5] the authors predicted trending and non-trending topics on twitter using machine learning model to identify features: like, lifetime, tweet count, user count, velocity and acceleration. Authors in [6] used Twitter API with python and analyzed it with Sentiment analysis to predict the business popularity of certain cosmetic products. In [7] the authors have worked on two issues regarding Twitter trend prediction using ARF, RF and LDS. In this work, the authors [8] have taken six parameters, e.g., specific changes, stemming length, total delays, hole diameter, spacing and analyzed them with two soft computing based models, Support Vector Machine (SVM) and Adaptive Neuro Fuzzy Inference System (ANFIS) and compared with Kuz-Ram method. The authors in [9] used two dimensional principle component analysis (2D-PCA) for feature extraction and vectors were applied to Adaptive Neuro Fuzzy Inference System (ANFIS) to get a 97.1% classification accuracy in predicting face recognition. Authors in [10] used multivariate image analysis through quantitative structure toxicity relation (MIA-QSTR) method coupled with System PCA-ANFIS to assess the toxicity of esters to Daphnia Magna. The authors of this paper [11] implemented an ANFIS based approach to estimate prices of residential properties.

3   Data Set Descriptions

Our main objective is to predict the live popularity of a presidential candidate before election. To do that, we have collected the tweets about election from Harvard Dataverse [12]. In brief, we have data of 101 days before US election and total number of tweets in our entire dataset is approximately 228 million. After the Tweet IDs have been collected, we have downloaded the tweets from twitter using Twarc Library and Twitter API. When the tweets are downloaded, a dictionary is created for arranging the attributes on a day basis. A python code is written to check the attributes from the tweets and if the tweets' conditions are filled according to our written codes, the values for every attribute are sorted according to a particular day. In addition to that, there are some attributes which show how much positive or negative any particular tweet is, for a candidate. Therefore, we use sentiment analysis to judge whether a tweet is positive or negative and the sentiment analysis score has been added to the dictionary for a particular day. The workflow of our data collection has been represented in Fig. 1 which will display how the data collection and other processes have been carried out. As we are predicting the live popularity of a presidential candidate, we need to choose a candidate. Without any specific candidate we cannot train our system and also it is not possible to validate the result. As a consequence, we elect Donald Trump for our work and the motivation to select him is that he was a candidate of the most recent of US Presidential Election. In addition to that, Donald Trump was also a famous candidate in the last US Election and he is also very active on twitter. For his popularity, people tweeted about him and showed their opinion about him as a candidate. That is what we need for prediction. Consequently, we work with 19 attributes for predicting the popularity of a presidential candidate. The 19 attributes for our work have been explained below:

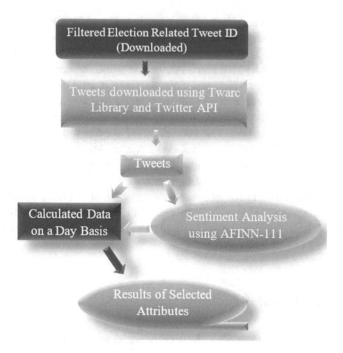

Fig. 1 Architecture of the data collection process

Number of Tweets: It represents total number of tweets about election on that day. As Trump was a popular candidate, tweets about election will definitely affect his popularity.

Tweets about Candidate with URL: It conveys total number of tweets that contains URL and also the word Trump in it for a particular day. As we all know when a tweet has an URL, people tend to show more curiosity about that tweet. We searched for those tweets which has the word 'Trump' and also an URL in it.

Hashtags per Tweet: This attribute illustrates the Number of hashtags in per tweet about trump. Hashtags tends to represent the summary of the whole tweet and sometimes the hashtags are so powerful to represent emotion. Some tweets only contain a single or multiple hashtags. In addition to that, positive hashtags like #Trump2016, #TrumpForPresident and negative hashtags like #AntiTrump can immensely impact on his popularity.

Any State in USA: It represents how many times any tweet was tweeted using location of any state. The reason behind electing this attribute is that, when a name of state is given, we can analyze the candidate's position on that particular state. It also represents how well a candidate like Trump is doing during his campaign on that particular state.

Presidential Debate: Any presidential debate on that day? Yes or no. Yes was represented by 1 and No was represented by 0.

Retweet of Candidate's official tweets: Total number of times tweets from Trump's official ID (@realDonaldTrump) was retweeted on any particular day. As retweet generally shows strong support or strong objection about the candidate.

Followers: This attribute illustrates total number of followers till that day following Trump's official Twitter ID (@realDonaldTrump). The more followers a person has, the more popular he is among the people.

Likes: Total number of likes on Trump's tweets on that day will definitely indicate a portion of people supporting Donald Trump, as likes are an indication of support.

Political Parties' name: It shows total number of times the words 'Republican', 'Republicans' or 'Republic Party' were mentioned across the twitter. Generally, a candidate's popularity insanely depends on his Political Party and Donald Trump is no exception.

Candidate's Name: Candidate's Name represents total number of times Trump's name was mentioned across the twitter on that day. It is a clear sign of popularity on any particular day.

Political Party and Candidate's Name: It narrates total number of times "Republicans" and "Trump" was mentioned together across the twitter on that day. Mentioning the political party and candidate's name together is a clear sign of a specific opinion which could impact the popularity.

Total Retweets about Candidate: It represents total number of times tweets about Trump were retweeted across the twitter on that particular day. Retweeting someone's tweet who has tweeted about a candidate definitely influences his popularity.

Total Emoji: This attribute represents total types of emoji used in particular tweet and then added to generate total emoji. As emoji is a popular way to express our emotions, analyzing it will help us to understand the popularity.

Positive Tweets: It shows the number of positive tweets about trump on that day. For getting the value of this attribute we did sentiment analysis using a library AFINN-111. If the total sentiment value of a tweet remains 0 or more than 0, that particular tweet was regarded as a positive tweet. The positive tweets will definitely increase popularity.

Negative Tweets: It shows total number of negative tweets on that day about trump. Same as before, we got the value of this attribute using sentiment analysis and the library AFINN-111. If the total sentiment value of the tweet is less than 0, that particular tweet was regarded as negative tweet. Same as positive tweets, negative also affects popularity.

Official Twitter ID Mention: Total number of times Trump's official ID (@realDonaldTrump) was mentioned across the Twitter on that day. If any tweet mentions Trump's official ID, that will indeed impact Trump's popularity.

Vice President Mention: It illustrates total number of times Trump's Vice President (Mike Pence) was mentioned across the twitter. If the Vice Presidential Candidate is popular that will definitely increase the Popularity of Presidential Candidate.

Table 1 Sample data set of a single day

Variables	Values	Variables	Values
Number of Tweets	2066966	Political Parties' name	6030
Tweets about Candidate with URL	73320	Candidate's Name	484689
Hashtags per Tweet	1.0527	Political Party and Candidate's Name	2906
Any state in USA	42192	Total Retweets about Candidate	379873
Presidential Debate	0	Total Emoji	2622138286
Retweet of Candidate's official tweet	51299	Positive Tweets	423221
Followers	35508692	Negative Tweets	470222
Likes	121289	Official Twitter ID Mention	325301
Number of Candidates Tweet	1	Vice President Mention	1360
Number of Users Tweeted about Candidate	272359		

Number of Candidate's Tweet: How many times candidate tweeted on any particular day is represented by this attribute. A person's popularity enormously depends how much he is active on social media and the same goes for Donald Trump.

Number of Users Tweeted about Candidate: Total number of users tweeted on that particular day about Trump and it is needed to generate the estimation rules, which is the objective of our research. As for our work, it represents how many people talked about Donald Trump on any particular day and the estimation result will be generated based on this attribute.

Among 19 attributes, 17 of them have been found directly by checking the tweets and if the condition matched for that attribute, the value was added in the dictionary for a particular day. Remaining two attributes, positive tweets and negative tweets have been done by sentiment analysis using AFINN-111. A sample value for every attribute has been given in Table 1.

4   Methodology

Our popularity prediction system has been partitioned into three different parts named as, Feature Scaling, PCA (Principle Component Analysis) and ANFIS (Adaptive Neuro Fuzzy Inference System). The architecture has been given in Fig. 2.

A. Feature Scaling

Feature Scaling is a formula, which is used to scale a data, when different attributes have high differences in their respective values. For example, the number

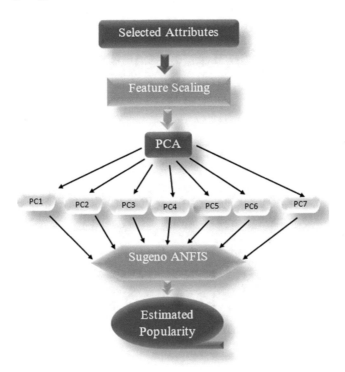

Fig. 2 Architecture of the prediction system

of tweets on a particular day is around 1.5 million and some other day the same number is around 3.5 million.

So the difference of the values between the same attribute for different days is enormous and it is not very harmonious for a system to generate result with such differences. To give a solution we came out with Feature Scaling. For feature scaling, we used the famous feature scaling formula mentioned in Fig. 3. After feature scaling has been done the different values of different attributes stays between −2 and +2. However, Feature Scaling has not been done on the output result which is 'Number of user tweeted about Candidate'. As it turns out it would create even more complexity to go back from scaled value to the authentic value.

B. Principal Component Analysis (PCA)

PCA is a dimensionality reduction algorithm, which is widely used in Machine Learning, Data Mining, and Fuzzy Theory. Except the output value, we have 18 attributes and not all 18 attributes are equally responsible for generating estimated popularity of the candidate. We want to take those variables, which are most responsible for generating the output.

We considered data of 101 days in our research. Among all of the days, we divided our data into 2 parts containing 80 days for training the system and 21 days

$$x' = \frac{x - \min(x)}{\max(x) - \min(x)}$$

Fig. 3 Feature scaling formula

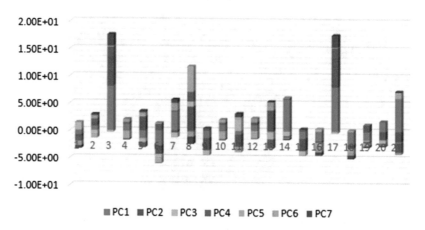

PCA contribution for the test set

Fig. 4 Stacked column chart of PCA attributes

for testing the system. According to the calculated variances, the 1st 7 variables of PCA contain 92.3% of total variance of the whole variance. Figure 4 illustrates 21 column charts of 7 PCA variables which basically represent the test set, where PC1, PC2, PC3, PC4, PC5, PC6 and PC7 have been represented by different colors.

C. Adaptive Neuro Fuzzy Inference System (ANFIS)

ANFIS has been widely used for prediction system for years and as we are predicting a presidential candidate's popularity, we also took ANFIS in our fuzzy approach to predict the popularity. However, there are multiple variants of ANFIS used for predictions and according to the nature of our data; the Sugeno ANFIS suits the most. 7 PCA variables have been passed through the ANFIS and the corresponding result on that particular day is generated. For the prediction using PCA variables, we used MATLAB Neuro Fuzzy Toolbox which is incredibly effective for generating predictive value. After generating the training data and implementing the FIS, the 128 rules based ANFIS structure is generated and the generated structure has been shown on Fig. 5a.

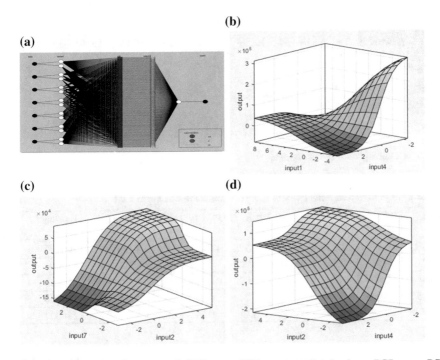

Fig. 5 a Generated Anfis structure. **b** PC1 versus PC4 versus predicted value. **c** PC7 versus PC2 versus predicted value. **d** PC2 versus PC4 versus predicted value

The generated surface graphs based on estimated popularity have been given in Fig. 5b–d where the graphs were generated on the basis of PC1, PC2, PC3, PC4, PC5, PC6, PC7 and output.

5 Results and Analysis of Results

Among all the days, we divided our dataset into two parts, which is 80% for training set and 20% for the test set. In other words, 80 days was included into training set and 21 days were considered as test set. We have done PCA on both training and testing data with 92.3% variance accomplished by 7 PCA variables. PCA values of the training set have been given into ANFIS system and the error of the training data was matched with the FIS output, from the Fig. 6a we can clearly visualize that there was hardly any error. Apart from that, we have trained our training set using 1000 epoch. But, when the system was put on a test set, the system shows more error as what we can visualize from Fig. 6b. There can be several reasons. One of the reason is the number of days in the training set was not huge. One solution

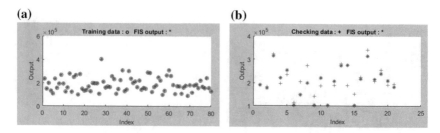

Fig. 6 a Training data. **b** Testing data

Table 2 PCA variables, actual values and predicted values of the test set

PC1	PC2	PC3	PC4	PC5	PC6	PC7	Actual value	Predicted value
−0.831	−0.363	1.398	−0.831	−0.325	−0.471	−0.493	190137	190000
0.771	0.584	−0.687	0.714	−0.705	0.587	0.258	183266	177000
8.041	5.434	−0.303	1.783	0.072	0.115	2.132	322918	313000
1.068	−0.061	−0.157	0.380	−1.279	0.570	−0.231	196096	225000
−1.448	2.476	−0.731	−0.916	0.441	0.343	0.278	235909	254000

could be that we could have designed our system based on hourly data but that would have shown much fluctuations. Another reason the error is high on the test set is that the way we have collected our data. As mentioned before, we have collected our data from Harvard archive and the ratio of the total number of tweets was not same for 101 days. That leads to an unbalanced number of people who tweeted about Trump. There could be another way we could have collected our data, which is manually downloading twitter feeds from twitter. But twitter does not allow downloading past feeds after a specific amount of time. In addition to that, the number of tweet was download in the permitted time was around 300–400 tweets for a day which was no way near to the number of tweets we needed to conduct in our study.

After calculating %RMSE (Root Means Square Error) we came up with 16.97% error which leads to 83.03% accuracy. As, from Table 2, we can visualize that, for different values of PCs, the real popularity is 190137 and the estimated popularity is 190000 which is almost equal. All the actual values and predicted values from the test set which has been represented in Table 2, also has been represented respectively in Fig. 7. The actual value was represented with blue bars and the predicted value was represented with the red bars.

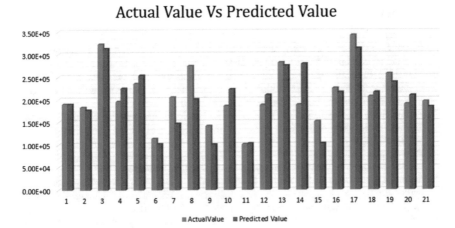

Fig. 7 Actual value versus predicted value for all 21 days from test set

6 Conclusions and Future Works

Downloading 228 millions of tweets, working with them and sorting them based on our attributes for a single day was indeed one of the biggest challenges. When there is massive dataset and prediction work has been done on that dataset, a simple mistake on choosing the attributes can have effect on the prediction system. As no work has been done before on predicting the popularity of a presidential candidate, choosing the attributes was one of the main challenges. Apart from that, implementing the appropriate fuzzy system on the basis of those attributes, choosing the membership functions and fuzzy rules were undoubtedly the biggest challenges of our work. How the enormous data from twitter can be used to predict someone's popularity, our work is a perfect example for that. In addition to that, there was no work has been done prior to us to predict someone's personality using ANFIS and we took it to another level by predicting the popularity of a presidential candidate. As it took enormous amount of time for compressing the data for just only one candidate, we could not verify our prediction system for multiple candidates. In our future work, we plan to use our current prediction system for predicting multiple candidates' popularity and will compare the result among different candidates. In addition to that, our system is only based on one prediction formula which is ANFIS. Our future implementation will have more fuzzy and machine learning approaches for predicting a candidate's popularity.

References

1. Pedro, S., Carlos, S.: Learning from the news: predicting entity popularity on twitter. In: International Symposium on Intelligent Data Analysis (IDA), pp. 171–182 (2016)
2. Tom, M., Shoumik, R., Fang, Z., Zoran, O.: Predicting poll trends using twitter and multivariate time-series classification. In: International Conference on Social Informatics SocInfo 2016: Social Informatics, pp. 273–289 (2016)
3. Prabhsimran, S., Sawhney, R.S., Kahlon, K.S.: Predicting the outcome of Spanish general elections 2016 using twitter as a tool. In: Advanced Informatics for Computing Research, pp. 73–83 (2017)
4. Hridoy, S.A., Ekram, M.T., Islam., M.S., Faysal, A., Rahman, R.M.: Localized twitter opinion mining using sentiment analysis. Decis. Anal. **2**, 8 (2015)
5. Anubrata, D., Moumita, R., Soumi, D., Saptarshi, G., Das, A.K.: Predicting trends in the twitter social network: a machine learning approach. In: International Conference on Swarm, Evolutionary, and Memetic Computing SEMCCO, pp. 570–581 (2015)
6. Pajaree, Y., Sukanlaya, L., Bundit, T., Ne, P.: Business popularity analysis from twitter. In: International Conference on Computing and Information Technology IC2IT 2017, pp. 337–348 (2017)
7. Peng, Z., Xufei, W., Baoxin, L.: Evaluating important factors and effective models for twitter trend prediction. In: Online Social Media Analysis and Visualization, pp. 81–98 (2015). https://doi.org/10.1007/978-3-319-13590-8_5
8. Mohammad, E., Alireaza, S., Carsten, D., Maliheh, A., Bazzazi, A.A.: Application of PCA, SVR, and ANFIS for modeling of rock fragmentation. Arab. J. Geosci. **8**(9), 6881–6893 (2014)
9. Hitesh, S., Rahul, K., Ketan, P.: Face recognition using 2DPCA and ANFIS classifier. In: Proceedings of Fourth International Conference on Soft Computing for Problem Solving, pp. 1–12 (2015)
10. Asadollahi-Baboli, M.: Aquatic toxicity assessment of esters towards the Daphnia magna through PCA-ANFIS. Bull. Environ. Contam. Toxicol. **91**(4), 450–454 (2013)
11. Jian, G., Jozef, Z., Levitan, A.S.: An adaptive neuro-fuzzy inference system based approach to real estate property assessment. Real Estate Res. (JRER) **30**(4), 396–421 (2008)
12. Harvard Dataverse: http://pages.jh.edu/jrer/papers/pdf/past/vol30n04/01.395_422.pdf. Accessed 12 Dec 2017

DNA Sequences Representation Derived from Discrete Wavelet Transformation for Text Similarity Recognition

Phan Hieu Ho, Ngoc Anh Thi Nguyen and Trung Hung Vo

Abstract Recognizing text similarity, also known as duplicated documents, is considered as the most important solution for plagiarism detection which is a rising dramatically in the era of digital revolution recently. With the aim to contribute an efficient plagiarism system, we investigate a new approach for in text similarity mining via DNA sequences representation derived from Discrete Wavelet Transformation (DWT). Consequently, the contribution of the paper is classified as threefold. Firstly, we convert the raw source materials into a unique set of floating-number series called a DeoxyriboNucleic Acid (DNA) sequences using DWT. The DNA-based structure then is also required for the testing documents input at the second step. Lastly, text similarity discovery algorithm is performed for those given input DNA strings via computing the Euclidean distance. The experimental result demonstrates the advantages of the proposed method with very high precision for detecting text similarity on standard dataset of PAN, known as Plagiarism Analysis, Authorship Identification, and Near-Duplicate detection.

Keywords Text similarity · Discrete Wavelet Transformation
Text analysis and mining · Plagiarism system · Euclidean measurement

1 Introduction

Text similarity mining has been extensively considered as one of the most well known problems in the field of data mining, exactly due to its prevalence in many real high-impact text research and applications such as information retrieval,

P. H. Ho (✉) · N. A. T. Nguyen · T. H. Vo (✉)
The University of Danang, 41 Leduan St., Danang City, Vietnam
e-mail: hophanhieu@ac.udn.vn

T. H. Vo
e-mail: vthung@dut.udn.vn

N. A. T. Nguyen
e-mail: ngocanhnt@ued.udn.vn

© Springer International Publishing AG, part of Springer Nature 2018   75
A. Sieminski et al. (eds.), *Modern Approaches for Intelligent Information and Database Systems*, Studies in Computational Intelligence 769,
https://doi.org/10.1007/978-3-319-76081-0_7

text summarization, supervised and unsupervised learning, topic tracking and many more [1, 2]. As the evolved digital age, the large of digitalized sources such as free online libraries or social media posts are wide-spread available in the internet. This issue has reinforced the need of text similarity detection since the positive effects of the digital revolution lead to the negative impact; especially, in the academic institutions area whereby documents are easier to be plagiarized by students [3–5]. Therefore, one of prevention is to setup the anti-duplicated text system to detect the student's works having similar parts with others. The automatic text similarity detection not only helps educators in academic and education environment reduce illegal copying and distribution of information but also helps students to check and improve their work.

Typically, the function of measuring text similarity will be assigned a real number in range from 0 to 1. The zero value denotes the documents are completely dissimilar with others. In contrast, the one value presents that the documents are identical well [6]. Therefore, to detect the similarity of the documents, we need to calculate their similarity measures. Many researchers have been proposed different techniques for measuring the similarity between documents such as Brute-Force, Morris-Pratt, Knuth-Morris-Pratt (KMP), Boyer-Moore, Karp-Rabin, Horspool, … [7]. One of the well-known approaches is vector-based mathematical models that are used for computing the similarity in document. As such, all of the features formulated in the documents are performed under vector-based models via measures of Euclidean distance, Cosine similarity, Jaccard, Dice, … [8]. Another popular alternative measuring approaches are finding similarity based on lexical and semantic, database-based, knowledge-based approaches, and natural language processing [9, 10].

Since text similarity analysis strongly relying on data representation whereby a proper mathematical modeling is one of prime concern in order to boost performance and computing processing. It means that a raw document under string-based representation should be converted into other structures in order to facility text analysis. As mentioned above, the most popular model is vector-based approach that represent document under formulation of the features of terms/words and then the weight of these features are calculated via TF-IDF measurement [11].

In this paper, we propose a new representation for text similarity mining based on the idea of converting source document into a unique set of floating-number series, called a DeoxyriboNucleic Acid (DNA) sequences by applying DWT technique [12]. Consequently, text similarity detection will be carried out via term of detecting similarity in DNA sequences. To the best of our knowledge, there are no literature researches that have applied DWT for textual document similarity detection. Therefore, this paper has original and novelty characteristic. Our method is conducted via three steps as following: (1) Transformation: the available materials of document source are converted into a set of the unique floating-number sequences, named as source DNAs, which are derived by applying DWT; (2) Testing input representation: to check the similarity for an arbitrary document, we also apply DWT to derive its own DNAs to which the smallest Euclidean distances from the source DNAs are computed; (3) Similarity detection:

As compared to a threshold level, the values of these distances indicate whether any piece of the checked document is duplicated from another source. Our simulation is perform on real standard dataset of PAN in order to corroborates the advantages of the proposed method with very high precision for detecting text similarity.

The rest of paper is structured as follows: In Sect. 2, we provide fundamental of the Haar Wavelet Filtering and a new approach for generating DNA sequences representation based on Haar Wavelet Filtering. In the upcoming Sect. 3, the details of the proposed system setup for text similarity detection is described. Experimental results and discussion of the proposed system is given in Sect. 4. The last section provides the summary of the paper and point out our oriented research works in future.

2 Fundamental of Haar DWT and Proposed Algorithm for Generating DNA Sequences

It is realized that Haar DWT applied to both original text and suspicious text is a key to the text comparison and decision through DNA sequences. In fact, among methods of DWT, Haar DWT is popularly used since it is easy to transform and to invert back to the original signal. Accordingly, we propose an algorithm to efficiently utilize Haar DWT so that we obtain their own DNA sequences.

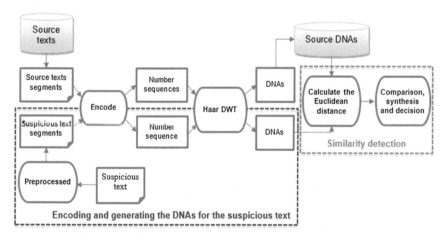

Fig. 1 General diagram for the text-similarity detection system

As illustrated in Fig. 1, the input data fetched into the Haar DWT is a sequence of floating number, and the length of the sequence is $N = 2^K$. The Haar DWT executes K iterations, and the output sequence at the k-th iteration is expressed as

$$\mathbf{x}^{(k)} = \left[\mathbf{x}_{low}^{(k)} \ \mathbf{x}_{high}^{(k)} \ \mathbf{x}_c^{(k-1)} \right] \tag{1}$$

where the approximation-coefficient vector $\mathbf{x}_{low}^{(k)}$ and detail-coefficient vector $\mathbf{x}_{high}^{(k)}$ are given as

$$\mathbf{x}_{low}^{(k)} = \left(\mathbf{x}_a^{(k-1)} * \mathbf{f}_L \right) \downarrow 2, \tag{2}$$

$$\mathbf{x}_{high}^{(k)} = \left(\mathbf{x}_a^{(k-1)} * \mathbf{f}_H \right) \downarrow 2, \tag{3}$$

with $\mathbf{f}_L = [1 \ 1]$ and $\mathbf{f}_H = [-1 \ 1]$ being low-pass and high-pass filter, respectively; $\mathbf{x}_a^{(k-1)}$ and $\mathbf{x}_c^{(k-1)}$ are the approximation-coefficient vector at the $(k-1)$-th step and the concatenation of detail-coefficient vectors from 1-st to the $(k-1)$-th step, respectively.

At the initialization, $\mathbf{x}_a^{(0)}$ and $\mathbf{x}_c^{(0)}$ are set to

$$\mathbf{x}_a^{(0)} = \mathbf{x}^{(0)}, \tag{4}$$

$$\mathbf{x}_c^{(0)} = [], \tag{5}$$

where $\mathbf{x}^{(0)}$ is the initial sequence after text encoding and [] is an empty vector. The vector $\mathbf{x}_a^{(k)} \in \mathbb{R}^{1 \times N_a(k)}$, with $N_a(k) = 2^{K-k}$, $\mathbf{x}_c^{(k)} \in \mathbb{R}^{1 \times N_c(k)}$ and $N_c(k) = \sum_{i=1}^{k} 2^{K-i}$, $k = 1, 2, \ldots, K$ are updated by:

$$\mathbf{x}_a^{(k)} = \mathbf{x}_{low}^{(k)}, \tag{6}$$

$$\mathbf{x}_c^{(k)} = \left[\mathbf{x}_{high}^{(k)} \ \mathbf{x}_c^{(k-1)} \right]. \tag{7}$$

It can be proved that $N_a(k) + N_c(k) = 2^{K-k} + \sum_{i=1}^{k} 2^{K-i} = 2^K = N$, $k = 1, 2, \ldots, K$. Therefore, the length of number sequence after K iterations is still N as that of the initial sequence. Since each of transformed sequences is unique as corresponding to its input sequence, they are called DNAs.

In summary, we develop an algorithm for calculating DNAs as described in Algorithm 1.

Algorithm 1: Calculating DNAs

Input: The sequences of the floating numbers, generated by text encoding.
Output: The K-th sequence is as the DNA for text.
  1: **Initialization:** The vectors as in (4) and (5)
  2: **For** $k:= 1 \rightarrow K$
  3: Calculate the sequence at the k-th step as (1), (2) and (3)
  4: Update the values of vectors as in (6) and (7)
  5: **End**

---

3   Text Similarity Recognition: Proposed System Setup

In this section, the whole system of text similarity recognition is presented. Specifically, the task of every block in the diagram as previously illustrated in Fig. 1. In the preprocessing block, the system first removes all special character and splits a long text into some segments. Then, a certain word is encoded as a specified number, and thus a sequence of the integer numbers related to a segment of text is fetched to the input of Haar DWT. The Haar DWT block samples the sequence and generates the corresponding DNAs. For convenience, in the first subsection we present the preprocessing block which is applied to both source and suspicious texts, while their main processes are independently detailed in two remaining subsections. The detail of each step can be classified into three phases including: (1) Preprocessing; (2) DNAs representation via generating source DNAs and (3) Text similarity detection.

3.1 *Preprocessing*

The set of collected source texts is fetched to the preprocessing. For each text, this block is designed to recognize a segment as a sentence in the text, and thus the periods (.) are used for splitting into the segments. The special characters, e.g., (! ?, []…), are removed from the segments, and then these preprocessed segments are stored to provide the detail of text comparison. Similarly, a suspicious text is preprocessed to generate the segments, but these segments are used for real-time similarity detection. The segments belonging to the source texts are called source segments, while those of the suspicious texts are called suspicious segments.

3.2 *Generating Source DNAs*

The DNAs corresponding to source texts are called source DNAs. To generate the source DNAs, we encode the source segments so that a unique sequence of the integer number stands for a certain source segment. However, splitting a source text into the sentences makes the length of segments unequal due to the various number of words in the sentences. Therefore, we use a shifting window to sample the sequences with a given length. To further reduce the complexity for encoding the source segment, we encode all words for a segment before sampling the segment, and then generate the source DNAs.

Encoding and aligning data. First, we transform all words in a segment into a sequence of the floating numbers such that one word is represented by a unique value. To do this, we use Unicode to encode all characters of a word and then a concatenation of these codes makes a specific number for the word. However, the number of encoding digits for a character is different from that for the others. By setting the number of encoding digits following a maximum value in Unicode, all characters are denoted by the same-length integer number, with containing some padding zeros on the left if any. The maximum value for the number of encoding digits is denoted by m. In particular, the specific number of the i-th word, denoted by s_i, is expressed as

$$s_i = \sum_{l=1}^{L_i} s_{i,l} \times 10^{m \times (L_i - l)} \tag{8}$$

where L_i is the number characters of the i-th word and $s_{i,l}$ stands for the code of the l-th character ($l = 1, 2, \ldots, L_i$) in the i-th word.

Second, we align the words by defining a maximum number of characters for a word as L_{max}. It is obvious that the lengths of the words are various, and thus we align all words by padding s_i with zeros as a suffix such that the length of an arbitrary word is equal to the maximum number of digits $M = m * L_{max}$. Since the zero-padded value of an integer number is quite large, we scale this value using the *common logarithm*. Finally, at the input of Haar DWT, a floating number representing a word is calculated as

$$x_i^w = M - m \times L_i + \log_{10}(s_i) \tag{9}$$

Sampling the segments and generating DNAs. After encoding and aligning data, a segment is now denoted by a sequence of floating numbers (SFN). Suppose that a certain segment has W words and the i-th word is represented by x_i^w as in (9), meaning that a corresponding SFN contains W values. To generate the DNAs for the segment, we sample the SFN such that the length of a sample, called a *sampling window*, is equal to $N = 2^K$. For a shift of window, an N-length sample is extracted from the SFN. If $N > W$, it is obvious that a unique sample from SFN is generated and that we can pad some random number into the SFN to make its length equal to N, and thus the similarity is dependent on the values contained in the original SFN. If $N > W$, we obtain a group of samples for an SFN. The number of samples in a group follows the length of SFN and how much the sampling window is right-shifted. Simply, we set the shift interval to one element of SFN, and then we have a group of $W-N+1$ samples for a W-length SFN. Summarily, a sample in a group, which is extracted from an SFN, is expressed as

$$\mathbf{x} = \mathbf{x}^{(0)} = [x_1 \, x_2 \ldots x_N] = \left[x_i^w x_{i+1}^w \ldots x_{i+N-1}^w \right] \tag{10}$$

These samples are fetched to Haar DWT to generate a group of source DNAs for the segment as in Algorithm 1.

Data structure for source DNAs. After obtaining the set of source DNAs through two previous steps, we sort these DNAs as the ascending values of the first element. This structure enables the binary search on all database of source DNAs to reduce the complexity. It is realized that the first element of a DNA is the sum of all values of original sequence at the input of Haar DWT. Therefore, this value is called approximation coefficient after K steps of subsampling, and then we can find a source DNA, which is closest to a suspicious DNA from the suspicious text, through the first element.

Algorithm 2: Generating the source DNAs

Input: All source text collected.
Output: Source DNA database in arrangement.
 1: **Initialization:** Length of DNA (N).
 2: Preprocessing, segmenting and storing the preprocessed source text.
 3: **For** each segment:
 4:    Encoding and aligning data as in section 3.2.
 5:    Sampling the segment and generating the group of DNAs as in section 3.2.
 6:    **For** each DNA in current group:
 7:       Binary searching to insert the DNA into the source DNA database such that the first elements of DNAs are sorted in ascending order.
 8;   **EndFor** // *end for loop starting at the line 6*
 9: **EndFor** // *end for loop starting at the line 3*

3.3 Text Similarity Recognition

This phase detects text similarity matching via two following steps, shown in the last block of the Fig. 1.

Encoding and generating the DNAs for the suspicious text. As mentioned earlier, the suspicious text is preprocessed as same as the source text, and thus we also collect the suspicious segments. Encoding the suspicious segments is similar to encoding the source segments as in Sect. 3.2. Differently from processing the source segments, the suspicious segments needn't to be stored in library, but rather they are successively processed in real-time. In particular, the suspicious segments are fetched in one-by-one to be encoded before being sampled at the Haar DWT block. As a result, for each segment a group of suspicious DNAs is obtained at the output of Haar DWT. Therefore, at a certain moment, one group of DNAs is processed to make the comparison and decision.

Comparison and Decision. The final block of system executes three tasks: DNA comparison, synthesis and decision. First, by searching the source database the comparison block determines the group of source DNAs which are closest to group of suspicious DNAs. As a result, one suspicious DNA in its group is only matched to one source DNA in library. To measure the similarity between two DNAs, we use Euclidean distance as given as

$$d(\mathbf{x}, \mathbf{y}) = \|\mathbf{x} - \mathbf{y}\|_2^2, \tag{11}$$

where $\mathbf{x} \in \mathbb{R}^{1 \times N}$ and $\mathbf{y} \in \mathbb{R}^{1 \times N}$ are the source and suspicious DNAs, respectively. The Euclidean distance is compared to a given threshold ε. If $d(\mathbf{x}, \mathbf{y}) < \varepsilon$, two DNAs are same and their positions in the segments are marked. Note that the proposed algorithm finds the closest DNA $\mathbf{x}$ to the DNA $\mathbf{y}$ with respect to $d(\mathbf{x}, \mathbf{y}) < \varepsilon$. Since each segment has a unique DNA as encoded by Unicode in alignment, an ε-radius ball containing two points $\mathbf{x}$ and $\mathbf{y}$ indicates that two segments corresponding $\mathbf{x}$ and $\mathbf{y}$ contain a large of the common letters. Second, the synthesis is to aggregate all marked positions to analyze and connect DNAs in the group. Finally, the decision task is to detect the similarity through determining how much similar the source and suspicious segments are, and then to show the result of detection. The algorithm for text similarity detection is summarized in Algorithm 3.

Algorithm 3: Text-similarity detection

Input: Suspicious text.
Output: Show the result of detection, i.e., the percentage of similarity…
  1: **Initialization:** the length of DNA (N) and threshold (ε).
  2: Preprocessing, segmenting and storing the data for output.
  3: **For** each segment:
  4:       Encoding and generating a group of DNAs.
  5:       **For** each DNA $\mathbf{y}$ in the group:
  6:             Binary searching on source DNA database to find a DNA $\mathbf{x}$ such that the first element of $\mathbf{y}$ is closest to that of $\mathbf{x}$.
  7:             Calculate the Euclidean distance $d(x, y)$ as in (11).
  8:             **If** $d(\mathbf{x}, \mathbf{y}) < \varepsilon$ **then**
  9:                   Mark DNAs $\mathbf{x}$ and $\mathbf{y}$.
  10:           **Endif**
  11:     **Endfor** // *end for loop starting at the line 5*
  12:           Synthesize all DNAs marked if any and connect them to reconstruct the segment.
  13:           Detect some strings of the suspicious segment which are similar to some of source one (if any).
  14:**Endfor** // *end for loop starting at the line 3*

---

4 Experimental Results

4.1 Datasets Acquisition

In this paper, a real dataset taken from PAN is carried out in order to evaluate the efficiency and effectiveness of the proposed system. PAN is a well-known series of scientific events and shared tasks on digital text forensics. One of the most important tasks of PAN is plagiarism detection [13, 14]. In this task, a set of source documents and a set of suspicious documents are given. Plagiarism detention algorithm is tasked to find all text passages in the suspicious documents which have been plagiarized and the corresponding text passages in the source documents. PAN proposed two datasets for doing performance evaluation on plagiarism detection algorithms: training Scopus and testing Scopus. The training Scopus has a set of suspicious documents and a set of source documents. A suspicious document is generated by adding passages from one or more source documents. Each suspicious document has a corresponding xml file which contains plagiarism information such as which source documents it plagiarized from, offsets of the plagiarized passages, etc. In this work, we use these xml files to calculate performance measures. The testing Scopus has similar structure of training Scopus which is also published on PAN website.

4.2 Evaluation Method and Experimental Results

In this work, we calculate two measures to evaluate proposed algorithm: *prec* (precision) and *rec* (recall). Calculation steps follow PAN's instructions to generate un-biased results. In detail, we define the set of plagiarized passages from source text and detected passages in the suspicious texts as

$$\mathcal{S} = \{S\} \text{ and } \mathcal{D} = \{D\}, \tag{12}$$

respectively. The values of *prec* and *rec* measures are respectively formulated as

$$prec = \frac{1}{|\mathcal{D}|} \sum_{D \in \mathcal{D}} \frac{|D \cap (\cup_{S \in \mathcal{S}} S)|}{|D|} \text{ and } rec = \frac{1}{|\mathcal{S}|} \sum_{S \in \mathcal{S}} \frac{|S \cap (\cup_{D \in \mathcal{D}} D)|}{|S|}, \tag{13}$$

where $|\mathcal{S}|$ and $|\mathcal{D}|$ are the numbers of elements of $\mathcal{S}$ and $\mathcal{D}$ respectively, while $|S|$, $S \in \mathcal{S}$, and $|D|$, $D \in \mathcal{D}$ are the lengths of strings, respectively.

In this work, we use 2009 training Scopus which is published on PAN website to evaluate the proposed algorithm. The training Scopus comprises 7214 source documents and 7214 suspicious documents[1]. The testing configurations are shown in the Table 1.

---

[1]http://www.uni-weimar.de/medien/webis/corpora/corpus-pan-labs-09-today/pan-09/pan09-data/pan09-external-plagiarism-detection-training-corpus-2009-03-30.zip.

Table 1 Testing configurations

Parameters	Setting values
Maximum number of digits used to encode a character (m)	5
Maximum number of characters of a word (L_{max})	45
Length of a DNA (N)	8
Maximum decomposition level of Haar (K)	3
Euclidean threshold (ε)	10^{-16}–10^{-5}

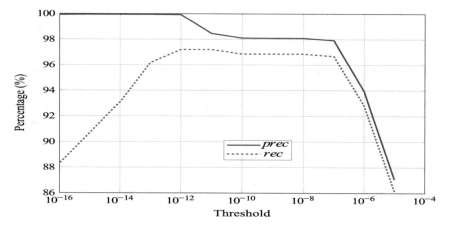

Fig. 2 The precision and recall versus threshold

Figure 2 indicates the change of *prec* and *rec* when modifying the value of the threshold. As can be seen, both *prec* and *rec* are quite low, 87% and 86% respectively, as the high level of threshold, 10^{-5}. This is because when the threshold increases, the more suspicious sentences are detected. However, only some words in these sentences are similar to the source sentences. As compared to the length of the sentences, the similar words contribute a low ratio. Therefore, when the threshold increases, both $|\mathcal{S}|$ and $|\mathcal{D}|$ strongly increase while the similarity ratios in most of them are very low. This leads *prec* and *rec* decrease. On the other hand, when the level of threshold decreases under 10^{-12}, the detected sentences must be almost similar to the source sentences. Hence, *prec* continuously rises while *rec* surprisingly decreases. In fact, there exists many sentences in high similarity, but they aren't detected since the very low threshold requires the suspicious sentences being much more similar to or perfectly same as the source sentences. Due to time limitation, we will simulate and demonstrate the superiority of the proposed method compared with other existing methods in our future work.

5 Conclusion

In this paper, we have developed a novel approach for text-similarity detection. Our main contribution lies on converting textual document into unique sequences of numbers, called as DNA series that boost performance of text matching detection through the efficiently storage, update and usage of dataset. In addition, the advantage of the proposed system also emphasized with a proposed approach of recognizing and evaluating the text similarity through binary search and Euclidean-distance comparison. To verify the success and effectiveness in text similarity detection, our proposed algorithm is implemented with a real dataset of PAN that shows high percentages of detection on both precision and recall with a given threshold.

Acknowledgements This research is funded by Funds for Science and Technology Development of the University of Danang under grant number B2017-DN01-07.

References

1. Hotho, A., Nürnberger, A., Paaß, G.: A brief survey of text mining. Ldv Forum **20**(1), 19–62 (2005)
2. Cer, D., Diab, M., Agirre, E., Lopez-Gazpio, I., Specia, L.: SemEval-2017 Task 1: Semantic Textual Similarity-Multilingual and Cross-Lingual Focused Evaluation (2017). arXiv:1708.00055
3. Meuschke, N., Gipp, B.: State-of-the-art in detecting academic plagiarism. Int. J. Educ. Integr. **9**(1), 50–71 (2013)
4. Brinkman, B.: An analysis of student privacy rights in the use of plagiarism detection systems. Sci. Eng. Ethics **19**(3), 1255–1266 (2013)
5. De, T.C., et al.: Developing plagiarism detection system for Vietnamese University. In: 12th Vietnam—Japan International Joint Symposium, Can Tho (2014)
6. Reddy, G.S., Rajinikanth, T.V., Ananda Rao, A.; Clustering and classification of text documents using improved similarity measure. Int. J. Comput. Sci. Inf. Secur. **14**, 39–54 (2016)
7. Wahlstrom, S.: Evaluation of string searching algorithms. In: IDT Mini-conference on Interesting Results in Computer Science and Engineering (2004)
8. Lin, Y.S., Jiang, J.Y., Lee, S.J.: A similarity measure for text classification and clustering. IEEE Trans. Knowl. Data Eng. **26**(7), 1575–1590 (2014)
9. Gomaa, W.H., Fahmy, A.A.: A survey of text similarity approaches. Int. J. Comput. Appl. **68**(13), 13–18 (2013)
10. Nawab, R.M.A., Stevenson, M., Clough, P.: An IR-Based approach utilizing query expansion for plagiarism detection in MEDLINE. IEEE/ACM Trans. Comput. Biol. Bioinf. **14**(4), 796–804 (2015)
11. Mountassir, A., Berrada, I., Benberahim, H.: Representing text documents in training document spaces: a novel model for document representation. J. Theor. Appl. Inf. Technol. **56**(1), 30–39 (2013)
12. Vidakovic, B.: Statistical Modeling by Wavelets, vol. 503. Wiley (2009)
13. Stein, B., zu Eissen, S.M., Potthast, M.: Strategies for retrieving plagiarized documents. In: Proceedings of the 30th Annual International ACM SIGIR Conference on Research and Development in Information Retrieval, pp. 825–826 (2007)
14. Potthast, M., et al.: Overview of the 1st international competition on plagiarism detection. In: Stein, B., et al (eds.) PAN'09, pp. 1–9 (2009)

Tweet Integration by Finding the Shortest Paths on a Word Graph

Huyen Trang Phan, Dinh Tuyen Hoang, Ngoc Thanh Nguyen and Dosam Hwang

Abstract Twitter is a well-known social network service. Every second, users post a large number of tweets on different topics, which leads to a significant problem-it is time-consuming for users to get useful information for their individual purposes. It is difficult for a user to receive necessary information from all topics with high accuracy. Thus, integrating the tweets to create summaries is very convenient solution for users. There are some previous works trying to solve the problem of tweet integration. However, they did not consider automatic grouping tweets into small clusters according to topic. Moreover, the tweets have not analyzed for sentiment mining before summarization. In this study, we propose an approach to integrate tweets by taking into account techniques such as topic modeling to automatically determine the number of topics as well as the tweets inside each topic, plus sentiment analysis to classify the attitudes of the users. The experimental results show that the proposed model achieves promising results.

Keywords Data-integration · Tweets-summarization · Word-graph
K-shortest-paths

H. T. Phan · D. T. Hoang · D. Hwang (✉)
Department of Computer Engineering, Yeungnam University,
Gyeongsan, South Korea
e-mail: dosamhwang@gmail.com

H. T. Phan
e-mail: huyentrangtin@gmail.com

D. T. Hoang
e-mail: hoangdinhtuyen@gmail.com

N. T. Nguyen
Faculty of Computer Science and Management, Wroclaw University
of Science and Technology, Wroclaw, Poland
e-mail: Ngoc-Thanh.Nguyen@pwr.edu.pl

© Springer International Publishing AG, part of Springer Nature 2018
A. Sieminski et al. (eds.), *Modern Approaches for Intelligent Information
and Database Systems*, Studies in Computational Intelligence 769,
https://doi.org/10.1007/978-3-319-76081-0_8

1 Introduction

A social network is a service that helps users, share ideas and interests, respond to each other, rate information, and make new friends quickly and more efficiently. Social networks become not only familiar but also occupy a prominent place in the Internet user community. They become more and more popular communication channels for people. There are about 3.66 billion Internet users in 2017,[1] of which social network users are more than 2.51 billion.[2] They are potential virtual communities that businesses and service providers always want to reach in order to promote their products. Twitter is one of the most well-known online social networks, which has enjoyed extreme popularity in recent years. Twitter has increased in both number of users and service quality. Every second, about 6,000 tweets are posted on Twitter, which corresponds to over 350,000 tweets per minute, 500 million tweets per day, and about 200 billion tweets per year.[3] The service they provide is referred to as microblogging. That is a variant of blogging where, on Twitter, the pieces of content are extremely short (limited to 140 characters for each tweet. Twitter is different from other social media platforms; it is not bidirectional, meaning that the connections do not have to be mutual, so you can follow users who do not follow you, and the other way round. Also, Twitter offers a series of application programming interfaces (APIs) which are open source programs. They allow users access to Twitter data, including reading tweets, accessing user profiles, and posting content on behalf of a user. That helps users gather a large amount of public data related to specific topics or events. This data can be sliced, split, integrated, and visualized in a variety of ways in order to serve different purposes. Moreover, Twitter provides one of the best social media platforms for publishing and following content in real time. It is a major platform for real-time information. Twitter is a potential gold mine for data miners and researchers.

Many people use Twitter as the main communication channel where they can find and exchange useful information. Twitter creates both opportunities and challenges for data miners. The biggest challenge is how users can capture information they are interested in following in the fastest, most concise, and the most complete way. In other words, from the large number of tweets that are posted by many users, how can you know the number of topics, the attitudes of the posters, as well as the main topic in the content? Therefore, we have to integrate all tweets that have the same topic. The problem of integrating tweets from different sources is an essential task. It becomes significant in a variety of fields, which includes commercial, political, and scientific domains.

Some methods have been proposed to solve this problem [1–3]. In [1] the authors have proposed a new method for summarizing tweets with documents based on conversations. The main idea of this method is based on user-to-user interaction on Twitter tending to revolve around a few related topics. So summarizing tweets

---

[1]http://www.internetlivestats.com/internet-users/.

[2]https://www.statista.com/statistics/278414/number-of-worldwide-social-network-users/.

[3]http://www.internetlivestats.com/twitter-statistics/.

by conversation can create more coherent document aggregation and more relevant topic extraction. Other authors have considered summarizing topics by using subtopics along the timeline [3]. This method helps to quickly capture topic evolution on Twitter. They ranked and selected salient tweets as a summary of each subtopic. Then, they modeled and formulated the tweet ranking in a unified mutual reinforcement graph. Yet other authors developed a method that finds a trending phrase or any phrase specified by a user [2]. Next, this method collects a large number of posts containing the phrase and provides an automatically created summary of the posts related to this phrase. The previous methods focused on generating the tweet summarization with available topics. However, most of them did not automatically determine the topics. That led to some tweets having content unrelated to the available topics, but they were still put into one of the corresponding groups. It means for those clusters, such tweets are outliers. That affects the accuracy of the achieved results. Moreover, a sentiment analysis technique was not considered. The users will not know the attitudes of others for each topic. That leads to users who do not obtain full useful information.

In this paper, we propose a new method for integrating tweets by taking into account the following steps. First, a large number of tweets from many different users is collected. Next, the number of topics is automatically determined, and tweets having the same theme are grouped into a cluster. Then, a sentiment analysis technique is employed to specify the positive, negative, and neutral tweets. Finally, we build word graphs corresponding to topics and a finding-shortest-paths technique is used on the word graphs to make out the results of the tweet integration process. The finding-shortest-paths technique on the word graphs is employed to integrate tweets from various resources because it fits into short data integration. It allows fast updates, as well as quick interrogations when generating the summary.

The rest of the paper is organized as follows. Section 2 contains a review of related works. Section 3 presents the methodology of our proposed model. The implementation of the proposed method is shown in Sect. 4 The conclusion and future works are in Sect. 5.

2 Related Works

Data integration on Twitter associates the content of the tweets from these different users. These tweets must have the same topic. In other words, this process generates short texts from various different tweets. It becomes very necessary in many domains. For example, it helps travelers find feedback from people about the place (traffic, hotel, food) they will visit via comments from Twitter users. It helps a business know the opinions of users about the product or service they provided. It helps famous people know the attitudes of people toward them. From that, they may adjust their behavior toward the public.

Tweet integration is a process that consists of many steps, and tweet summarization is the major step in that process. Until now, the published methods just stopped at the tweet summary. There is no method that mentions the problem of creating a model to integrate tweets. A summary on Twitter or a microblog is a particular kind of multi-document summarization [4], where each tweet corresponds to one document. Sharifi et al. [2] proposed a method to take a trending phrase or any phrase specified by a user, and a huge number of tweets were found. These tweets have to contain the phrase specified. Then, a summary is generated by the tweet integration. Other researchers described two approaches to microblog summarization [5]. They also collected a large number of posts on the trending topics, and then automatically generated a short post of all the posts on these themes with two algorithms as follows. First, the phrase reinforcement method collected the most common phrase on one side of the search phrase. Then, it found posts that consist of the most common phrase on the other side as well. Second, the hybrid term frequency inverse document frequency (TF-IDF) method ranked tweets using the TF-IDF scheme and gave better results. There is research about tweet summarization that relies on the topic published. Yajuan et al. [3] detected topics on Twitter based on a framework of the timeline. They detected topics by subtopics along the timeline to capture topics on Twitter in a rapid, full way. In particular, they ranked and selected salient and diversified tweets as a summary of each subtopic. They modeled and formulated the tweet ranking by using a reinforcement graph. The experimental results showed their approach achieved improvements that are better than the LexRank and phrase graph.

The previous methods created a summarization of tweets. However, they were based on available topics for the summary without generating topics automatically. Besides, they did not implement sentiment analysis of the tweets on each subject. That leads to information received after tweet summarization that is not complete. Also, the attitudes and opinions of the poster for each topic were not analyzed.

3   Tweet Integration Method

This section presents how to build the tweet integration model on Twitter by finding the shortest paths on a word graph. The model focuses on finding a summarization of tweets, which is the result of integrating tweets from many different users. We chose the find-shortest-paths method on a word graph to implement the proposed model because it combines advantages from sentence clustering methods and word frequency. Besides, this method provides a formal model for computing the importance of sentences. Details of the proposed model are presented in Fig. 1 and Algorithm 1.

3.1   Tweet Preprocessing

A tweet is a short text, informal style of writing. It can include noise data, special characters, symbols, and emojis. This phase aims to remove all the unnecessary

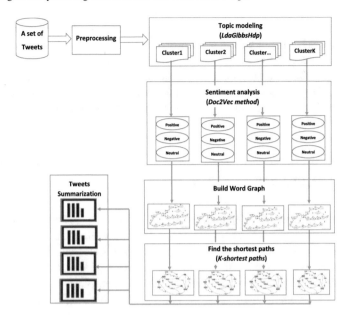

Fig. 1 The tweet integration model

elements in the tweets, such as non-English timelines, URLs, hashtags, punctuation, or tweets composed of only one or two words. Then, the tokenize method is used to break the tweet into words. This step is crucial because it affects the results of the method proposed.

3.2 Topic Modeling

Topic modeling is used to find a group of words that represent the content of the tweets. Then, the topics are identified by using a process of finding similarities in the found words. The topic modeling method is a convenient approach to analyzing and determining the corresponding topics for each tweet. A topic contains a group of words that frequently occur together. A topic model can use related words with similar meanings and can distinguish words with multiple meanings.

There are several methods for tweet topic modeling, such as TF-IDF [6], latent Dirichlet allocation (LDA) [7], Word2Vec [8], LDACLM [9]. Because a tweet can have several topics, we use a Gibbs sampling algorithm for the hierarchical Dirichlet process (the acronym is LdaGibbsHdp [10] in our work, where Hdp stands for hierarchical Dirichlet process). (Lda stands for Latent Dirichlet Allocation; Hdp stands for Hierarchical Dirichlet Process). The LdaGibbsHdp algorithm is extended from the latent Dirichlet allocation algorithm, which is a widely used topic modeling approach. The LdaGibbsHdp algorithm can automatically determine the number of topics, the number of tweets on each topic, and the names of those topics. It uses a Dirichlet process to capture the uncertainty in the number of topics. Then, one base distribution is selected to represent the countably infinite set of possible topics in the tweets, and the finite distribution of topics for each tweet is sampled from this base distribution. The advantage of the LdaGibbsHdp algorithm is the maximum number of topics that are determined by learning from the data; it does not need to be specified in advance.

3.3 Sentiment Analysis of Tweets

Sentiment analysis is the computational study of opinions, attitudes, and emotions of people toward a topic. After implementing topic modeling, the received result is a set of clusters where each cluster contains a set of tweets, which can be positive, negative, or neutral. For the task of sentiment analysis, given a tweet, x, determine whether x expresses a positive, negative, or neutral opinion. Currently, there are many algorithms for sentiment analysis on Twitter that are effective: Nave Bayes, Support Vector Machines, K-nearest Neighbor, Doc2Vec.

Algorithm 1 Generality algorithm of tweet integration

Input: A set of tweets.
Output: A set of summaries corresponds to the topics.
 1: Using Streams API to collect tweets;
 2: Preprocessing;
 3: **For** each tweet in a set of tweets
 4:     Determining the number of topics and content of each particular topic;
 5: **For** each topic in a set of topics
 6:     **For** each tweet in a set of tweets
 7:        Using Doc2Vec model to implement sentiment analysis;
 8: **For** each tweet in topic
 9:     Adding into tweet two special words: S and F
10:     **For** each bigram in tweet
11:        **If** bigram does not exist on graph
12:           Creating a new node;
13:           Mapping bigram into the new node;
14:           Creating a new edge from new node to the previous adjacent node;
15:        **If** bigram existed on graph
16:           Mapping bigram into the corresponding node;
17: **For** each node in a set of nodes
18:     **For** each edge in a set of edges
19:        Using the K-shortest paths algorithm to find k shortest paths;
20: **return** A set of k_shortest paths are the result of tweets integration on Twitter;

Doc2vec [11] is an unsupervised algorithm to create vectors for sentences, paragraphs, or documents. The Doc2Vec model was generated by Quoc Le and Tomas

Mikolov [12]. The algorithm is a further development of the Word2Vec algorithm. The vectors created by Doc2Vec can be employed for classifying sentiment.

In this work, the Doc2Vec model is employed to implement sentiment analysis, because the performance of this method is better than other methods. The averaging word vector method ignores the order of words, and the parse tree method just works for sentences, whereas the Doc2Vec method can overcome these issues. The Doc2Vec model has an accuracy that is high (about 0.87%).[4]

3.4 Word Graph Construction

Definition 1 A word graph is a triple consisting of nodes, edges and weights of edges. It is denoted $G = (V, E, w)$. Where, $V = \{S, v_1, v_2, ..., v_n, F\}$ is a set of nodes. S and F denote the start and end of each tweet. v_i ($i \in \{1 ... n\}$) corresponds to a bigram of words. $E = \{e_1, e_2, ..., e_m\}$ is a set of edges. e_j corresponds to a trigram of words. e_j ($j \in \{1 ... m\}$) presents relations between the adjacent bigrams, and w is a set of weights of edges.

The word graph is built with the following steps. First, add into each tweet two special words, which are S and F, to tick the beginning and finish of one tweet. Second, browse through the tweets, and with each tweet, map the bigrams sequentially into a graph, where each bigram is a node. When a new node is added, a directed edge from this new node to the previous adjacent node is created. This edge is a trigram. If a bigram is where its equivalent node has existed on the graph, then no new node is generated, but instead, this bigram is mapped onto the corresponding available node.

3.5 Find-Shortest-Paths Method on a Word Graph

The result of tweet integration is returned by finding the paths beginning at node S, going through bigrams and trigrams, finishing at node F. The paths should be short, but they must have at least eight words [13]. The k-shortest paths algorithm is employed to find the shortest paths and the expressions used to compute the weight of edges, defined as follows:

$$weight(edge_{i,j}) = \frac{count(i) + count(j)}{\sum_{t \in T} dist(t, i, j)^{-1}} \qquad (1)$$

where i and j are nodes; $edge_{i,j}$ is an edge connecting i and j; $count(i), count(j)$, $count(edge_{i,j})$ respectively are the frequencies of i, j, and $edge_{i,j}$; and $dist(t, i, j)$ is the distance between nodes i and j in tweet t.

---

[4]http://linanqiu.github.io/2015/10/07/word2vec-sentiment/.

$$dist(t, i, j) = \begin{cases} pos(t, i) - pos(t, j) & \text{if } pos(t, i) < pos(t, j) \\ 0 & \text{otherwise} \end{cases} \tag{2}$$

where $pos(t, i)$ and $pos(t, j)$ are positions of i and j in tweet t.

To generate a summary concerning the topic, we force the path to go through the most frequent nodes by decreasing edge weight of the connections between the most common nodes. The edge weight is redefined as follows:

$$weight'(edge_{i,j}) = \frac{weight(edge_{i,j})}{count(i) \times count(j)} \tag{3}$$

4 Experiment

4.1 Dataset

We gathered 90.000 raw tweets on Twitter from 20 users by using the available Streams API for Python. For each Twitter's home page, we collected the maximum number of tweets. After the tweets were processed, we obtained 70.000 tweets which we used to evaluate the performance of the model.

4.2 Performance Evaluation

There is no standard method to assess the tweet integration model. For evaluation of the proposed model, we relied on evaluation of the final result the model obtained. That is an evaluation of the summaries the model created. We used a suite of automatically measured Recall-Oriented Understudy for Gisting Evaluation (ROUGE) metrics. They are widely used to liken the similarity between the summaries of another system and the proposed model. The simplest ROUGE metrics are the ROUGE_N metric [14] and ROUGE_L metric [14]. ROUGE_N relies on the similarity of n_grams. ROUGE_L is based on the longest common subsequence (LCS).

$$ROUGE\_N = \frac{\sum_{s \in MS} \sum_{n\_grams \in s} match(n\_gram)}{\sum_{s \in MS} \sum_{n\_grams \in s} count(n\_gram)} \tag{4}$$

where MS is the set of summaries of another system; $count(n\_gram)$ is the number of n_grams in the summary by another system; $match(n\_gram)$ is the number of co-occurring n_grams between the summaries of another system and the proposed model; and s is a particular summary of another system.

The *ROUGE_L* metric is calculated as follow:

$$LCS(X, Y) = \frac{length(X) + length(Y) - dist(X, Y)}{2} \tag{5}$$

where X and Y represent sequences of words or lemmas; $length(X)$ is the length of string X; and $dist(X, Y)$ is the distance of X and Y.

In this study, we have evaluated our proposed model by implementing it step by step as follows: We employed the Sumbasic method [15] to create the reference summaries. The first hypothesis summaries are generated by using the proposed method. The hybrid TF-IDF method [5] was used to give the second hypothesis summaries. The third hypothesis summaries were created by using the LextRank method [16]. Finally, we compared the results of the three approaches via the ROUGE metrics.

4.3 Results and Discussion

Topic modeling method automatically grouped tweets into 22 topics. Then, the sentiment analysis technique determined attitude of other users for each subject as well as the information is not limited to the available topics.

The final result of the proposed method is the summaries of tweets corresponding to topics. According to Table 1, the proposed method obtained results that have a higher accuracy than the other methods. This is understandable for the following reasons: (i) Tweet topic modeling helped to group the related tweets into a cluster. That led to noise reduction. (ii) The summaries of tweets are created by finding the shortest paths on a word graph. On which, the nodes are bigrams and the edges are trigrams. That helps the summaries keep the semantics of the original tweets. Figure 2 presents the particular results of Table 1. We can see the ROUGE_N and ROUGE_L metrics consist of F_measure, Precision, and Recall of the proposed method which are greater than the other methods. The final values of *Precision*, *Recall*, and *F_measure* are calculated by averaging of 22 corresponding values.

Table 1 The average performance of three methods

Method	Rouge_N			Rouge_L		
	F_measure	Precision	Recall	F_measure	Precision	Recall
Proposed method	0.30	0.31	0.29	0.20	0.20	0.19
Hybrid TF-IDF	0.28	0.29	0.27	0.19	0.19	0.18
LextRank	0.22	0.23	0.21	0.16	0.16	0.15

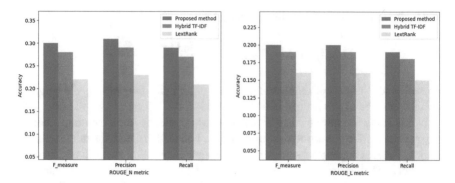

Fig. 2 Comparing the accuracy of the proposed method with other methods

5 Conclusion and Future Work

This work proposed a new model for tweet integration by finding the shortest paths on a word graph. A topic modeling technique is considered to automatically determine the number of the topics in the tweets. A sentiment analysis method is used to calculate the percentage of opinions that agree, disagree, or are neutral for each issue. For each topic, a weighted directed graph is built to present the results of the tweet integration process, which is the shortest path on a word graph. The experimental results show that the proposed method achieves promising results. It gives a higher accuracy, compared to other methods. In future work, we plan to integrate tweets by concatenating related tweets according to the evolution of time in the hope that the accuracy of integrating tweets will be improved.

Acknowledgements This research was supported by Basic Science Research Program through the National Research Foundation of Korea (NRF) funded by the Ministry of Science, ICT & Future Planning (2017R1A2B4009410).

References

1. Alvarez-Melis, D., Saveski, M.: Topic modeling in twitter: aggregating tweets by conversations. In: ICWSM, pp. 519–522 (2016)
2. Sharifi, B., Hutton, M.A., Kalita, J.: Summarizing microblogs automatically. In: Human Language Technologies: The 2010 Annual Conference of the North American Chapter of the Association for Computational Linguistics, pp. 685–688. Association for Computational Linguistics (2010)
3. Yajuan, D., Zhimin, C., Furu, W., Ming, Z., Shum, H.Y.: Twitter topic summarization by ranking tweets using social influence and content quality. In: Proceedings of the 24th International Conference on Computational Linguistics, pp. 763–780 (2012)
4. Zhang, R., Li, W., Gao, D., Ouyang, Y.: Automatic twitter topic summarization with speech acts. IEEE Trans. Audi Speech Lang. Process. **21**(3), 649–658 (2013)

5. Sharifi, B., Hutton, M.A., Kalita, J.K.: Experiments in microblog summarization. In: 2010 IEEE Second International Conference on Social Computing (SocialCom), pp. 49–56. IEEE (2010)
6. Salton, G., Buckley, C.: Term-weighting approaches in automatic text retrieval. Inf. Process. Manag. **24**(5), 513–523 (1988)
7. Blei, D.M., Ng, A.Y., Jordan, M.I.: Latent dirichlet allocation. J. Mach. Learn. Res. **3**(Jan), 993–1022 (2003)
8. Mikolov, T., Chen, K., Corrado, G., Dean, J.: Efficient estimation of word representations in vector space. arXiv:1301.3781 (2013)
9. Zhou, S., Li, K., Liu, Y.: Text categorization based on topic model. Int. J. Comput. Intell. Syst. **2**(4), 398–409 (2009). https://doi.org/10.1080/18756891.2009.9727671
10. Heinrich, G.: Infinite LDA implementing the HDP with minimum code complexity. Technical note, Feb 170 (2011)
11. Hoang, D.T., Tran, V.C., Nguyen, V.D., Nguyen, N.T., Hwang, D.: Improving academic event recommendation using research similarity and interaction strength between authors. Cybern. Syst. **48**(3), 210–230 (2017)
12. Le, Q., Mikolov, T.: Distributed representations of sentences and documents. In: Proceedings of the 31st International Conference on Machine Learning (ICM-14), pp. 1188–1196 (2014)
13. Filippova, K.: Multi-sentence compression: finding shortest paths in word graphs. In: Proceedings of the 23rd International Conference on Computational Linguistics, pp. 322–330. Association for Computational Linguistics (2010)
14. Steinberger, J., Ježek, K.: Evaluation measures for text summarization. Comput. Inf. **28**(2), 251–275 (2012)
15. Vanderwende, L., Suzuki, H., Brockett, C., Nenkova, A.: Beyond sumbasic: task-focused summarization with sentence simplification and lexical expansion. Inf. Process. Manag. **43**(6), 1606–1618 (2007)
16. Erkan, G., Radev, D.R.: Lexrank: graph-based lexical centrality as salience in text summarization. J. Artif. Intell. Res. **22**, 457–479 (2004)

Event Detection in Twitter: Methodological Evaluation and Structural Analysis of the Bibliometric Data

Musa Ibarhim M. Ishag, Kwang Sun Ryu, Jong Yun Lee
and Keun Ho Ryu

Abstract Twitter—a social networking service is increasingly becoming an important source of news and information for various aspects of our life. However, harnessing reliable sources is both tedious and challenging. Algorithms for mining and detecting events from Twitter have been developed. In this paper, event detection techniques are investigated. In essence, a theoretical comparison of the state-of-the-art event detection algorithms is performed along with highlights to the current issues and proper suggestions to mitigate them. In addition, a knowledge domain map analysis using CiteSpace is applied to the bibliometric data in the field in order to explore the structural dynamics of the research in this domain.

Keywords Twitter · Event detection · Bibliometric analysis
CiteSpace

1 Introduction

Twitter—a social networking service has become an integral part of human's daily life. The rapid development in smart devices has enabled almost every adult to stay connected on the web and hence own his own Twitter account. The short messages

M. I. M. Ishag · K. S. Ryu · K. H. Ryu (✉)
College of Electrical and Computer Engineering, Chungbuk National University,
Cheongju, South Korea
e-mail: khryu@dblab.chungbuk.ac.kr

M. I. M. Ishag
e-mail: ibrahim@dblab.chungbuk.ac.kr

K. S. Ryu
e-mail: ksryu@dblab.chungbuk.ac.kr

J. Y. Lee
Department of Software Engineering, Chungbuk National University, Cheongju
South Korea
e-mail: jongyun@chungbuk.ac.kr

© Springer International Publishing AG, part of Springer Nature 2018
A. Sieminski et al. (eds.), *Modern Approaches for Intelligent Information
and Database Systems*, Studies in Computational Intelligence 769,
https://doi.org/10.1007/978-3-319-76081-0_9

communicated via this medium, bring about issues such as trusting the source and the communicated message.

Researchers have adopted twitter in order to harness useful information from it. Their attempts varied from adopting it to discover emerging disease outbreaks [1] and wellbeing [2], earthquakes [3], to predicting stock prices [4].

The main goal behind these applications is to detect groups of Tweets that behave differently; event detection [5], which distinguishes the process from the similar task of outlier detection [6] that focusses merely on detecting an anomalous data point. An event is defined as a real world occurrence over a specific place and time [7] and it can be considered significant if the number of people discussing it is large enough [7, 8].

Although a number of techniques have been developed to achieve this task, those techniques lack a benchmarking in order to decide which one suits what application domain. Therefore, the goal of this paper is to investigate the state-of-the-arts techniques for event detection from twitter streams, along with knowing the significant contributions, institutions, and their collaboration patterns. Accordingly, the paper puts forward the following contributions:

- Theoretical comparison and evaluation of the Twitter event detection techniques
- Highlight open research issues and present possible solutions
- Domain map analysis to the bibliometric data in the field using CiteSpace

The remainder of this paper is structured in a logical way that will clarify the contributions.

2 Problem Definition

An event is defined as a real world occurrence over a specific place and time [7]. Within the context of categorical data, event detection is carried out through a process in which the distribution of the data in a previous observation is compared against the distribution within the current stream and if the difference is deemed significant, an event will be declared along with the portion of the data that was affected the most [9]. Within the context of Twitter textual streams however, the same definition holds with the addition of significance [7]. Therefore, the problem of event detection in Twitter is concerned with finding real world occurrence over specific place and time that is discussed by a considerable amount of Twitter users.

3 Related Work

3.1 Comparing Event Detection Technique

Although surveys on topic detection and tracking [10] from traditional text data are available [11], the recent literature shows few attempts [5, 12–14] to organize the work on event detection techniques focusing on Twitter data. In essence, the literature can be broadly classified into two categories: studies that have tried empirical comparison of the techniques, and those that tried theoretical evaluations. An example of the first category is the work of Aiello et al. [5] where they have compared two base-line works against four other techniques developed by the authors using three different data sets. Unlike their work, in this paper, a theoretical comparison is followed.

A representative work in theoretical comparisons, however, is the recent work by Farzinar and Khreich [12]. The authors compared sixteen papers according to the event type, detection task, detection methods, and the features used. Although our work follows similar concepts as [12–14] to compare Twitter event detection techniques, we evaluate the most cited work in the literature which are selected by a bibliometric analysis tool; CiteSpace [15].

3.2   Bibliometric Analysis

Bibliometric analysis—also known as mapping a knowledge domain is about discovering knowledge, tracing its flow and change [16]. It comprises of visualizing, and mining a research field to identify institutions, expertise, and social networks [17]. Bibliometric analysis has been applied to various research fields ranging from Security Informatics [18], Medical Informatics [19], Drug delivery [20], and Big Data [21]. To the best of our knowledge, except for the work of Silva et al. [22] on the dynamics of publication venues in computer science, our work is the first to consider Domain Map Analysis to bibliometric data of the event detection in twitter. Although a number of tools are convenient to analyze the citation data [23], in this paper, CiteSpace [15, 24–28] is used to accomplish the task.

4   Research Methods

In this paper, a synergistic view of the research in Twitter event detection is considered within which both theoretical evaluation of the state-of-the-art techniques is given along with an exploratory analysis of the literature in the field. Figure 1 explains the framework followed in this paper wherein, a key-word search query was posted to the Thomson Reuter's Web of Science [29] to retrieve all the ISI papers related to the query. Thereafter, two branches of tasks were followed; one to download the citation data and carry out the Domain Map Analysis [30], whereas in the other branch, the most cited methodological approaches were theoretically evaluated.

As explained in Fig. 1, the methodology can be viewed as comprised of three main steps; data collection, analysis, and reports. In the collection part, two datasets

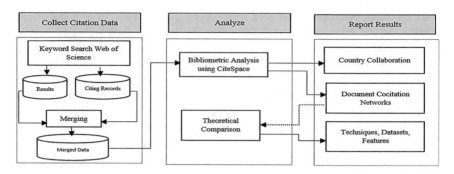

Fig. 1 Research framework

were considered for analysis. The search result in addition to the citing articles. In the analysis part, although CiteSpace was used, only two kinds of analysis were performed.

4.1 Bibliometric Analysis

Citation Dataset. Bibliographic records were collected from Thomson Reuters' Web of Science Core Collection (WoSCC) [29]. The records were collected from the beginning until 2016. The search query was; "Event Detection from Twitter" posed for the topic. The core dataset [28] returned consisted of 128 records of SCI and SCIE publications. For more comprehensive set of references, additional records citing the search results were also retrieved and combined with the core dataset.

Document Co-Citation Analysis. Document Co-Citation analysis was applied to the citation dataset using CitSpace and the resulting clusters were labeled by the Mutual Information (MI) measure [31].

The top clusters are identified by one digit ID followed by the number of documents in the cluster which resembles it's size. Afterwards, references to articles that cite most of the elements in each cluster are given followed by a Silhouette [32] value that shows the cohesion of the documents within each cluster. The clusters are labeled by terms extracted from the abstract portion of the publications using measures of Term Frequency Inverse Document Frequency (TFIDF) [33], Log Likelihood Ratio (LLR) [34], and MI respectively. Table 2 shows the top ten highly cited articles related to event detection in twitter for the time period of the analysis sorted according to their citation count. It also shows the IDs of the clusters in which those articles appear. In the table, the most cited article is by Sakaki et al. [35] in which they have developed an earthquake reporting system that notifies about earthquakes. They have analogized Twitter users as sensors and made correspondences between the process of event detection and object detection in a

a - Countries ranked by citation count		b- Collaboration network between countries
Country	**Citation Count**	
USA	25	
China	15	
England	15	
India	13	
Italy	10	
Japan	9	
Australia	8	
Canada	5	
France	5	
Singapore	5	

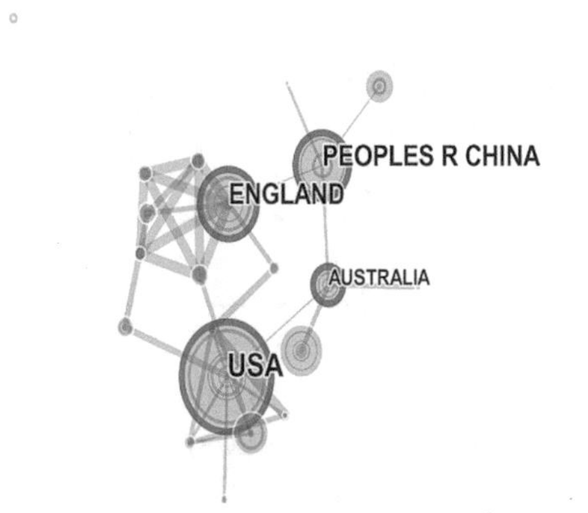

Fig. 2 Collaboration network

ubiquitous environment. These top cited papers are used for the purpose of comparing the techniques in this paper.

Institutional Collaborations. In order to understand the collaboration among research institutions across countries, a collaboration network analysis was conducted in CiteSpace and the results are shown in Fig. 2. In essence, Fig. 2b illustrates a network consists of research institutions by countries as nodes and links are added if at least a research article was collaboratively published by authors affiliated to more than one research institution in different countries. From Fig. 2b, it is clear that few collaborative works exist between various universities and countries. Particularly, countries like China, The United States of American, England, and Australia are the most salient. On the other hand, Fig. 2a shows the top ten countries with the most cited articles. The United States of America is topping the list with 25 citations followed China and England both with 15 citations. Although Japan has the most cited article by Sakaki et al. [35] and other significant studies, surprisingly, it is ranked the fifth in the table.

4.2 Comparing the Methods

In this paper, we have considered the most popularly cited works and tools available in the literature which were revealed by our analysis of the citation data as explained in Table 2. These include TwitterStand [36], Toretter [35], HotStream

Table 1 Distributions of the papers considered

Methodological papers	Surveyed previously by		
	Frazindar et al. [12]	Muhammad Imran et al. [13]	Cordeiro Mário et al. [14]
[35–38]	✓	–	✓
[39, 40]	–	✓	✓
[41]	✓	–	–
[42]	–	✓	–
[43]	–	–	–

[37], and the study of Lee et al. [38] all of these four systems were also surveyed in previous studies by Atefe and Khreich in 2012 [12], and by Cordeiro and Gama in 2016 [14]. Two other systems; TEDAS [39], and TwitterMonitor [40] which have previously been studied and compared from different prospective by Imran et al. [13] in 2015 and by Cordeiro and Gama [14]. The last three studies are: Petrović et al.'s [41] which was included in [12], TwitInfo [42] which was studied in [13], in addition to Jasmine [43] which is added for comparison for the first time in this study. Table 1 explains the distribution of the methods compared in this manuscript compared to the current literature.

The methods considered for the compare and contrast task have been applied to various application domains ranging from finding events related to Breaking News [36, 37], Natural Disasters/Earthquake [35], General Interest Events [41–43], Geo-Social Events [38], Trend Detection [40] and Crimes and Disasters [39]. They are compared according to the following three dimensions:

Data Sources. Although some datasets crawled and made publicly available by some researchers, these datasets violate the need for urgency in event detection from twitter. Therefore, most of the systems developed for the purpose of event detection in twitter, rely on Application Programming Interfaces (APIs) created by the online social networking service providers like Twitter. These APIs provide two ways to obtain data from twitter [13]. One way to allow researcher to access archived records through queries known as—search APIs. Examples of methods using this approach from the methods considered are Toretter [35] where queries containing keywords like "earthquake" are used to retrieve previously occurred twitter messages related to natural disaster events, and TEDAS [39]. The other way is used to obtain twitter messages as they are posted in a streaming scenario [13]. All of the remaining systems compared use the Twitter streaming API. This classification was explained in column 5 of Table 3 entitled collection. After obtaining data for analysis, researchers decide to represent and preprocess the data in order to proceed with the analysis. In essence, the event detection process can be described as using features like keywords related to a specific event and group the other twitters related to it in an approach called feature pivot approach [5], or considering the whole tweet as a document and group related tweets based on a similarity measure in an approach termed as document pivot [14]. This analogy was borrowed

Table 2 Top 10 most cited papers

Paper				Rank		Cluster ID
Year	Title	Venue	Author	Citation counts	Centrality	
2010	Earthquake shakes Twitter users: real-time event detection by social sensors	ACM WWW	Sakai et al.	39	0.16	0
2010	Twittermonitor: trend detection over the twitter stream	ACM SIGMOD	Mathioudaki et al.	12	0.22	0
2011	Twitinfo aggregating and visualizing microblogs for event exploration	ACM SIGCHI	Marcus et al.	8	0.10	9
2010	Breaking news detection and tracking in Twitter	ACM/IEEE WIC	Phuvipadawat et al.	8	–	0
2013	Tweet analysis for real-time event detection and earthquake reporting system development	IEEE TKDE	Sakai et al.	8	–	1
2012	Tedas: a twitter-based event detection and analysis system	IEEE ICDE	Li et al.	8	0.08	1
2010	Streaming first story detection with application to twitter	ACL HLT	Petrović et al.	8	–	0
2009	TwitterStand: news in tweets	ACM SIGSPATIAL	Sankaranarayanan et al.	7	–	0
2011	Jasmine: a real-time local-event detection system based on geolocation information propagated to microblogs	ACM CIKM	Watanabe et al.	7	0.08	0
2010	Measuring geographical regularities of crowd behaviors for Twitter-based geo-social event detection	ACM SIGSPATIAL	Lee et al.	7	0.09	6

Table 3 Comparing the methods in terms of datasets and techniques

Year	System/Tool	References	Technique		Classification	Data	Publicly available
			Clustering			Collection	
2009	TwitterStand	[36]	Online-Group-Burst		Naïve Bayes	Twitter Streaming API	No
2010	Toretter	[35, 51]	–		SVM	Twitter Search API	No
2010	TwitterMonitor	[40]	Online-leader-follower clustering		–	Twitter Streaming API	No
2010	HotStream	[37]	Online-TF-IDF based grouping		–	Twitter Streaming API	No
2010	–	[41]	Online-locality sensitive hashing (LSH)		–	Twitter Streaming API	No
2010	–	[38]	–		–	Twitter Streaming API	No
2011	Jasmine	[43]	–		–	Twitter Streaming API	No
2011	Twitinfo	[42]	–		–	Twitter Streaming API	No
2012	TEDAS	[39]	–		General Classifier	Twitter Search API	No

from the traditional Topic Detection and Tracking (TDT) [10] seeking to find topics and trends from stream of textual sources.

Features. As a processing step, the approaches can be classified as features related to the tweet content which is basically the text message conveyed, and features that are related to the way these messages are organized by twitter; meta-features. The meat features include the structure of the twitter network in terms of the relationships among the users, timestamps, retweets, and hashtags. Accordingly, most of the methods considered for comparison use keywords and term vector representation when it comes to representing the tweet contents. However, regarding the meta features, retweets, followers, favorites, and timestamps are considered by the methods which incorporate meta-features in addition to the contents of the tweet. These methods include, TwitterStnad, HotStream, TEDAS [38, 41]. On the other hand, Toretter, TwitterMonitor, Jasmine, and TwitInfo only utilize tweet related features in terms of keywords. This classification is illustrated by Table 4.

Event Types. In order to proceed with the process of detection and devising, or using existing appropriate algorithm, it is crucial to decide whether the events are known beforehand, or any unidentified events are also of interest. The literature then classifies the types of events into Specified and unspecified ones [12, 14].

According to our analysis, four of the methods discover unspecified events. Namely, Twittermonitor, HotStream, Jasmine, and [41]. Contrary, the remaining five methods search to find specified events.

Techniques. The techniques used to find events are generally classified as those that use clustering or classification or a combination of both. For example, Sankaranarayanan et al. in TwitterMonitor [36] use Naïve Bayes classifier to

Table 4 Comparing the methods in terms of the features considered

Year	System/Tool	References	Features		Real-time	Spatio temporal
			Tweets	Meta		
2009	TwitterStand	[36]	Term Vector	Hashtags Timestamp	Yes	Yes
2010	Toretter	[35, 51]	Keywords	–	Yes	Yes
2010	TwitterMonitor	[40]	Keywords	–	Yes	No
2010	HotStream	[37]	Term Vector	Hashtags Timestamp Retweet Followers	Yes	No
2010	–	[41]	Tweets Entropy	Users	Yes	No
2010	–	[38]	Tweets Crowd	Geo-tagged micro blogs	Yes	Yes
2011	Jasmine	[43]	Keywords	–	Yes	Yes
2011	Twitinfo	[42]	Keywords	–	Yes	Spatial
2012	TEDAS	[39]	Keywords	Retweets Favorites	No	Yes

identify tweets as news or junks and then carryout online clustering to group tweets about the same news together. Table 3 shows how the methods considered use the classification and clustering. In essence, techniques like Naïve Bayes, and Support Vector Machine are used for classification. From these contrasts performed in this paper, it is clear that our approach of mixing the bibliometric analysis and theoretical comparison is more effective in the sense that it can help the authors select the most relevant papers.

This claim is supported by the contents of Table 1, where it is evident that this paper has considered more approaches for comparison compared to recent surveys in the literature performed by subject matter experts.

5   Open Issues

Although the lack of benchmarking dataset, the difficulty of finding accurate spatial information and the difficulty in describing events along with the multilingualism of tweets have all been reported in a recent survey [14] we add the following two issues with bright suggestions to mitigate them.

- **Incorporating Meta-features**

Incorporating more meta-features is needed to boost the accuracy and reliability of the events. The recent work of Mario Cataldi et al. [44] where they have calculated the authorities of the users of twitter based on PageRank and used that to increase the reliability and trustworthiness of the discovered events can be thought of as a leading article in this direction.

- **Exploring Efficient Techniques**

Efficient techniques are needed to help in finding meaningful events in an efficient manner. Recently, Gaglio et al. [45] utilized frequent pattern mining as a new approach for real-time detection of events [46]. In essence, they have improved the Soft Frequent Pattern Mining (SFP) [47] for better performance. More recently, Huang et al. [48] have used the more sophisticated frequent pattern mining method; High Utility Pattern [49] to find top-k terms from the tweets, cluster them, and announce emerging topics in clusters. In this regard, the authors also foresee the potential of emerging pattern mining [50] as an additional methodology for efficient event detection.

6 Conclusions

This paper investigated the state-of-art techniques used to find events from the stream short text messages conveyed through Online Social Networking Services. In essence, the study considered event detection from twitter. Contrarily to the traditional way of classifying the literature manually, this paper used CiteSpace software to map the domain of the research by analyzing citation data from the starting of the research in event detection until 2016. To the best of the authors' knowledge, this is the first time a combined bibliometric and comparative analysis is performed in a single study. The results uncovered the methodologies mostly used, and the collaboration patterns between institutions and countries. The need to consider new methods for the task of event detection along with the necessity to use more features are identified as new challenges and bright suggestions were provided to tackle them.

Acknowledgements This research was supported by Basic Science Research Program through the National Research Foundation of Korea (NRF) funded by the Ministry of Science, ICT & Future Planning (No. 2017R1A2B4010826), and MSIT (Ministry of Science and ICT), Korea, under the ITRC (Information Technology Research Center) support program (IITP-2017-2013-0-00881) supervised by the IITP (Institute for Information & Communications Technology Promotion).

References

1. Thackeray, R., et al.: Using Twitter for breast cancer prevention: an analysis of breast cancer awareness month. BMC Cancer **13**(1), 508 (2013). https://doi.org/10.1186/1471-2407-13-508
2. Loff, J., Reis, M., Martins, B.: Predicting well-being with geo-referenced data collected from social media platforms. In: Proceedings of the 30th Annual ACM Symposium on Applied Computing, pp. 1167–1173. ACM, Salamanca, Spain (2015). https://doi.org/10.1145/2695664.2695939
3. Earle, P.S., Bowden, D.C., Guy, M.: Twitter earthquake detection: earthquake monitoring in a social world. Ann. Geophys. **54**(6), 708–715 (2012). https://doi.org/10.4401/ag-5364
4. Bollen, J., Mao, H., Zeng, X.: Twitter mood predicts the stock market. J. Comput. Sci. **2**(1), 1–8 (2011). https://doi.org/10.1016/j.jocs.2010.12.007
5. Aiello, L.M., et al.: Sensing trending topics in Twitter. IEEE Trans. Multimedia **15**(6), 1268–1282 (2013). https://doi.org/10.1109/TMM.2013.2265080
6. Rousseeuw, P.J., Leroy, A.M.: Robust Regression and Outlier Detection, vol. 589. Wiley (2005)
7. McMinn, A.J., Moshfeghi, Y., Jose, J.M.: Building a large-scale corpus for evaluating event detection on twitter. In: Proceedings of the 22nd ACM international conference on Information & Knowledge Management, pp. 409–418 ACM, San Francisco, California, USA (2013). https://doi.org/10.1145/2505515.2505695
8. Rei, L., Grobelnik, M., Mladenić, D.: Event Detection in Twitter With an Event Knowledge Base
9. Wong, W.-K., Neill, D.B.: Tutorial on Event Detection KDD 2009. Age 9 (2009)
10. Allan, J.: Topic Detection and Tracking: Event-Based Information Organization, vol. 12. Springer Science & Business Media (2012)

11. Hogenboom, F., et al.: A survey of event extraction methods from text for decision support systems. Decis. Support Syst. **85**, 12–22 (2016). https://doi.org/10.1016/j.dss.2016.02.006
12. Atefeh, F., Khreich, W.: A survey of techniques for event detection in twitter. Comput. Intell. **31**(1), 132–164 (2015). https://doi.org/10.1111/coin.12017
13. Imran, M., et al.: Processing social media messages in mass emergency: a survey. ACM Comput. Surv. (CSUR) **47**(4), 67 (2015). https://doi.org/10.1145/2771588
14. Cordeiro, M., Gama, J.: Online social networks event detection: a survey. In: Solving Large Scale Learning Tasks. Challenges and Algorithms, pp. 1–41. Springer International Publishing (2016). https://doi.org/10.1007/978-3-319-41706-6_1
15. Chen, C.: CiteSpace II: detecting and visualizing emerging trends and transient patterns in scientific literature. J. Am. Soc. Inf. Sci. Technol. **57**(3), 359–377 (2006). https://doi.org/10.1002/asi.20317
16. Balaid, A., et al.: Knowledge maps: a systematic literature review and directions for future research. Int. J. Inf. Manage. **36**(3), 451–475 (2016). https://doi.org/10.1016/j.ijinfomgt.2016.02.005
17. Chen, H., et al.: Terrorism Informatics: Knowledge Management and Data Mining for Homeland Security, vol. 18. Springer Science & Business Media (2008)
18. Liu, W., et al.: Collaboration pattern and topic analysis on intelligence and security informatics research. IEEE Intell. Syst. **29**(3), 39–46 (2014). https://doi.org/10.1109/MIS.2012.106
19. Qian, D., et al.: Mapping knowledge domain analysis of medical informatics education. In: Frontier and Future Development of Information Technology in Medicine and Education, pp. 2209–2213. Springer Netherlands (2014). https://doi.org/10.1007/978-94-007-7618-0_269
20. Lee, Y.-C., Chen, C., Tsai, X.-T.: Visualizing the knowledge domain of nanoparticle drug delivery technologies: a scientometric review. Appl. Sci. **6**(1), 11 (2016). https://doi.org/10.3390/app6010011
21. Singh, V.K., et al.: Scientometric mapping of research on 'Big Data'. Scientometrics **105**(2), 727–741 (2015). https://doi.org/10.1007/s11192-015-1729-9
22. Silva, T.H.P., Moro, M.M., Silva, A.P.C.: Authorship contribution dynamics on publication venues in computer science: an aggregated quality analysis. In: Proceedings of the 30th Annual ACM Symposium on Applied Computing, pp. 1142–1147. ACM, Salamanca, Spain (2015). https://doi.org/10.1145/2695664.2695781
23. Federico, P., et al.: A survey on visual approaches for analyzing scientific literature and patents. IEEE Trans. Visual Comput. Graph. **23**(9), 2179–2198 (2016). https://doi.org/10.1109/TVCG.2016.2610422
24. Chen, C., Leydesdorff, L.: Patterns of connections and movements in dual-map overlays: a new method of publication portfolio analysis. J. Assoc. Inf. Sci. Technol. **65**(2), 334–351 (2014). https://doi.org/10.1002/asi.22968
25. Chen, C.: Predictive effects of structural variation on citation counts. J. Am. Soc. Inf. Sci. Technol. **63**(3), 431–449 (2012). https://doi.org/10.1002/asi.21694
26. Chen, C., Ibekwe-SanJuan, F., Hou, J.: The structure and dynamics of cocitation clusters: a multiple-perspective cocitation analysis. J. Am. Soc. Inform. Sci. Technol. **61**(7), 1386–1409 (2010). https://doi.org/10.1002/asi.21309
27. Chen, C.: Searching for intellectual turning points: progressive knowledge domain visualization. Proc. Natl. Acad. Sci. **101**(suppl 1), 5303–5310 (2004)
28. Chen, C., Dubin, R., Kim, M.C.: Emerging trends and new developments in regenerative medicine: a scientometric update (2000–2014). Expert Opin. Biol. Ther. **14**(9), 1295–1317 (2014). https://doi.org/10.1517/14712598.2014.920813
29. Reuters, T.: Web of Science (2012)
30. Chen, H., et al.: Terrorism Informatics: Knowledge Management and Data Mining for Homeland Security. Springer Science & Business Media (2008)
31. Church, K.W., Hanks, P.: Word association norms, mutual information, and lexicography. Comput. Linguist. **16**(1), 22–29 (1990)

32. Campello, R.J.G.B., Hruschka, E.R.: A fuzzy extension of the silhouette width criterion for cluster analysis. Fuzzy Sets Syst. **157**(21), 2858–2875 (2006). https://doi.org/10.1016/j.fss. 2006.07.006

33. Aizawa, A.: An information-theoretic perspective of tf–idf measures. Inf. Process. Manag. **39** (1), 45–65 (2003),. https://doi.org/10.1016/s0306-4573(02)00021-3

34. Quackenbush, S.R., Barnwell, T.P., Clements, M.A.: Objective Measures of Speech Quality. Prentice Hall (1988)

35. Sakaki, T., Okazaki, M., Matsuo, Y.: Earthquake shakes Twitter users: real-time event detection by social sensors. In: Proceedings of the 19th International Conference on World Wide Web, pp. 851–860. ACM, Raleigh, North Carolina, USA (2010). https://doi.org/10. 1145/1772690.1772777

36. Sankaranarayanan, J., et al.: Twitterstand: news in tweets. In: Proceedings of the 17th ACM SIGSPATIAL International Conference on Advances in Geographic Information Systems, pp. 42–51. ACM, Seattle, Washington, USA (2009). https://doi.org/10.1145/ 1653771.1653781

37. Phuvipadawat, S., Murata, T.: Breaking news detection and tracking in Twitter. In: 2010 IEEE/WIC/ACM International Conference on Web Intelligence and Intelligent Agent Technology (WI-IAT), vol. 3, pp. 120–123. IEEE, Toronto, ON, Canada (2010). https:// doi.org/10.1109/wi-iat.2010.205

38. Lee, R., Sumiya, K.: Measuring geographical regularities of crowd behaviors for Twitter-based geo-social event detection. In: Proceedings of the 2nd ACM SIGSPATIAL International Workshop on Location Based Social Networks, pp. 1–10. ACM, San Jose, California, USA (2010). https://doi.org/10.1145/1867699.1867701

39. Li, R., et al.: Tedas: a twitter-based event detection and analysis system. In: 2012 IEEE 28th International Conference on Data Engineering, pp. 1273–1276. IEEE, Washington, DC, USA (2012). https://doi.org/10.1109/icde.2012.125

40. Mathioudakis, M., Koudas, N.: Twittermonitor: trend detection over the twitter stream. In: Proceedings of the 2010 ACM SIGMOD International Conference on Management of data, pp. 1155–1158. ACM, Indianapolis, Indiana, USA (2010). https://doi.org/10.1145/1807167. 1807306

41. Petrović, S., Osborne, M., Lavrenko, V.: Streaming first story detection with application to twitter. In: Human Language Technologies: The 2010 Annual Conference of the North American Chapter of the Association for Computational Linguistics, pp. 181–189. Association for Computational Linguistics, Los Angeles, California, USA (2010)

42. Marcus, A., et al.: Twitinfo: aggregating and visualizing microblogs for event exploration. In: Proceedings of the SIGCHI Conference on Human Factors in Computing Systems, pp. 227–236. ACM, Vancouver, BC, Canada (2011). https://doi.org/10.1145/1978942.1978975

43. Watanabe, K., et al.: Jasmine: a real-time local-event detection system based on geolocation information propagated to microblogs. In: Proceedings of the 20th ACM International Conference on Information and Knowledge Management, pp. 2541–2544. ACM, Glasgow, Scotland, UK (2011). https://doi.org/10.1145/2063576.2064014

44. Cataldi, M., Di Caro, L., Schifanella, C.: Emerging topic detection on twitter based on temporal and social terms evaluation. In: Proceedings of the Tenth International Workshop on Multimedia Data Mining. ACM, Washington, DC, USA (2010). https://doi.org/10.1145/ 1814245.1814249

45. Gaglio, S., Lo Re, G., Morana, M.: Real-time detection of twitter social events from the user's perspective. In: 2015 IEEE International Conference on Communications (ICC), pp. 1207–1212. IEEE, London, UK (2015). https://doi.org/10.1109/icc.2015.7248487

46. Lee, Y., Nam, K.W., Ryu, K.H.: Fast mining of spatial frequent wordset from social database. Spat. Inf. Res. **25**(2), 271–280 (2017). https://doi.org/10.1007/s41324-017-0094-6

47. Petkos, G., et al.: A soft frequent pattern mining approach for textual topic detection. In: Proceedings of the 4th International Conference on Web Intelligence, Mining and Semantics (WIMS14), p. 25. ACM, Thessaloniki, Greece (2014). https://doi.org/10.1145/2611040. 2611068

48. Huang, J., Peng, M., Wang, H.: Topic detection from large scale of microblog stream with high utility pattern clustering. In: Proceedings of the 8th Workshop on Ph.D. Workshop in Information and Knowledge Management, pp. 3–10. ACM, Melbourne, Australia (2015). https://doi.org/10.1145/2809890.2809894

49. Yun, U., Ryang, H., Ho Ryu, K.: High utility itemset mining with techniques for reducing overestimated utilities and pruning candidates. Expert Syst. Appl. **41**(8), 3861–3878 (2014). https://doi.org/10.1016/j.eswa.2013.11.038

50. Dong, G., Li, J.: Efficient mining of emerging patterns: discovering trends and differences. In: Proceedings of the Fifth ACM SIGKDD International Conference on Knowledge Discovery and Data Mining, pp. 43–52. ACM, San Diego, California, USA (1999). https://doi.org/10.1145/312129.312191

51. Sakaki, T., Okazaki, M., Matsuo, Y.: Tweet analysis for real-time event detection and earthquake reporting system development. IEEE Trans. Knowl. Data Eng. **25**(4), 919–931 (2013). https://doi.org/10.1109/TKDE.2012.29

Combination of Inner Approach and Context-Based Approach for Extracting Feature of Medical Record Data

Van-Minh Le, Quang-Ngu Truong and Tu-Thien Huynh

Abstract Changing from a legacy Health Information System (HIS) to a modern HIS creates a problem of migration. Particularly, it requires us to handle unstructured data. In this paper, we proposed a new approach which is used to detect keywords from textual documents, and it involves two stages. Firstly, we study extracting features from the words by exploiting the relation between their characters. The following stage is presenting a combination of inner approach and context-based approach in order to make this extraction. The method is tested with MIMIC-II dataset and, in our problem, it shows a better result compared to old methods. We believe that it can be applied into natural language processing problems in other fields.

Keywords Machine learning · Text extraction · Pattern recognition · Data migration · Neural network

V.-M. Le (✉)
School of Information and Communication Technology - The University
of Danang, Da Nang, Vietnam
e-mail: vanminh.le246@gmail.com

V.-M. Le
IRD, UMI 209, UMMISCO, Bondy, France

Q.-N. Truong
Danang University of Science and Technology - The University
of Danang, Da Nang, Vietnam
e-mail: nguqtruong@gmail.com

T.-T. Huynh
ABX Software Services, Ltd., Da Nang, Vietnam
e-mail: marsch.huynh@gmail.com

© Springer International Publishing AG, part of Springer Nature 2018   113
A. Sieminski et al. (eds.), *Modern Approaches for Intelligent Information
and Database Systems*, Studies in Computational Intelligence 769,
https://doi.org/10.1007/978-3-319-76081-0_10

1   Introduction

1.1   Context of Problem

Health Information System integrates different modules such as patient management, appointment management, financial management and so on. It is clear that implementing HIS can bring many benefits. For instance, it improves patient's satisfaction as they can check appointment or access their medical records online. HIS also improves hospital administration by keeping a record of patients, necessary information and providing a secured data backup. Briefly, HIS can improve the quality of healthcare services and remove operational inefficiencies.

The company sponsoring us has already launched a HIS which conform to the FHIR HL7 standard [1, 2]. In detailed, FHIR (Fast Healthcare Interoperability Resources) is a draft standard for data formats, elements and application programming interfaces (APIs) for the exchange of electronic health records. An interaction between existing health care systems makes it easier to provide medical information to providers. The FHIR data format is based on HL7.

Turning into the term of Electronic Health Records (EHR), it is an electronic version of a patients medical history, and is maintained by the provider over time. [3–5]. The EHR helps improve accuracy and clarity of storing medical records and capture patients' condition. It automates access to information through a networked system.

1.2   Problem of Migration in Electronic Health Records

The company sponsoring us has successfully developed a new HIS based on the FHIR HL7 standard. Deploying the new HIS faces a large task of data migration. The Fig. 1 presents this task of every Electronic Health Records.

In the context of a healthcare system, the migration task faces two main problems. Firstly, the old system and the new system are both running; therefore the transforming is required to be automatic and has a fast recognition ability. Secondly, privacy is a major concern because the system manages Electronic Health Records which are very sensitive. It is clear that transforming records in a paper is definitely a time-consuming task as we need to read the paper and then type it into the database. As to protect the privacy, we cannot outsource the task in order to keep the information secret. Briefly, the way forward to solve these problems involves building an intelligent system helping us doing that manual work.

The plugin has to convert PDF files to text files and then recognize meaningful keywords in the text. Particularly, the plugin takes a PDF file as input and return

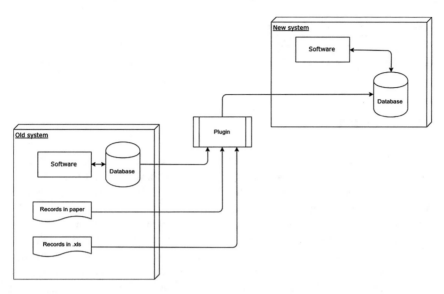

Fig. 1 General schema of migration task

NGUYEN, VAN A

70 Y old Female, **DOB:** 06/28/1926

Account Number: 34777

11 ABC CENTER DR, TEXAS, TX-77072

Home: 478-789-2278

Guarantor: NGUYEN, VAN A Insurance: VUN LEN IPA

External Visit ID: 37240993

Appointment Facility: Center for Integrated Medicine

Vital Signs:

Ht 60, Wt 142, BP 107/69, Temp 98.2, HR 71, BMI 27.73.

Fig. 2 A progress note sample

a record containing words/phrases and their meanings. This words/phrases will be inserted into a database with the same metadata and the same schema of the new system (Fig. 2).

The result should looks like this (Table 1):

Table 1 A sample of record in database

Name	Birthday	BMI	W	H	BP	Facility
Nguyen Van A	06/18/1946	27.73	142	60	107/69	Center for Integrated Medicine

1.3 Problem to Solve

The problem is how to build a machine learning model in order to automatically recognize words/phrases and their meanings. The idea of this work is finding a good approach to extract features of data (in this case, data is in form of text), after that, choosing the right machine learning model.

In this paper, we propose a new method combining inner approach and context-based approach for extracting feature for medical data. This method is tested in MIMIC-II dataset, and, in our problem, the method shows a better result compared to many old feature extraction methods. Therefore, we believe that the method can be applied to other natural language processing problems. This paper is organized as follow: Sect. 2 presents related work, Sect. 3 is our proposed method, Sect. 4 shows the experiment, Sect. 5 is the evaluation and Sect. 6 is our conclusion.

2 Related Work

Since the essence of the problem is to extract necessary information from medical records, our theoretical studies indicate that the problem is in the field of data mining. Data mining is researching field where we learn the relationships and general patterns in big data, such as the relationship between patient data and diagnoses [6–9].

According to the work presented in [10], we consider our problem as the process of classifying text/words into different classes. That is the reason for learning about word sorting methods in the text. In the paper [11, 12], the authors explain how to solve a problem of recognition. In our case, we focus on the important part of this problem, the feature extraction.

The solution we selected for the problem of pre-processing is Named-entity recognition [13, 14]. Named-entity recognition (NER) is a subtask of information extraction that seeks to locate and classify named entities in text into pre-defined categories such as the names of persons, organizations, locations, expressions of times, quantities, monetary values, percentages, etc.

In fact, the NER issue requires extracting features of a word. Fortunately, there are already several methods such as Word2vec, Glove, and local feature. These models can be categorized into two main model families: (1) global matrix factorization methods, such as latent semantic analysis (LSA) of Deerwester et al. [15] and (2) local context window methods, such as the skip-gram model of Mikolov et al. [16].

Word2vec was created by a team of researchers led by Tomas Mikolov at Google and it is described in paper Efficient Estimation of Word Representations in Vector Space [17]. The vector representations of words learned by word2vec models have been shown to carry semantic meanings and are useful for various NLP tasks [18].

3 Proposition

In this section, we propose the methods to extract features from word so that we can you in the learning model. This learning model which recognition phrases/words play the most important role in data migration.

3.1 Naive

First, we begin with the most direct approach (or the naive approach) to extract features. We were inspired by the image processing method with the MNIST dataset when many studies use directly all pixels of an image in feature vectors. In this case, we use all character of a word to build the feature vector. Each word has many characters, and we fix this word's length by 10 characters. In detail, if the length of a word is bigger than 10, we concentrate the word to its 10 first characters; in case of the length is smaller than 10, we fill out the feature vector with number 0. At this time, the character is encoded based on the ASCII encoding and each word is converted to a 10-dimensional vector.

This method brings a poor performance as many classes of words/phrase are fully not recognized. Details of this performance will be represented in the evaluation section

3.2 Re-using Word2vec

In natural language processing, a context of a word plays a major role in finding and determining the meaning of this word, the synonyms, and the antonyms. Especially, English is a language that a word can be understood simply by collecting meaning from the words around it. Thus, we want to build a word representation method that considers the co-occurrence between this word and its context.

In our study, we reuse the word2vec model to create the vector feature. All words in the dataset are trained by a skip-gram model and the dimension of feature $N\text{-}dim$ is set to 100.

3.3 Character2vec

Initially, inspired by the performance of word2vec, we believe that a character and its context characters can contain meaning, too. We reuse the skip-gram model from word2vec which is proved that it performs very well in some natural language processing tasks. In our case, the target character is encoded at the input layer, and the context characters are at the output layer.

Our proposed character2vec is similar to word2vec. The first layer represents the character set containing all the characters in a corpus. Each character is in one-hot format. Therefore, the length of this one-hot vector is equal to the size of the character set. The matrix $\mathbf{W}_{V \times N}$ is the embedding matrix of characters. The size of hidden layer N is set to 16. Finally, the output layer illustrates the context of the character. To define the context of a character, we choose a window size of 1. In detail, each character is considered in the context of 2 words next to it.

Like naive approach, each word is fixed the length by 10 characters. Afterward, we concatenate vectors which are corresponding to each character in order to build the feature vector for the word. When we test this character2vec model in MIMIC-II dataset, it performs better than the Naive approach a lot.

3.4 Combination of Inner Approach and Context-Based Approach

We understand that the semantics of the word is expressed by its context and also its characters. Therefore, we suggest a new method combining the two methods of word2vec and character2vec. This is simply a combination of two vectors. We have a 100-dimensions vector from word2vec and a 160-dimensions vector from character model. We concatenate it into a 260-dimensions vector representing a word. The Fig. 3 illustrates an example of this combination of the word "interested".

We conduct Word2vec model training on MIMIC-II data by using the library Gensim [19, 20]. Gensim is a Python open source library for vectorizing and modeling word.

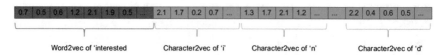

| 0.7 | 0.5 | 0.6 | 1.2 | 2.1 | 1.9 | 0.5 | ... | 2.1 | 1.7 | 0.2 | 0.7 | ... | 1.3 | 1.7 | 2.1 | 1.2 | ... | ... | 2.2 | 0.4 | 0.6 | 0.5 | ... |

Word2vec of 'interested   Character2vec of 'i'   Character2vec of 'n'   Character2vec of 'd'

Fig. 3 Example about of combine extract feature method. The part in red color is the 100-dimensional word2vec embedding vector of word "interesting". Each character has a 16-dimensional embedding vector which is represented in blue color; we subsequently concatenate embedding vectors of the word's characters to the word embedding in order to get the final feature vector

4 Experiments

4.1 Datasets

We train our models on MIMIC-II dataset [21]. Particularly, we exploit nearly 32.500 event notes in this dataset. The dataset is already de-identified and classified manually; we then modify the dataset with mock-up data. In fact, we build a dataset in which all words are classified into 6 classes: Clip number, hospital ward name, human name, location, hospital and uninteresting class.

4.2 Implementation Details

Initially, we choose Multi Hidden Layer Neural Network as the method used to evaluate the effectiveness in recognition of each feature selection method. We do tests in 5 cases which are Word2vec (w2v), Character2vec (c2v), naive approach (naive), a combination of Word2vec and Character2vec (w2v_c2v), and a combination of Word2vec and naive approach (w2v_naive). In each case, we calculate recall and precision regarding 10 Multi Hidden Layer Neural Network models from no hidden layer to 9 hidden layers. Then, we use f1 [22] score measure to compare a solution to each others.

5 Evaluation

In this case, we run a general evaluation for machine learning. We divide our dataset into 2 parts: one for training, other for testing.

5.1 Compare Naive with Character2vec

First, we compare our proposed method with the naive one. In this case, we compare the accuracy of prediction of these two methods. The Fig. 4 shows that the character2vec better than naive method.

For observing confusion matrix, we find that naive method cannot recognize some classes. The reason for this problem is that naive method consider exactly every character of a word. Then, this method cannot recognize the new word which is not in the train data set. On the other hand, character2vec studies the relation between a character and its neighbors. In the top side of Table 2, naive method cannot recognize "Hosp. Ward Name" class which describe the place where the hospital locates. But in the bottom side, this class can be recognized by character2vec method.

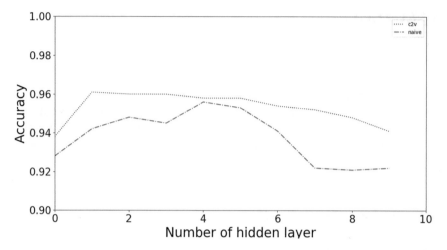

Fig. 4 Accuracy in 2 cases naive and character2vec

Table 2 Confusion matrix in 2 cases naive (above) and character2vec (below)

	Clip number	Hosp. ward name	Human name	Location	Hospital
Clip number	3692	0	0	0	0
Hosp. ward name	0	0	0	0	0
Human name	0	0	355	33	0
Location	0	0	3	552	1
Hospital	0	0	47	0	5
	Clip number	Hosp. ward name	Human name	Location	Hospital
Clip number	2692	0	0	0	0
Hosp. ward name	0	208	104	0	0
Human name	0	0	5870	459	1
Location	0	0	139	1909	0
Hospital	0	0	431	324	863

From observing result, we notice that more hidden layers do not give better result. The explanation for this phenomenon is the overfitting issue because the train data set does not provide enough data. Once more notice is that there are a lot of words belonging to uninterested class. This is normal for a problem of text mining or information retrieval. We have a lot of data but we need to find some of them which interest us. Then, we focus on other measures which are suitable for this case: precision and recall. In Fig. 5 which shows precision and recall in both cases, we can find that the character2vec prevail naive method in most cases.

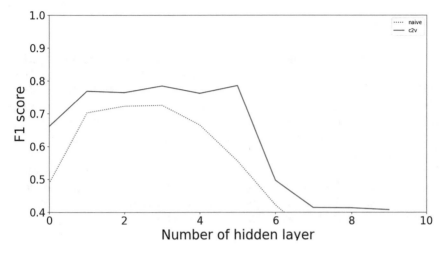

Fig. 5 F1 score in two cases: naive and chracter2vec

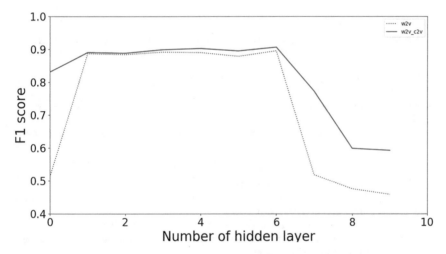

Fig. 6 F1 score in 2 cases word2vec and word2vec combines character2vec

5.2 Compare Word2vec with Word2vec Combines Character2vec

In this test case, we compare existing word2vec with our proposed method (combination of word2vec and character2vec). Figure 6 shows the result of 2 methods in precision and recall. From the observation, we can see that our proposed method is better. The main reason for this result is that the combination benefits both inner approach and context-based approach.

6 Conclusion

In this paper, we propose character2vec and also combination of character2vec and word2vec which are useful for extracting feature of word in order to extract text from medical record documents. Our proposition is tested with MIMIC-II dataset and gives the good result. In practice, we implement the proposed method with LSTM model and the model performs very well. Therefore, we believe that the method is really useful and we can apply it to extract features of words in other natural language processing tasks.

In future work, we propose to integrate another approach of extraction which focuses on subword of a word. The approach is inspired by study in [23]. Besides, we propose to test the extraction with other complicated neural networks such as recurrent network and long short-term memory (LSTM).

References

1. Braunstein, M.L.: Patientphysician collaboration on FHIR (Fast Healthcare Interoperability Resources). In: 2015 International Conference on Collaboration Technologies and Systems (CTS), pp. 501–503. IEEE (2015)
2. Franz, B., Schuler, A., Kraus, O.: Applying FHIR in an integrated health monitoring system. EJBI **11**(2) (2015) en61–56
3. Habib, J.L.: EHRs, meaningful use, and a model EMR. Drug Benefit Trends **22**(4), 99–101 (2010)
4. Kierkegaard, P.: Electronic health record: wiring Europes healthcare. Comput. Law Secur. Rev. **27**(5), 503–515 (2011)
5. Gunter, T.D., Terry, N.P.: The emergence of national electronic health record architectures in the United States and Australia: models, costs, and questions. J. Med. Internet Res. **7**(1) (2005)
6. Chakrabarti, S., Ester, M., Fayyad, U., Gehrke, J., Han, J., Morishita, S., Piatetsky-Shapiro, G., Wang, W.: Data mining curriculum: a proposal (version 1.0). Intensive Work. Gr. ACM SIGKDD Curric. Comm. **140** (2006)
7. Han, J., Pei, J., Kamber, M.: Data mining: concepts and techniques. Elsevier (2011)
8. Hastie, T., Tibshirani, R., Friedman, J.: The elements of statistical learning: data mining, inference, and prediction. Biometrics (2002)
9. Pujari, A.K.: Data mining techniques. Universities press (2001)
10. Paulus, D.W., Hornegger, J.: Applied pattern recognition: a practical introduction to image and speech processing in C++. Morgan Kaufmann Publishers (1998)
11. Burges, C.J.: A tutorial on support vector machines for pattern recognition. Data Min. Knowl. Disco **2**(2), 121–167 (1998)
12. Hopfield, J.J., et al.: Pattern recognition computation using action potential timing for stimulus representation. Nature **376**(6535), 33–36 (1995)
13. Marsh, E., Perzanowski, D.: Muc-7 evaluation of IE technology: overview of results. In: Seventh Message Understanding Conference (MUC-7): Proceedings of a Conference Held in Fairfax, Virginia, 29 April–1 May 1998. (1998)
14. Lagerweij, R., Bron, M., Monz, C.: A Joint Classification Approach to Slot-Filling (2012)
15. Deerwester, S., Dumais, S.T., Furnas, G.W., Landauer, T.K., Harshman, R.: Indexing by latent semantic analysis. J. Am. Soc. Inf. Sci. **41**(6), 391 (1990)
16. Mikolov, T., Yih, W.t., Zweig, G.: Linguistic regularities in continuous space word representations. In: HLT-NAACL. vol. 13, pp. 746–751 (2013)

17. Mikolov, T., Chen, K., Corrado, G., Dean, J.: Efficient estimation of word representations in vector space. arXiv:1301.3781 (2013)
18. Rong, X.: Word2vec parameter learning explained. arXiv:1411.2738 (2014)
19. Rehurek, R., Sojka, P.: Gensim-python framework for vector space modelling. NLP Centre, Faculty of Informatics, Masaryk University, Brno, Czech Republic (2011)
20. Rehurek, R., Sojka, P.: Gensim–statistical semantics in python (2011)
21. Goldberger, A.L., Amaral, L.A.N., G.L.H.J.I.P.M.R.M.J.M.G.P.C.K.S.H.: Components of a new research resource for complex physiologic signals. Circulation **101**(23):e215–e220 [Circulation Electronic Pages; http://circ.ahajournals.org/content/101/23/e215.full]; 13 June 2000
22. Powers, D.M.: Evaluation: from precision, recall and F-measure to ROC, informedness, markedness and correlation (2011)
23. Bojanowski, P., Grave, E., Joulin, A., Mikolov, T.: Enriching word vectors with subword information. arXiv:1607.04606 (2016)

A Novel Method to Predict Type for DBpedia Entity

Thi-Nhu Nguyen, Hideaki Takeda, Khai Nguyen, Ryutaro Ichise
and Tuan-Dung Cao

Abstract Based on extracting information from Wikipedia, DBpedia is a large scale knowledge base and makes this one available using Semantic Web and Linked Data principles. Thanks to crowd-sourcing, it currently covers multiples domains in multilingualism. Knowledge is obtained from different Wikipedia editions by effort of contributors around the world. Their goal is to manually generate mappings Wikipedia templates into DBpedia ontology classes (types). However, this cause makes the type inconsistency for an entity among different languages. As a result, the quality of data in DBpedia can be affected. In this paper, we present the statement of type consistency for an entity in multilingualism. As a solution for this problem, we propose a method to predict the entity type based on a novel conformity measure. We also evaluate our method based on database extracted from aggregating multilingual resources and compare it with human perception in predicting type for an entity. The experimental result shows that our method can suggest informative types and outperforms the baselines.

Keywords DBpedia · Ontology · Mappings · Conformity · Consistency

T.-N. Nguyen (✉)
Hai Phong University, Haiphong, Vietnam
e-mail: nhunt@dhhp.edu.vn

H. Takeda · K. Nguyen · R. Ichise
National Institute of Informatics, Tokyo, Japan
e-mail: takeda@nii.ac.jp

K. Nguyen
e-mail: nhkhai@nii.ac.jp

R. Ichise
e-mail: ichise@nii.ac.jp

T.-N. Nguyen · T.-D. Cao
Hanoi University of Science and Technology, Hanoi, Vietnam
e-mail: dungct@soict.hust.edu.vn

© Springer International Publishing AG, part of Springer Nature 2018
A. Sieminski et al. (eds.), *Modern Approaches for Intelligent Information
and Database Systems*, Studies in Computational Intelligence 769,
https://doi.org/10.1007/978-3-319-76081-0_11

1 Introduction

Semantic Web is the development of the World Wide Web to represent semantical information in a machine-understandable way. Currently, the potential of Semantic Web is mostly enriched by Linked Data (LD) and Linked Open Data (LOD) so that amount of data accessed in a simple way on the Web has rapidly increased in short time. DBpedia—is one of the most remarkable examples of LD, it is the central LOD repository [1, 5]. It is built upon the community effort to extract the knowledge from Wikipedia [7]. Wikipedia is known as an encyclopedic knowledge of mankind which covers multiples domains [8]. However, Wikipedia itself only offers very limited querying and searching capabilities while needs of human about information searching are very diverse. DBpedia with publishing structured data on the Web in LD principles seems to be a solution to fill this gap. It was only based on English Wikipedia at the first [5, 6] and now it provides structured data for more than 100 Wikipedia language editions. It becomes the huge, cross-domain knowledge base in multilingualism.

Taking the advantage from Wikipedia knowledge base of DBpedia project opens up many opportunities. However, the issue of data quality needs to be the concern when this data is used in many different applications. Currently, the structured data in DBpedia is maintained by its community based on the mapping project [8]. Contributors from many countries have joined this project and their target is to populate this knowledge base by mapping the Wikipedia templates into the corresponding types (e.g., Species, Person, and Place) in DBpedia ontology [7]. These templates are particular pages that almost contain infoboxes—a set of subject-attribute-value to summarize a Wikipedia article. For example, infobox country in English Wikipedia is mapped to type country in ontology and etc. It is clear that the type information plays an important role as the backbone in knowledge base. The quality of mappings thus will affect directly to the quality of data in DBpedia. Despite the maturity of the DBpedia community, the lack of consensus between the contributors from different languages is still remaining as an issue.

In detail, a real-world entity is represented by multiple instances in DBpedia. Each instance is described in a specific language and its type is based on the mappings constructed for that language. There are only 32 languages having mappings currently, this means an entity will be typed in these ones. Due to independent hand-generated mappings for specific languages, the types of particular instances are different even those ones describe as the same entity. Concretely, considering an entity, some types may be different at the specific-level, correct, or incorrect. For example, the entity of Barack Obama is recognized as Person, Politician, President, Artist and Book in 29 languages. Here, there is an agreement between Person, Politician and President but still different at the specific levels. Meanwhile, Artist and Book are incorrect. Obviously, the mapped types are not consistent among different languages. Meanwhile, the type consistency is necessary to guarantee the quality of data and consolidation of DBpedia, especially for other languages if they want to join in this project. Therefore, choosing the most suitable type is an important task but it is difficult, too.

There are also some approaches in type inference such as SDType [9], Airpedia [10] and Hypernyms [4]. These ways are good at predicting higher-level types (as `Person`), while predicting more specific types (such as `President`) are limited. Thus, in this paper, we propose a novel conformity measure to predict an entity type. We combine the consideration of the specific-level of and the majority voting. Before coming to this issue, we provide the detailed statistic about the consistency of type for an entity based on analysis DBpedia dumps of all languages have mappings. This helps the community of editors to have an overview picture about the statement of type consistency.

Organization: The remainder of the paper is structured as follows. Section 2 reviews existing solutions related to this research. The statement of type consistency is described in Sect. 3. The novel idea of type prediction is presented in Sect. 4. The experiments, obtained results and theirs evaluations are showed in Sect. 5. Finally, a conclusion is given in Sect. 6 with some further suggested studies.

2   Related Work

Type inference is an issue that is concerned by many research, especially mappings are hand-generating currently. Airpedia [10] is to aim to automatically map Wikipedia templates to types. They infer types of entities based on 5 pivot languages by the basic algorithm with using the most specific ancestor and some other heuristic hypotheses. They also exploit cross-language links between different languages in DBpedia. Comparing with our work, this is the most related research.

With the same purpose is to predict type, the authors of [4, 11] have the other approaches. In particular, [11] proposes a link-based type inference called SDType. This method benefits links between resources as indicators for types. For each link, they used the statistical distribution of types in the subject and object position of the property for predicting the instances type. Kliegr and Zamazal [4] gives a method to infer a type statistically for dataset in Linked Hypernyms Dataset. This database is only used to describe articles of Dutch, English and German languages. Most of these types are inferred based on exited ones of ontology due to using the same data modeling approach in DBpedia. Thus, the results of this method completely based on assigning a type for an entity in DBpedia.

Meanwhile, [3] provides a tool for automatically typing DBpedia entities based on Tipalo algorithm. Tipalo interprets natural language definition of a DBpedia entity, which is extracted from its corresponding Wikipedia page abstract. Types are identified by means of a set of heuristics based on graph patterns, disambiguated to Word-Net and aligned to two top-level ontologies.

Fossati et al. [2] present a method called DBTax to learn automatically the taxonomy from the Wikipedia category system without supervising and assigning types to DBpedia entities. DBTax also propose a solution for the problem of wrong classification of entities and the overlapping among the four mostly populated DBpedia types, namely `Place`, `Person`, `Organization` and `Work`. They mainly based on Wikipedia category system to derive the taxonomy for the classification of DBpedia resources by using natural language processing.

Miao et al. [9] also infers this type automatically but it refers only to the case of Chinese. They used the inference model, which trained data and used both 2 elements: interpolation and external link from DBpedia. Their method received a positive result but the interpolation still had some drawbacks about time and the number of trained samples. For example, after training session type could be assigned to only one particular entity instead of a general entity. In conclusion, the entity in their research is the same as the instance which we have pointed out. In our research, the entity concept is used to indicate things in real world, not only in a specific subject.

3   The Consistency of Type

In this section, we present the consistency of types for an entity and provide the statistic of obtained results. Firstly, we brief the main information of DBpedia ontology and the statement of mappings in different languages currently. The figure, which we used, is the latest one and is updated from DBpedias website.

3.1   The Statements of Mappings in DBpedia

The latest DBpedia Ontology version encompasses 755 classes which has a form as the sub-sumption hierarchy and are described by more than 3000 different properties.[1] They characterize briefly the important features of million things in multilingualism. As mentioned above, mappings are created manually by the community to populate the shared ontology in different languages. The goal is to map Wikipedia templates to corresponding types in DBpedia ontology. Currently, the mapping communities grew up for 32 languages. However, this number is still small when comparing with 128 languages that are extracted by DBpedia. Thorough the statistics can be found on the mapping DBpedia website,[2] we draw out information of top languages, which have the ratio of template mapping is highest on the total number of languages having mappings as shown in Table 1. Obviously, Dutch has the maximum percentages of templates mapped with 95.82%. We also recognize that almost of them have their own chapters[3] in DBpedia. Recent survey shows that there are 19

---

[1] http://wiki.dbpedia.org/dbpedia-version-2016-04.

[2] http://mappings.dbpedia.org/server/statistics/.

[3] http://wiki.dbpedia.org/about/language-chapters.

Table 1 The statistic of mappings in 10 languages

Code	Language	Mapped templates (%)
nl	Dutch	95.82
sv	Swedish	92.62
ga	Irish	86.71
sr	Serbian	82.62
en	English	80.92
es	Spanish	80.01
pt	Portuguese	77.24
de	German	76.53
bg	Bulgarian	75.97
be	Belarusian	75.18

chapters with the newest Arab version. The growth of DBpedia community creates many chances in LOD area, which in turn gives birth to specific DBpedia languages chapter.

For this information, our aim is to provide the statement of mappings by different communities in particular and the development of DBpedia project in general. There-by, it also reflects the type assignment in multilingualism. There are only 32 languages that have been assigned types for instances. Therefore, it is the useful information utilized so that we can analyze the consensus of entity type in DBpedia. We also focus on mapped types of these languages to evaluate and compare the agreement among them. Furthermore, we especially concern about the languages that have the high percentage of mapped templates.

3.2 The Analysis of Type Consistency

Mappings are created manually by the different communities basing on using the uniform DBpedia ontology. As a sequence, this task is accomplished with dissimilar policies in multilingualism. Thus, the same template may be not mapped the same type in languages. Considering 32 languages have mappings, this disagreement is evident. The character of entitys type in the DBpedia ontology is a hierarchical classification. There will be the following cases occurred for an entity when it is mapped to classes in DBpedia ontology. An entity can be mapped to the top-level class owl:Thing and we ignored this case. We, therefore, may be have (1) the only type it is the perfect case and the percentage of consistency equals 100%; (2) the different types but these ones share an ancestor and have the same specific level it is also the good case; (3) the completely different types in specific languages and there is no the relationship among them it is the worst case but it occurs rarely; (4) the types mixed of (2) and (3) the complex types. Basing on these cases, we go analyze the

Table 2 The statistics of the agreement of types between two languages

Pairs		#instances have type in both languages	Instances have the same type (%)
nl	sv	308462	21.86
en	nl	201248	53.28
en	sv	158842	8.47
en	es	149234	30.92
it	en	144815	10.59
en	sv	158842	8.47
nl	es	143634	77.53
pt	en	132773	68.55
nl	it	130938	8.83
pl	it	118935	8.75

consensus of types in different aspects. This will provide the clear overview picture for editors, especially new beginners.

First, we compare all pairs of languages in multilingual database and we obtained the results of these mentioned cases. According to our analysis, the case (4) has the highest ratio with 47% pairs have more than 50% of resources assigned with the complex type among 476 language pairs. And the agreement of type assignment among different languages in (1) is not high. Table 2 illustrates the percentage of instances sharing the same type in 10 particular language pairs, the number of instances had a type in both languages are the most among all pairs. In general, only 37% pairs have more than 50% of instances with the same type.

Second, we also evaluate the type consistency for an entity by using the value of entropy based on the frequency of mapped types in different languages. In detail, we consider randomly 500 entities whose type is available in at least 5 languages. And we retrieve the results as follow as shown in Fig. 1.

Figure 1 shows that the type consistency values of 500 entities are not high and average about 0.79. Clearly, the maximum value does not belong to the entity mapped with the number of languages is the most. This value seemingly decreases when the number of languages increases.

Although the type information is very important in DBpedia, these statistics show that the unity among mappings communities is still weak. As a result, the consistency of mapped type among different languages remains many issues. These mappings were fixed and any updating or changing will be take place with the next DBpedia version. Meanwhile, DBpedia community is growing with more new editors who want to create mappings for their own languages. Giving a solution to increase the consensus among these communities is necessary in this context. Thus, taking the advantage of the share model—DBpedia ontology will be easier and type prediction for an entity is more suitable.

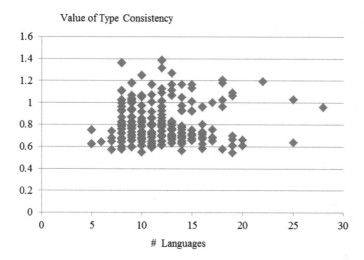

Fig. 1 The consistency of type for 500 entities in multilingualism based on the value of entropy

4 Type Prediction

Mappings are desirable about the type consistency. This not only affects directly the quality of data in DBpedia, but also makes many difficulties for new editors of mapping communities later. Therefore, the entity type prediction is considered as an important problem. It is helpful for the utility of DBpedia versions whose mapping communities are immature. In addition, it is also the core of automatic mapping creation. Since the type information is inconsistently described in different languages, it is difficult to recognize the most suitable type of an entity. In this section, we describe how to predict that type for an entity. Our idea is based on the combination of the specific-level and the majority voting. We propose prediction which use a heuristic search method to find out the most suitable type for an entity. The pseudo code of TPrediction is given in Algorithm 1.

The input is any entity in DBpedia dumps that is aggregated with multilingual resources. Thus, the most specific-level types assigned by different languages can be searched. Then, we define the conformity $Con(x)$ of a most specific-level type x. The conformity is a recursive value taking the sum of the frequency of x and the conformity of its parent.

$$Con(x) = frequency(x) + Con(parent(x)) \tag{1}$$

where the frequency of x is the number of languages that the entity are mapped with type x. For an entity, we select the most suitable type by picking the one with the highest conformity. Obviously, this chosen type will meet the condition that it is used in the most languages and also adequately specific.

Algorithm 1 Tprediction

Input: an entity e of in multilingualism L
Output: the most suitable type t of e
 Begin
 T $\longleftarrow$ Ø
 best $\longleftarrow$ 0
 for instance *ins* $\in e$ **do**
  T$\longleftarrow$ T $\cup x_i, l_i \in L$
  for x $\in$ T **do**
   frequency $(x) \longleftarrow$ is the total number of treating the entity as x
   if x= "owl:Thing" **then**
    $Con(x) \longleftarrow 0$
   else
    $Con(x) = frequency(x) + Con(parent(x))$
   end if
  end for
 end for
 while T$\neq$ Ø **do**
  for $x \in$ T **do**
   best $\longleftarrow Con(x)$
   $t \longleftarrow x$
  end for
 end while
 return t

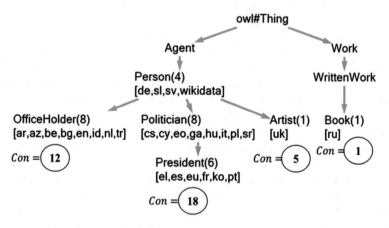

Fig. 2 The conformity of the most specific types

If there are many types that have the same highest conformity, we rank the type based on the conformity of their parent type. Lets consider the example in Fig. 2. In this figure, the entity of Barack Obama is assigned to 6 types in 29 languages. The conformity of the type President is the highest ($Con(President) = 18$). Therefore, it is selected as the prediction result.

Table 3 The accuracy of type suggestion methods

Our method	Majority voting
55%	44%

5 Results and Evaluation

We compare our method with a manually crafted dataset and two other baselines: (1) majority voting and (2) most specific ancestor. We build an entity database from all available language versions of DBpedia. This database contains 86,290,758 entities, which are constructed from 128,866,644 instances. An entity is the compilation of the instances interconnected via `owl:sameAs` links.

The difficulty of type suggestion is the diversity of types. Therefore, we employ the following procedure to create the dataset. We randomly select 500 entities whose type is available in at least 5 languages. After that, we pick up the 100 most inconsistent ones. Here, the inconsistency is estimated by the entropy of types frequency. Different from the conformity in Eq. 1, in order to guarantee the hierarchical relations, transitive types are counted. Meanwhile, transitive types are a set of ancestor types. After the selection, an expert is asked to assign the most suitable type among available of the entity (3). Finally, we compare the results of our method (1), (2) against (3).

Table 3 implies that our method gives the best result and even when we carry out the experiment with 500 entities, this remains unchanged. As entities of high inconsistency are selected, the most specific ancestor method always chooses the `owl:Thing`, which is the root of the DBpedia ontology. Thus, we do not enumerate its result here. Although majority voting is better than the most specific ancestor, in general, its result is not specific enough. This experiment demonstrates our prediction method is good but still 45% of the types are different from humans opinion. Most of them belong to types of place entity because among countries, the definitions of administrative region and residential area are different.

DBpedia ontology currently lacks types to represent all these dissimilarities. For example, Baug is a commune in France but there is no type for commune. Therefore, this entity should be mapped to `Settlement` type. However, our method returns the inaccurate type `City` because this type is more specific than `Settlement`. If we exclude this case, our method returns the result of accuracy type prediction with approximately 70% and the majority voting is 30%.

6 Conclusion and Future Work

In this paper, we present an overview on type consistency of an entity in DBpedia. In context of missing of documentation on mapped classes and properties, our statistics are helpful for the mapping community. Realizing the important of type prediction, we also proposed a new method to predict the most suitable type. For our approach, we combine the consideration of the specific-level and the majority voting to suggest the most suitable type of an entity. We also evaluated three methods and the results show that our method is the most promising one although it remains some weaknesses.

This work is just the first step in type prediction and much more work needs to be done to further explore the issue to construct an automatic DBpedia chapter for different languages. For future work, we will improve our method by also considering the conformity of the transitive types in order to give better predictions.

Acknowledgements This work was supported by NII (National Institute of Informatics) in Japan based on MOU agreement with Hanoi University of Science and Technology.

References

1. Bizer, C., Lehmann, J., Kobilarov, G., Auer, S., Becker, C., Cyganiak, R., Hellamnn, S.: DBpedia a crystallization point for the web of data. Web Semant. Sci. Serv. Agents World Wide Web **7**(3), 154–165 (2009)
2. Fossati, M., Kontokostas, D., Lehmann, J.: Unsupervised learning of an extensive and usable taxonomy for DBpedia. In: The 14th International Conference on Semantic Web, (ISWC), pp. 177–184 (2015)
3. Gangemi, A., Nuzzolese, A.G., Draicchio, F., Musetti, A., Ciancarini, P.: Automatic typing of DBpedia entities. In: The 11th International Conference on Semantic Web, (ISWC), vol. 7649, pp. 65–81 (2012)
4. Kliegr, T., Zamazal, O.: Towards linked hypernyms dataset 2.0: complementing DBpedia with hypernym discovery and statistical type inference. In: Proceedings of the 9th International Conference on Language Resources and Evaluation, (LREC2014), pp. 3517–3523 (2014)
5. Kobilarov, G., Bizer, C., Auer, S., Lehmann, J.: Dbpedia—A linked data hub and data source for Web applications and enterprises. In: The 18th International World Wide Web Conference (2009)
6. Kontokostas, D., Bratsas, C., Auer, S., Hellmann, S., Antoniou, I., Metakides, G.: Internationalization of linked data. The case of the Greek DBpedia edition. Web Semant. Sci. Serv. Agents World Wide Web **15**(3), 51–61 (2012)
7. Lehmann, J., Isele, R., Jakob, M., Jentzsch, A., Kontokostas, D., Mendes, P.N., Hellmann, S., Morsey, M., van Kleef, P., Auer, S., Bizer, C.: DBpedia—A large-scale, multilingual knowledge base extracted from Wikipedia. Semant. Web **6**(2), 167–195 (2015)
8. Mendes, P.N., Jakob, M., Bizer, C.: A multilingual cross-domain knowledge base. In: Proceedings of the 8th International Conference on Language Resources and Evaluation (LREC'12), pp. 1813–1817. European Language Resources Association (ELRA) (2012)
9. Miao, Q., Fang, R., Song, S., Zheng, Z., Fang, L., Meng, Y., Sun, J.: Automatic identifying entity type in linked data. In: Proceedings of the 30th Pacific Asia Conference on Language, Information and Computation: Posters, pp. 383–390. Springer (2016)
10. Palmero, A.A., Giuliano, C., Lavelli, A.: Automatic mapping of Wikipedia templates for fast deployment of localised DBpedia datasets. In: Proceedings of the 13th International Conference on Knowledge Management and Knowledge Technologies, pp. 1:1–1:8. ACM (2013)
11. Paulheim, H., Bizer, C.: Type inference on noisy RDF data. In: Proceedings of the 12th International Sematic Web Conference, pp. 510–525. Springer (2013)

Context-Based Personalized Predictors of the Length of Written Responses to Open-Ended Questions of Elementary School Students

Roberto Araya, Abelino Jiménez and Carlos Aguirre

Abstract One of the main goals of elementary school STEM teachers is that their students write their own explanations. However, analyzing answers to question that promotes writing is difficult and time consuming, so a system that supports teachers on this task is desirable. For elementary school students, the extension of the texts, is a basic component of several metrics of the complexity of their answers. In this paper we attempt to develop a set of predictors of the length of written responses to open questions. To do so, we use the history of hundreds elementary school students exposed to open questions posed by teachers on an online STEM platform. We analyze four different context-based personalized predictors. The predictors consider for each student the historical impact on the student answers of a limited number of keywords present on the question. We collected data along a whole year, taking the data of the first semester to train our predictors and evaluate them on the second semester. We found that with a history of as little as 20 questions, a context based personalized predictor beats a baseline predictor.

Keywords Written responses to open-ended questions · Online STEM platforms
Text mining · Context based predictors

R. Araya (✉) · A. Jiménez · C. Aguirre
Centro de Investigación Avanzada en Educación, Universidad de Chile,
Periodista Mario Carrasco 75, Santiago, Chile
e-mail: roberto.araya.schulz@gmail.com

A. Jiménez
e-mail: abjimenez@cmu.edu

C. Aguirre
e-mail: carlosaguirre@automind.cl

A. Jiménez
Department of Electrical and Computer Engineering, Carnegie Mellon University,
Pittsburgh, PA, USA

© Springer International Publishing AG, part of Springer Nature 2018
A. Sieminski et al. (eds.), *Modern Approaches for Intelligent Information
and Database Systems*, Studies in Computational Intelligence 769,
https://doi.org/10.1007/978-3-319-76081-0_12

1 Introduction

The study of the effect of the questions posed by the teacher is a subject of great importance in the teaching practice and in the preparation of teachers. Already in 1912 [1] emphasized the realization of questions as a fundamental component in teacher training. In recent years there has been an explosion of applications and educational platforms that ask questions of students. However, unlike the oral questions posed by the teacher in the classroom, the platforms in the summative and formative assessment modules mainly perform closed-ended multiple-choice questions. This is undoubtedly a great tool that allows the teacher in real time to know the progress and achievement of their students. The analysis of this type of answers has also made possible to construct computational models of the students. These models estimate the state of knowledge of each student and degree of mastery of the concepts and procedures to be taught. However, there is still little use of technology to analyze written responses to open-ended questions. While the platforms contain the possibility of introducing open answers, the analysis is still very basic. Written responses collected from students contain a wealth of information that cannot be captured by simply analyzing the answers to questions with multiple options. Here lies a huge potential for gaining a deeper understanding of student learning.

Written answers to open-ended questions depend on many factors. One of them is how the teacher formulates them. Recent advances in language technology open up a great opportunity to understand the impact of open teacher questions. So far the analysis is done mainly by hand. But this is very slow and represents a heavy workload for the teacher. The challenge then is to build tools that help the teacher analyze the answers students write on online platforms. This would allow us to get closer to knowing in real time what each student thinks and how he responds in written form. It will also allow the teacher to know how to ask to motivate their students so that they write answers that are really informative, and avoid generating extremely brief answers.

Written answers to open-ended questions also add a great value that is different from what oral answers deliver. Unlike oral answers, where typically the teacher only gets the answer of one or two students for each question, receiving written answers will get the answers of all the students. In addition, answers on a platform capture independent responses, as students do not listen to the other answers and therefore do not rely on what others say. That is, we capture personalized information that allows us to better estimate the distribution of learning. Additionally, written response reduces the problem of inhibition that complicates many students in having to speak and to respond publicly. Another advantage of written answers is a technical advantage. Greater fidelity is achieved to capture student response compared to oral responses, as no personal microphone or speech recognition system is required.

Moreover, written answers are a powerful tool that promotes learning. Students need to think more carefully about what they will respond to. The written response requires a planning process and then a writing process. According to [2], writing, unlike speech has several characteristics that differentiate it and are very important in learning. Writing is a learned and artificial behavior, speech is natural and spontaneous. The written response is a technological tool, much slower (paused), and also more rigid than the oral response. Writing does not have an audience because one responds alone, no one is listening, and since it generates a tangible and permanent product, it is also much more responsible. Since writing is a representation that makes one's ideas more visible, writing requires much more work and therefore, according to [2], it achieves more learning than verbal responses. Moreover, according to [3] "Written speech is considerably more conscious, and it is produced more deliberately than oral speech .... Consciousness and volitional control characterize the child's speech from the very beginning of its development." The Common Core Standards in Mathematics contains Standards for Mathematical Practice "Construct viable arguments and critique the reasoning of others". It specifies that students should justify their conclusions, communicate them to others, and respond to the arguments of others. Writing is not only a powerful communication tool. It is a powerful reasoning tool as well. According to [4], "until I read what I have written, I do not see the holes in my logic, the missing steps, or the rambling thoughts" p. 4.

There are several reasons why it is very interesting to estimate the length of students' written responses. First, the teacher needs to know if his question is being answered and the impact he is achieving. It is ideal for the teacher to have this information in real time and thus allows him to adjust the next question. A central question is whether the question provides information that allows knowing the reasoning of the student. To achieve that, the teacher should avoid short answers such as Yes-No, True-False, one word answers or very short answers like "I do not know". For this there are several indicators that the teacher could obtain immediately: detection of very short answers, comparisons with the length of the answer to other questions, if it promotes more writing than the previous questions, comparisons of the length of the answers of students who are academically strong with those of the academically weak, if different impacts are observed according to gender, comparisons with the answers of the same question in other courses and/or levels. For example, [5] records and studies the length of student responses, concluding that with sufficient waiting time the length may increase between 300% and 700% [6].

Second, when preparing questions, it would be very useful for teachers to have a simulator tool that allows them to predict the effect they will have on the course. It would be ideal to have a predictor as accurate as possible. This would help the teacher examine alternatives and choose more effective questions.

Third, estimating the effect of questions on the length of responses is a basic component in estimating other indicators of the effects of teacher questions on student responses. For example, to estimate rates such as the number of key concepts per word, and the number of positive or negative words per word. In the literature of automatic text analysis, length is also critical feature of text complexity. For example, readability algorithms, like the Flesch-Kincaid Grade Level metric, use length of sentences [7]. Moreover, predictive indices of essay quality are also related to length, such as syntactic complexity (as measured by number of words before the main verb) [7].

Fourth, predicting the length of answers is a powerful tool for teacher improvement. It allows the teacher to make a retrospective analysis of the type of questions he or she has been asking. According to [8] "By analyzing your questioning behavior you may be able to decrease the percentage of recall questions and increase the percentage that requires students to think". Having the length of the answers could be an essential component in building a system that automatically pre-classify the questions. There are several other taxonomies for questions [9], and for each of them it would also be interesting to predict the length of their answers for different categories.

Fifth, having a predictor is a tool for testing teachers' beliefs. According to [8] "Teachers sometimes think that if they begin to question why, explain, compare, or interpret, they are automatically encouraging their students to perform divergent or evaluative thinking operations." Is this really so or are there other factors involved?

Sixth, longer answers are indicative of a dialogical discourse of the teacher, and shorter answers are indicative of a more authoritative discourse. For example, Chin [10] says that "Classroom discourse can be analyzed in terms of its authoritative and dialogic functions [11]. In authoritative discourse, the teacher conveys information; thus, teacher talk has a transmissive function. Teacher talk often involves factual statements, reviews, and instructional questions; and students' responses to the teacher's questions typically consist of single, detached words. On the other hand, in dialogic discourse, the teacher encourages students to put forward their ideas, explore and debate points of view."

What does the teacher gain by obtaining a predictor of the length of the response per student and not only for a whole class? The study of written responses of whole classes provides important information on the impact of the teacher's questions [12]. However, by capturing each student's data and their previous behavior on similar questions, the estimator can be much more accurate and personalized. The student's historical behavior may give more information than generic information such as belonging to a course, the subject matter or content of the question, the student's gender and academic performance. There are many other components that influence the length of the responses that are not captured in those variables. These are factors that belong to the student. Knowing how the student responded when there are certain keywords can be very important and different from the response that other students of the same course, same gender and academic performance have had. For example, if in several previous questions where the word "airplane" was written a student writes long answers, then surely that same student will write a long answer in a next question with the word "airplane". This may not be the case in another student who is not motivated by the theme "airplanes". This greater accuracy and personalization is an important gain if at one point the teacher wants to stimulate more a certain student or a certain specific group of students. For example, a group that is falling behind, or a disinterested group, or a group of students with particular interests. On the other hand, when preparing their questions, the teacher could have an estimate per student of the effect that will have.

It is also a tool that allows the teacher to predict not only the average behavior of the class but the entire spectrum. That is, this type of tool could predict the dispersion of responses and, even more, predict the distribution of class responses. It is also important to distinguish between predicting the distribution of response of a generic class and the distribution of response of a specific class. By having a custom model, you can predict the distribution of personalized responses per class. This ability to estimate and predict considering the personal history of each student and even more so in real time is exclusive of computer systems. It is something that requires the use of a platform.

2 Methods

In this article, we report the analysis of a year's use of the open-ended question feature of Conecta Ideas, an online cloud-based STEM platform [13]. This platform was used in 13 low-SES elementary schools in Santiago, Chile. Students attended lab classes twice a week. The platform was used both for math and science classes. Teachers and lab coordinators tracked the students' progress in real time using their smartphones. A real time, early warning system highlights which students are having most difficulty completing the tasks. The Conecta Ideas platform also includes features that encourage the more advanced students to cooperate with and help their peers. Students who receive support from the teacher, lab coordinator or their peers rate the quality of the help they receive. If the majority of students are having difficulties, the teacher can freeze the system and explain certain key concepts.

In each session, the students solved between 10 and 30 multiple-choice questions. The schools had been using this platform for 5 years. Since the beginning of 2015 the teacher has also been able to pose one or two open-ended questions during each session, in addition to the multiple-choice questions. The goal of this was to have students reflect on the contents and to encourage student metacognition. The open-ended question feature was implemented in 2015 and was completely new for teachers. However, not all teachers started using it immediately. Even though it was introduced as a metacognitive tool, the teachers initially started by asking mainly calculation questions or questions that only required the students to identify a fact. Following a couple of meetings reviewing questions together and discussing alternative ways of promoting reasoning and argumentation, the teachers started to change the type of questions they asked, in favor of truly open-ended questions. An example of a question is "Pedro has to buy 4 pencils, each one costs 150. How much money did he spend? Explain how you arrived at the result". And an example of an answer is "I summed to get to the result and the result is 600". We present the analysis of the data gathered during 2016. We present evidence from 25 classes. These classes ranged between fourth and eighth grade and were all from 12 low-SES schools based in Santiago, Chile. We define several models to predict the length of each student's responses. All models use the Q & A information for the first semester to predict the length of answers to questions to be asked in the second semester. The Baseline model for a student in a subject is the average length of all that student's responses in that subject in the first semester.

The following models are introduced with the goal to reduce the universe of questions characterizing them with only 200 words. The predictors are based on the list L of the 200 most frequent non-stop words in the questions posed on first semester. For example, the 20 most frequent words are: apple, tens, which, Pedro, chocolates, explain, why, obtain, answer, how many, hundreds, flowers, María, how much, subtract, candies, cover, trays, mathematics, Francisco. The list L defines the context component of the predictors. For each word w of the list L and for each student s, the average length of all answers to questions in the first semester containing that word is calculated. We call this number the *impulse response* of that word for that particular student and we write it down as $h(w, s)$. If for a word a student did not answer any questions in the first semester that included that word, then the value of the impulse response of that word for that student is No_Info. Thus, every student is characterized by 200 impulse responses.

The Context Based-1 model $CB1(q, s)$ predicts for each student s the length of the answer to each question q posed on the second semester to the student. This prediction is obtained by computing the impulse response $h(w, s)$ for each of the words w within the subset $L(q, s)$ of the words that are in the question q that belong to L, and that are in questions answered on the first semester by s. Then Context Based-1 model $CB1(q, s)$ is defined as the average of these impulse responses. That is

$$CB1(q,s) = \frac{1}{Card(L(q,s))} \sum_{w \in L(q,s)} h(w,s) \tag{1}$$

If for a question all the impulse responses are No_Info, then $CB1$ is No_Info. The $CB1$ model could be better than Baseline since it uses more historic data. It is important to emphasize that this predictor does not use any data for the semester 2 at all.

The second context based model is the Context Based-2 model $CB2$. For each student s and question q, $CB2(q, s)$ also uses the impulse responses of all the words on the question that are within the list of the 200 most frequent words on the first semester. Context Based-2 is defined as the weighted average of these impulse responses. The weights used are the same as the frequency of those words in the first semester. It is then divided by the sum of the active weights. That is, for each student s and every question q, $CB2$ is always a convex combination of impulse responses. Thus if we denote $f(w)$ the frequency of the word w on all responses on the first semester, then

$$CB2(q, s) = \frac{\sum_{w \in L(q, s)} f(w) h(w, s)}{\sum_{w \in L(q, s)} f(w)} \tag{2}$$

If for a question all the impulse responses are No_Info, then $CB2$ is No_Info. $CB2$ could be better than Baseline and Context Based-1 since it uses more historic information. This predictor does not use any data from the second semester.

The third model is Context Based-3 model $CB3$. This estimator is similar to $CB2$, but the weights are different. They are the Kolmogorov-Smirnov (KS) statistics of the impulse responses of the words w within the list of the 200 most frequent words on the first semester. They are obtained by classifying students according to their average length of response on the second semester. Two classes are defined: students with long answers and students with short answers. Students with long answers are those with average length response above the median of all students' average lengths. The rest are students with short answers. For a given word w, the Kolmogorov-Smirnov statistic $KS(w)$ is the maximum distance between the empirical cumulative distribution of the impulse response $h(w, s)$ of the students s with long answers and the empirical cumulative distribution of the impulse response $h(w, s)$ of the students s with short answers. $KS(w)$ ranges from 0 to 1. Words w with $KS(w)$ close to 1 are words w in questions that can discriminate between students whose answers are long from those that are short. As in the previous predictor, for each student and for each question, $CB3$ is a convex combination of impulse responses. Thus,

$$CB3(q, s) = \frac{\sum_{w \in L(q, s)} KS(w) h(w, s)}{\sum_{w \in L(q, s)} KS(w)} \tag{3}$$

If for a question all the impulse responses are No_Info, then $CB3$ is No_Info. $CB3$ might be better than Baseline, $CB1$ and $CB2$. This predictor does use data from semester 2, so that to build it there would have to be separate basis of construction and testing. However, for years to come the model could use the KS of previous years.

The fourth model is the Context Based-4 model $CB4$. This model is obtained by searching for weights $p(w)$ of the words w for the linear combination of impulse responses and an intercept b that minimizes the mean square error between the model prediction and the responses of all students in a sample of questions from the second semester. Thus

$$CB4(q, s) = b + \sum_{w \in L(q, s)} p(w) h(w, s) \tag{4}$$

It is important to stress that weights $p(w)$ and intercept m do not depend on each student or question. This model uses information from the second semester, and therefore the weights and intercept are calculated in a construction sample. It then sets its performance in an independent test sample.

To evaluate the performance of the models we use three metrics. First, we use the mean quadratic error to the questions of the second semester belonging to the test base. This is the average of the squares of errors in the length of responses between what the model predicts and the actual length. Second, we also use a less demanding metric that only considers the average length of each student's responses in the second semester. This means that we evaluate the model according to whether it is able to approach the average length of response, and no response per response. This metric is the square root of the average of the quadratic errors of the students, and where for each student the error is the difference between the predicted average of the lengths of the student's answers and the average of the real lengths of his answers. Third, we use the Kolmogorov-Smirnov distance metric. In this case we only measure the ability of each model to classify students according to their average length of response in the second semester will be below or above the median of the students averages. That is, if it predicts that the student will be a student with short answers or long answers.

3   Results

In each of the three metrics we will analyze the behavior of the metrics in the subpopulation of students who have answered at least a certain amount of questions in the first semester. The reason for this is that we expect that for those students with more historical information, that is to say that they have answered more questions in the first semester, then the models should make better predictions and thus errors should fall.

Figure 1 shows the square root of the mean square error for all the second semester questions in the test sample. It is observed that for students with the highest number of questions answered in the first semester the errors of the models improve to less than half of the error for all the population. These differences are in some cases statistically significant, but due to the low number of students who have answered many questions in the first semester then it is necessary to consider a larger population. Since the Baseline, Context Based-1, and Context Based-2 predictors do not use second-semester information, we can use the full base to make error estimates.

Figure 2 shows the mean square error in the total base and shows that Context Based-2 has smaller error than Baseline for students with more than 20 questions.

Figure 3 shows the evolution of the second metric. This metric is the prediction error in the averages of responses per student. It is again observed that for students with the highest number of questions answered in the first semester, the errors of the models improve over the Baseline-1 model, but now the error of the optimal model does not improve over others. This is because we are now observing a different metric. We do not measure the difference response to response but the difference of averages. These differences are in some cases statistically significant, but due to the

Fig. 1 Square root of the mean square error by question on the testing set

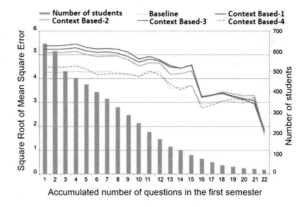

Fig. 2 Square root of the mean square error by question on the complete data base

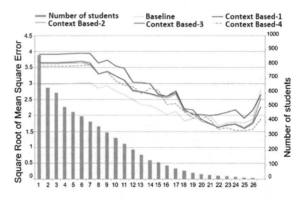

Fig. 3 Square root of the mean square error by student on the testing set

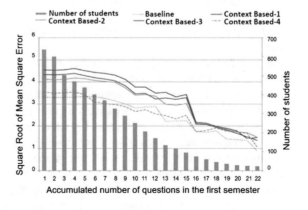

Fig. 4 Kolmogorov-Smirnov
(KS) distance by student on
the testing set

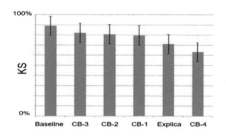

low number of students who have answered many questions in the first semester
then it is necessary to consider a larger population.

Figure 4 shows the distances KS for the different predictors. Context Based
predictors are abbreviated by CB. The KS of predictors based on a single word are
also included. It is observed that the predictor based on the impulse responses of
"explain" has a large KS, although less than the Baseline and Context Based-3,
Context Based-2 and Context Based-1 predictors.

4   Conclusions

In this paper we have analyzed written responses of students from various elementary school classes to open-ended questions. These are questions that teachers
pose to students on an online STEM platform. In each session the students answer
one or two open questions. One of the teacher's goals is for students to reflect on
the multiple-choice exercises they have been doing and respond by writing justi-
fications of their results and strategies. A critical problem is to get students to write
as long as possible, since at that age they are beginning to write and still write very
brief sentences. Research on science and mathematics learning [8, 4, 14] suggests
that writing helps students to learn. Moreover, the longer the students write, then
more opportunities students have to reflect and learn. For example, it is known that
the waiting time is key and that there is an optimal time that gets students to write
more [5].

In a previous paper [12] we have analyzed the impact in complete classes on the
length of the answers due to the presence of key words in the teacher's questions.
We analyzed the effect on the average of the entire class. However, in each class
there are many differences in the length of the answers between students, and in
calculating the average of the class long answers are compensated with short
answers. We had found that certain words have an important impact on the average
response length. In this work, unlike the previous one, we study the individual
behavior of each student. By knowing the history of the lengths of a student's
responses, it is possible to be more precise in predicting the length of answers to
future questions. For this purpose we first calculate stimulus response per word. It is
the length of that student's response when the word is in a question. This is the

effect size of each word when it appears in the questions. If we think of a word as an input or impulse, the computed average is the response to that impulse for that student. It is similar to the stimulus-response or impulse response analyzes in control systems. Based on the stimulus response for each student of each of the 200 selected non stop-words, we explore several predictors.

We use the history of 865 elementary school students from 25 classes exposed to open questions posed on an online STEM platform. In particular, we analyze four different personalized models with context information about the question, characterized by a list of 200 keywords. We collected data along a whole year, taking the data of the first semester to train our models and evaluate them on the second semester. We found that for students that have answered more questions on the first semester, the performance of one of the proposed context based models beats a baseline predictor. Moreover, given the obtained trend of error as function of the number of question on the first semester, it is probable that with a history of more questions, the performance of the context based predictors will be much better than the baseline predictor. With more questions we could also include more keywords and study the robustness of the predictors.

These results suggest that if there is sufficient history of a student's responses, then predictors based on a reduced set of keywords compete favorably with a baseline predictor. This baseline is the historical average of student response lengths in the previous semester. The baseline contains only general information, not information about the effect on the student of each word. By incorporating the impulse response of each word the prediction can be improved. The results achieved here are very preliminary because they are based on data from a few hundred students of elementary schools. As a next step we are gathering more information to increase the number of students and to increase the number of historic questions answered per student. We are also planning to explore the impact of clustering similar words [15] and reduce further the list of 200 key words.

Acknowledgements Funding from PIA-CONICYT Basal Funds for Centers of Excellence Project FB0003 is gratefully acknowledged and to the Fondef D15I10017 grant from CONICYT.

References

1. Stevens, R.: The question as a measure of efficiency in instruction: a critical study classroom practice. Teach. Coll. Contrib. Educ. **48** (1912)
2. Emig, J.: Writing as a mode of learning. Coll. Compos. Commun. **28**, 122–128 (1977)
3. Vygotssky, L.: Thought and Language. MIT Press (1986)
4. Urquhart, V.: Using Writing in Math to Deepen Student Learning. McREL (2009)
5. Rowe, M.: Wait time: slowing down may be a way of speeding up! J. Teach. Educ. 43–49 (1986)
6. Shahrill, M.: Review of effective teacher questioning in mathematics classrooms. Int. J. Human. Social Sci. **3**(17) Sept (2013)

 7. McNamara, D.; Graesser, A.: Coh-Metrix: an automated tool for theoretical and applied natural language processing. In: Applied Natural Language Processing: Identification, Investigation and Resolution, pp. 188–205. IGI Global (2011)
 8. Blosser, P.: How to ask the right questions. The National Science Teachers Association (2000)
 9. Tofade, T., Elsner, J., Haines, S.: Best practice strategies for effective use of questions as a teaching tool. Am. J. Pharma. Educ. **77**(7) Article 155 (2013)
10. Chin, C.: Teacher questioning in science classrooms: what approaches stimulate productive thinking? J. Res. Sci. Teach. **44**(6), 815–843, Aug (2007)
11. Scott, P.: Teacher talk and meaning making in science classrooms: A Vygotskian analysis and review. Studies in Science Education, 32, pp. 45–80 (1998)
12. Araya, R., Aljovin, E.: The effect of teacher questions on elementary school students' written responses on an online STEM platform. In: Andre, T. (ed.) Advances in Human Factors in Training, Education, and Learning Sciences, vol. 596, pp. 372–382. Springer, Cham (2017)
13. Araya, R., Gormaz, R., Bahamondez, M., Aguirre, C., Calfucura, P., Jaure. P., Laborda, C.: ICT supported learning rises math achievement in low socio economic status schools. LNCS, vol. 9307, pp 383–388 (2015)
14. Winograd, K.: What fifth graders learn when they write their own math problems. Educ. Leader. **64**(7), 64–66 (1992)
15. Pennington, J.; Socher, R.; Manning, C.: GloVe: global vectors for word representation. In: Empirical Methods in Natural Language Processing (EMNLP), pp. 1532–1543 (2014). http://www.aclweb.org/anthology/D14-1162

Part III
Machine Learning and Data Mining

Robust Scale-Invariant Normalization and Similarity Measurement for Time Series Data

**Ariyawat Chonbodeechalermroong
and Chotirat Ann Ratanamahatana**

Abstract Classification is one of the most prevalent tasks in time series mining. Dynamic Time Warping and Longest Common Subsequence are well-known and widely used algorithms to measure similarity between two time series sequences using non-linear alignment. However, these algorithms work at its best when the time series pair has similar amplitude scaling, as a little adjustment of scale can actually double the error rates. Unfortunately, sensor data and most real-world time series data usually contain noise, missing values, outlier, and variability or scaling in both axes, which is not suitable for the widely used Z-normalization. We introduce the Local Feature Normalization (LFN) and its Local Scaling Feature (LSF), which can be used to robustly normalize noisy/warped/missing-valued time series. In addition, we utilize LSF to match time series containing multiple subsequences with a variety of scales; this algorithm is called Longest Common Local Scaling Feature (LCSF). Comparing to the usage of Z-normalized data, our classification results show that our proposed LFN is impressively robust, especially on high-error and noisy datasets. On both synthetic and real application data for wrist strengthening rehabilitation exercise using a mobile phone sensor, our LCSF similarity measure also significantly outperforms other existing methods by a large margin.

Keywords Dynamic time warping · Longest common subsequence
Time series normalization · Time series features

A. Chonbodeechalermroong · C. A. Ratanamahatana (✉)
Department of Computer Engineering, Chulalongkorn University,
Bangkok 10330, Thailand
e-mail: chotirat.r@chula.ac.th

A. Chonbodeechalermroong
e-mail: 6070375521@student.chula.ac.th

© Springer International Publishing AG, part of Springer Nature 2018   149
A. Sieminski et al. (eds.), *Modern Approaches for Intelligent Information
and Database Systems*, Studies in Computational Intelligence 769,
https://doi.org/10.1007/978-3-319-76081-0_13

1 Introduction

Dynamic Time Warping (DTW) [1, 2] and Longest Common Subsequence (LCSS) [3] are the two well-known similarity measures for time series data with the ability to warp on the X-axis using non-linear alignment. However, the main obstacle of these algorithms is the variability of scales in the amplitude; they work at its best when the time series pair has similar amplitude scaling. Therefore, time series normalization becomes critical as a little adjustment of scale can actually double the error rates [4]. Currently, Z-normalization is one of the most widely used techniques to normalize time series data. Unfortunately, Z-normalization can produce incorrect scaling on data containing noise, missing data, outlier, subsequence scaling, and even the variability in the time axis. As demonstrated in Fig. 1a–d, each of the sequences contains a sine wave followed by a square wave with different variability added to the sequences, i.e., (a) Y-axis scaling (amplitude), (b) X-axis scaling (warping), (c) noise, and (d) combination of all three; Z-normalization produces inaccurate scaling, which in turn causes inaccuracies in classification. Finding similarity between two Z-normalized sequences in Fig. 1d using DTW or LCSS can lead to incorrect alignment. As shown in Fig. 1e, due to incorrect scaling, high-magnitude points of the top sine wave cannot be matched correctly through DTW, and the rest of the square wave is wrongly matched, overestimating the

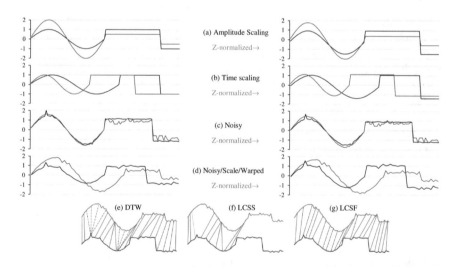

Fig. 1 Incorrect scaling caused by Z-normalization in sequences containing **a** amplitude scaling, **b** time scaling (warping), **c** noise, and **d** combination of three types of variability. **e** Due to incorrect scaling, high magnitude points of the top sine wave graph cannot be matched correctly using DTW; however, it cannot skip points, and therefore matches the rest of square wave part although the distances are high. **f** Though LCSS can skip data points, the square waves of the two sequences are too different for a relatively small ε parameter to match, and cannot be correctly aligned. **g** our proposed LCSF can correctly align the sequences

cumulative distances. Figure 1f, on the other hand, though LCSS can skip data points, the square waves of the two sequences are too different for a relatively small ε parameter to match, and cannot be correctly aligned.

To resolve this problem, we introduce a Local Feature Normalization (LFN) that first discovers a Local Scaling Feature (LSF) to normalize time series data more accurately, especially for time warped and noisy data. The Longest Common Local Scaling Feature (LCSF) similarity measurement is then proposed to effectively match time series sequences that contain subsequences with a variety of scales, as demonstrated in Fig. 1g.

2   Background and Related Works

Regardless of extended usage of DTW and LCSS in the past few decades, many researchers may not realize that DTW [1, 2] and LCSS [3] only work at their best when the data have similar amplitude scaling. Normalization has played a big role in trying to resolve this scaling issue [4]. However, as shown in Fig. 1, the most commonly used Z-normalization [4] technique for time series data can still cause DTW and LCSS to produce inaccurate results on data with such variability.

Nowadays, time series have been used in many applications such as body movement recognition from video [5], prosthesis control and rehabilitation [6, 7], and hand gesture recognition [8]. Most of the applications obtain data from sensors such as electromyography (EMG) or accelerometers, which are noisy and contain various types of scaling, and sometimes a single sensed time series sequence can have many subsequences with variety of scales. To match these varied-scale sub-sequences using the traditional DTW or LCSS, the training set needs to contain all possible combinations of scaling of all subsequences, which may not be possible to obtain. To the best of our knowledge, no current normalization technique is specially designed for such variability in the data. Our proposed LCSF can effectively solve this problem with only limited amount of training data. This section will first give a quick review of LCSS, which is the basis of our proposed work.

2.1   Longest Common Subsequence Similarity (LCSS)

Given two Time Series A and B with length n and m, $A = \{a_1, a_2, a_3, ..., a_n\}$, $B = \{b_1, b_2, b_3, ..., b_m\}$ where a_i and b_j are real numbers, LCSS score of A and B measuring the similarity between the two time series is:

$$LCSS(A, B) = f(n, m) = \begin{cases} 0 & \text{if } n = 0 \text{ or } m = 0 \\ \max \begin{cases} s(a_n, b_m) + f(n-1, m-1) \\ f(n, m-1) \\ f(n-1, m) \end{cases} & \text{otherwise} \end{cases}$$

(1)

where $s(a_i, b_j)$ is the similarity between a_i and b_j defined as $s(a_i, b_j) = 1$ if $|a_i - b_j| < \varepsilon$ and $|i - j| \leq l$, and $s(a_i, b_j) = 0$ otherwise; ε is a given small arbitrary value; l is a constraint window size. This discrete similarity has a drawback that if ε is too small, many points are considered as noise. If ε is too large, too many points can match with the others such that LCSS may produce just one-to-one matching (no warping/skipping). LCSS has non-duplicate alignment with score $\in [0, 1]$ for each alignment that makes output range $[0, \min(m, n)]$. LCSS' main advantage is its ability to skip noisy data points while DTW has to match every single data point.

3   Our Proposed Algorithm

This section explains our proposed Local Scaling Feature (LSF), which is then utilized in our Local Feature Normalization (LFN) and Longest Common Local Scaling Feature (LCSF) similarity measure.

3.1   Local Scaling Feature (LSF) and Local Feature Normalization (LFN)

Our LSF is based on LCSS concept, as we hold the assumption that the data is particularly noisy and LCSS can skip some noisy data points. However, our LSF still works well on non-noisy data.

Given two time series A and B with length n and m, $A = \{a_1, a_2, a_3, ..., a_n\}$, $B = \{b_1, b_2, b_3, ..., b_m\}$ where a_i and b_j are real numbers, we will normalize the time series A with some value a_i and normalize B with some value b_j. If the two time series are similar, but having different scaling in the Y-axis, then there must exist $a_i \approx b_j$, if A and B are in the same scaling. Hence, if we normalize A by a_i and B by b_j, we will get the correct scale of A and B called $A' = A/a_i$ and $B' = B/b_j$.

In order to find the correct A' and B', one might compare all possible combinations of a_i and b_j to normalize and select the best LCSS/DTW score among all matched normalized data; however, the time complexity is excessively high: $O(n^2m^2)$, nm for trying all possible A' and B', and another nm for doing LCSS/DTW on each of A' and B'. However, we can reduce this huge amount of combination by finding some potential candidates called **Local Scaling Features** (LSFs).

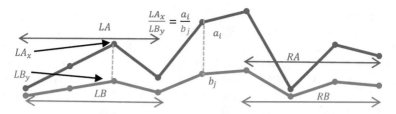

Fig. 2 An illustration of potential score calculation of a_i and b_j; the left/right subsequences will be scored using LCSS such that each two points are matched if their scale is equal to a_i/b_j

If a_i and b_j are potential candidates, then some neighboring points of a_i should match with some neighboring points of b_j. LCSS is used to match the left subsequences of length w: $LA = \{a_{i-w}, \ldots, a_{i-1}\}$ with $LB = \{b_{j-w}, \ldots, b_{j-1}\}$ as the Left Score (LS), and the right subsequences length w: $RA = \{a_{i+1}, \ldots, a_{i+w}\}$ with $RB = \{b_{j+1}, \ldots, b_{j+w}\}$ as the Right Score (RS), then the potential score $PS = LS + RS$; PS measures the local similarity. Then the best c candidates (c highest PS) are selected as the LSFs. Because the data are assumed to be noisy, we should not use Z-normalization in calculation of LS and RS. Instead, this local LCSS uses our special similarity function that requires no normalization.

To match the left subsequences LA and LB that is assumed to be in different scales and we are assuming that a_i and b_j are matched, there must be indices x and y that LA_x/LB_y is equal to a_i/b_j where $LA_x \in LA$ and $LB_y \in LB$ as well as matching RA and RB, as illustrated in Fig. 2.

However, another important problem is when a_i and b_j are much larger than LA_x and LB_y. We use the fact that if LA_x and LB_y are matched, then $(a_i - LA_x)/(b_j - LB_y)$ is also equal to a_i/b_j. For example, given $a_i = 10$, $b_j = 20$, $LA_x = 0.2$, $LB_y = 0.1$, we see that $a_i/b_j = 0.5$ while $LA_x/LB_y = 2$, but $(a_i - LA_x)/(b_j - LB_y) = 0.49$. This too little value of LA_x and LB_y comparing to a_i and b_j can be seen as noise. Therefore, we select the closest value to a_i/b_j from LA_x/LB_y or $(a_i - LA_x)/(b_j - LB_y)$. According to this principle, our similarity function given the candidate a_i and b_j is as follows:

$$s2(X, Y) = 1 - \frac{\min\left\{\begin{array}{l} \left|\frac{X}{Y} - \frac{a_i}{b_j}\right| \\ \left|\frac{a_i - X}{b_j - Y} - \frac{a_i}{b_j}\right| \end{array}\right.}{\varepsilon'\left|\frac{a_i}{b_j}\right|} \tag{2}$$

where $X \in \{LA, LB\}$; $Y \in \{RA, RB\}$; ε' is a parameter similar to ε in the original similarity function. The best c candidates are used to normalize the time series; then LCSS is applied to these c normalized series. In this LCSS part, to distinguish finer dissimilarity that helps solve the one-to-one matching problem of the original LCSS with large ε, we prefer the continuous similarity score instead of the old binary score. Therefore, Eq. (1) in Sect. 2.1 will be replaced by another similarity function, $s3(a_i, b_j) = 1 - |a_i - b_j|/\varepsilon$. The idea behind this function is that LCSS will not match a_i to b_j when the difference is larger than ε because $s3(a_i, b_j)$ is less than 0; however, if the difference is smaller than ε, the score will be a continuous value from 0 to 1. Similar idea also applies to $s2$ in LSF. To allow maximum flexibility of matching, window constraint is not applied. The sequence with the best LCSS score is chosen to be the output of our **Local Feature Normalization** (**LFN**). Note that the flipping candidates $(a_i/b_j < 0)$ needs not be calculated to save some time and improve overall accuracy as it rarely happens to be in the same class.

The time complexity of finding LSFs is $O(w^2 mn)$; w^2 is the complexity of using LCSS to calculate each candidate's PS; $O(mn)$ is for finding all possible candidates. Applying LCSS on the normalized series c times uses $O(cmn)$. Therefore, the overall complexity is $O(w^2 mn + cmn)$.

3.2   Longest Common Local Scaling Feature (LCSF)

As shown in Fig. 1a, some time series may contain several subsequences, each of which may also have different scaling. We could utilize LSF to handle multiple scaling in subsequences. An LFS is the local similarity between two points, such that ordered LSFs can match multiple-scaling subsequences. For each LSF, the maximum possible value of PS is $2w$ (LCSS(RA, RB) $\leq$ min($|RA|$, $|RB|$) $= w$, Sect. 2.1). Given a threshold t, a candidate whose PS score is larger than or equal to $2wt$ is called the potential candidate; t is the ratio of the maximum possible value of PS; $t \in (0, 1]$.

After finding the best c potential candidates $PS \geq 2wt$, each candidate consists of the indices i, j (denoting the position in A and B) and the score PS; however, there can be two or more candidates with the same index i such as $(i, j, PS1)$ and $(i, k, PS2)$: $j \neq k$; this multiple matching occurs because each potential candidate is found locally, or the threshold t is too small. In order to find a proper longest LSF matching, we will sort the potential candidates in an ascending order of i, together with a descending order of j, then find the longest increasing subsequence based on j on the sorted potential candidates. The output of the longest increasing subsequence is the LCSF score. For example, if the potential candidates are $(1, 2, 0.5)$, $(1, 3, 0.6)$, $(1, 4, 0.9)$, $(2, 3, 0.6)$ and $(3, 4, 0.6)$. After sorting, we will get $(1, 4, 0.9)$, $(1, 3, 0.6)$, $(1, 2, 0.5)$, $(2, 3, 0.6)$ and $(3, 4, 0.6)$ as the candidates. If we look only on the list of j, we will see $\{4, 3, 2, 3, 4\}$. Then the proper longest increasing subsequences on this set will be $\{2, 3, 4\}$, where $|\{2, 3, 4\}| = 3$ is the output of our LCSF similarity score.

Discovering all potential candidates consumes $O(w^2 mn)$ time complexity. Sorting c candidates needs $O(c\log(c))$, and finding the longest increasing subsequences using the efficient algorithm [9] requires $O(c\log(c))$. Hence, the overall complexity is $O(w^2 mn + c\log(c))$. LCSF has four parameters: w, ε, t and c to tune.

$LSF(A,B,w, t, c)$	$LFN(A, B, w)$
$pcs =\{\}$ #set of potential candidates	$pcs = LSF(A, B, w, 0, c)$
for i in $|A|$:	$max= 0$
for j in $|B|$:	for f in pcs:
$LS = lcss'(A[i\text{-}w,i\text{-}1], B[j\text{-}w,j\text{-}1], A[i], B[i])$	$A' = A/A[f.i]$
$RS = lcss'(A[i+1,i+w], B[j+1,j+w] , A[i], B[i])$	$B' = B/B[f.j]$
$PS = LS + RS$	$l = lcss''(A', B')$
if $PS >= 2wt$:	if $l > max$:
if $|pcs| < c$:	$max = l$
$pcs.add((i,j,ps))$	$bestA',bestB' = A',B'$
else if $PS > pcs.getMinPS()$:	return $max,bestA',bestB'$
$pcs.removeMin()$	# $lcss''$ is LCSS using $s3$ as a similarity function
$pcs.add((i,j,ps))$	
return pcs	
#$lcss'$ is LCSS using $s2$ as a similarity function, given $A[i]$ and $B[i]$	

$LCSF(A, B, w, t, c)$
$pcs = LSF(A, B, w, t, c)$ #c is set to ∞ in our experiment
$sort(pcs)$ in ascending order of i and in descending order of j on equal i
return $longest\_increasing\_subsequence(pcs)$ based on j

Fig. 3 Pseudocodes for LSF, LFN, LCSF

However, to make sure that we do not miss any potential candidates, we can ignore the c parameter and accept all potential candidates ($PS \geq 2wt$). The pseudocodes for LSF, LFN, and LCSF are shown in Fig. 3.

4 Experiments and Results

4.1 Local Features Normalization (LFN)

We evaluate the performance of LFN using 1-nearest neighbor (1-NN) classification [10] on 21 UCR's public datasets [11]. Note that every dataset on this archive is already Z-normalized, labeled, and split into training and test sets. The datasets are selected based on their high-noise and relatively small size criteria.

To find an optimal set of parameters w, ε' and ε, we use grid search on a given training set. Leave-one-out cross-validation is used to calculate the accuracy for each parameter setup. The lowest-error setup is used to evaluate the test set. To avoid a small floating point inaccuracy problem, we define integers $e' = 1/\varepsilon'$ and $e = 1/\varepsilon$, then we iterate the multiplier $e(e')$ instead of the divider $\varepsilon(\varepsilon')$. Based on our empirical results, e and e' are chosen from a set $\{1, 2, 3, 6, 10\}$, w from a set $[4, 10]$, and $c = 3$ The higher c value gives more opportunity for better accuracy, but it consumes more time; however, $c = 3$ empirically appears to be the smallest c value that does not sacrifice much accuracy. In Table 1, we compare classification error rates of our method with well-known similarity measures: (1) Euclidean Distance, (2) DTW with the best global constraint window reported on the UCR repository, (3) DTW with no global constraint, (4) the original LCSS, and (5) the LCSS using $s3$ similarity function with the same LFN's ε parameter that shows the comparison between Z-normalization and our normalization.

Our proposed normalization method produces the lowest error rates in 16 out of 21 datasets, and exclusively outperforming others on noisy and high-error (DTW error ≥ 0.3) datasets, e.g., MiddlePhalanxTW, Wine, Herring, BeetleFly, and

Table 1 Classification error rates on UCR datasets

Name	Euclidean distance	DTW (best global constraint)	DTW	LCSS	LCSS (s3)	LFN (w, e′, e)
BeetleFly[a]	0.250	0.300	0.300	0.300	0.300	*0.150(4, 10, 10)*
BirdChicken[a]	0.450	0.300	0.250	0.250	0.150	*0.150(3, 10, 10)*
DistalPhalanxOutlineAgeGroup[a]	0.218	0.228	0.208	0.240	0.242	*0.188(7, 10, 10)*
DistalPhalanxOutlineCorrect[a]	0.248	0.232	0.232	0.253	0.240	*0.223(6, 2, 2)*
DistalPhalanxTW[a]	0.273	0.272	0.290	0.273	0.273	*0.268(6, 2, 2)*
ECG	0.120	0.120	0.230	0.120	0.120	*0.110 (10, 3, 1)*
ECGFiveDays	0.203	0.203	0.232	0.230	0.230	*0.199(10, 1, 1)*
Herring[a]	0.484	0.469	0.469	0.422	0.515	*0.406(7, 6, 3)*
ItalyPowerDemand	0.045	0.045	0.050	0.170	0.155	*0.039(4, 1, 1)*
MedicalImages	0.316	*0.253*	0.263	0.359	0.293	0.297(8, 1, 1)
MiddlePhalanxOutlineAgeGroup[a]	0.260	0.253	0.250	0.250	0.275	*0.250(5, 10, 1)*
MiddlePhalanxOutlineCorrect	*0.247*	0.318	0.352	0.352	0.395	0.258(7, 1, 1)
MiddlePhalanxTW[a]	0.439	0.419	0.416	0.424	0.416	*0.409(7, 10, 10)*
MoteStrain	0.121	0.134	0.165	0.103	*0.097*	0.118(2, 6, 6)
ProximalPhalanxOutlineAgeGroup	0.215	0.215	*0.195*	0.254	0.223	*0.195(5, 6, 3)*
ProximalPhalanxOutlineCorrect	*0.192*	0.210	0.216	0.230	0.244	0.199(4, 10, 1)
ProximalPhalanxTW[a]	0.292	0.263	0.263	0.265	0.273	*0.255(5, 10, 10)*
SonyAIBORobot Surface[a]	0.305	0.305	0.275	0.358	0.393	*0.181(4, 1, 1)*
SonyAIBORobot SurfaceII	*0.141*	*0.141*	0.169	0.209	0.158	0.158(4, 1, 1)
TwoLeadECG	0.253	0.132	0.096	0.057	0.100	*0.045(6, 6, 6)*
Wine[a]	0.389	0.389	0.426	0.500	0.500	*0.259(6, 10, 1)*

[a]Observed as noisy dataset

BirdChicken. For small-error datasets, our algorithm performs better or not much worse as we observe that datasets is already in the correct scale. Our algorithm outperforms LCSS (original and s3) on 19/21 datasets, which shows significant improvement of our normalization. If c or w are too small, obtained LSFs might not cover true important features such that some errors may occur. LFN impressively outperforms DTW with global constraints in 19/21 datasets.

4.2 Longest Common Local Scaling Feature (LCSF) on Synthetic Data

To emphasize the advantage of the proposed LCSF, we evaluate it with a synthetic dataset. The generated dataset has four classes; each class has two parts selected from the three patterns: a sine wave, a square wave, or a triangle wave; each part has variety of lengths from 30 to 60 data points, and its amplitude is varied from 0.1 to 8.0. The dataset was generated with 5–20% noise multiplier (each sampling has been added by noise that is random between zero and the amplitude multiplied by the noise multiplier), 5–20% missing data (each sampling has this probability to be set to zero) and 5–20% combination of missing data and noise. The challenge of this dataset is that each time series sequence within each class contains multiple sub-sequences of different lengths and scaling. Each sequence has a total length of 100, and each class contains 25 sequences (100 instances/dataset). Examples of the datasets are shown in Fig. 4.

We evaluate the performance using 10-fold cross-validation and 1-NN classifier by reporting the misclassified rates. To tune the parameters, each training set of each fold is trained using Leave-one-out cross-validation with the setups (w, e', t); w and e' are the same as previous experiments; t is selected from {0.5, 0.6, 0.7, 0.8, 0.9}. We also compare LCSF with Derivative Dynamic Time Warping (DDTW) [12] because DDTW has the similar idea to LCSF that they both look at relationships among neighboring points. As shown in Table 2, our LCSF significantly outperforms all other methods by giving almost all perfect classification, demonstrating robustness of our proposed method. DDTW performs poorly because the slopes of different-scale sequences are different; for example, the derivative of a function $f(x)$ is half of the derivative of $2f(x)$.

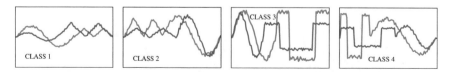

Fig. 4 Examples of the four classes of our synthetic data

Table 2 Classification error rates on the synthetic dataset

Missing chance%: noise multiplier%	Euclidean distance	DTW	DTW (best global constraint)	DDTW	LCSS	LCSF
5:0	0.20	0.17	0.18	0.35	0.16	0.000
10:0	0.14	0.15	0.14	0.29	0.16	0.000
15:0	0.20	0.23	0.20	0.32	0.22	0.000
20:0	0.18	0.19	0.18	0.37	0.18	0.000
0:5	0.17	0.17	0.16	0.19	0.18	0.000
0:10	0.14	0.12	0.12	0.28	0.17	0.000
0:15	0.17	0.17	0.17	0.34	0.23	0.000
0:20	0.13	0.15	0.13	0.38	0.20	0.001
5:5	0.17	0.17	0.16	0.19	0.14	0.000
10:10	0.14	0.12	0.12	0.28	0.15	0.000
15:15	0.17	0.17	0.17	0.34	0.12	0.001
20:20	0.13	0.15	0.13	0.38	0.11	0.011

4.3 LCSF on Real World Data and Applications

We apply our proposed LCSF to classify wrist strengthening rehabilitation exercises for wrist injuries, according to Dr. Steve Lucey, an orthopedic surgeon at Sports Medicine and Joint Replacement of Greensboro [13]. This wrist strengthening exercise has three classes as shown in Fig. 5. As a typical rehabilitation exercise usually contains multiple repetitive moves, the speed and force made by human in each repetition generally vary. This can cause a signal to have multiple subsequences with variability in scales, both in X and Y axes. We wrote an android application to collect accelerometer data (magnitude channel) with 20 Hz sampling rate from a mobile phone (Android 5.1.1, 2.0 GB RAM) due to its easy access and availability. We collected 70 samples for each class (210 samples in total), each with two periods of the routine to make the movement more apparent on the sampled signals. The maximum length of the sample is 48 data points.

We perform the experiment using 10-fold and 5-fold cross-validation with 1-NN classifier. We train the parameters as follows: $t \in \{0.5, 0.6, 0.7, 0.8, 0.9\}$, $w \in [4,$

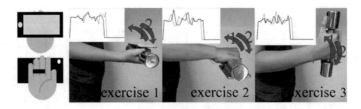

Fig. 5 The 3 wrist rehabilitation exercises for wrist injuries with sampled time series sequences showing for each class; the mobile phone is tied and taped on middle and ring fingers

Table 3 Classification error rates on wrist strengthening rehabilitation exercise data. Classification time is the average time consumed for classifying an instance in 5-fold cross-validation

Fold	Euclidean distance	DTW	DTW (best window global constraint)	DDTW	LCSS	LCSF
10	0.375	0.158	0.143	0.199	0.244	*0.080*
5	0.276	0.120	0.106	0.177	0.207	*0.063*
(5-fold) classification time (ms)	0	5	4	6	5	35

10], $e \in [1, 10]$. We use larger set of e' because the shorter length data consumes less time so that we can spend more time in the bigger searching space that we may get better results. As shown in Table 3, our approach has impressively better accuracy; although the classification time is higher, it is still feasible in real-time execution. More importantly, the obtained LCSF score can then be used as a crucial step to detect and correct the rehabilitation move as incorrect alignment or incorrect speed of movement can slow down the impairments [14].

5 Conclusion and Future Works

We propose a Local Feature Normalization (LFN) and a Local Scaling Feature (LSF) to normalize noisy/scaled data, and a Longest Common Local Scaling Feature (LCSF) similarity measure for time series data containing multiple subsequences with a variety of scales. Our classification results show that our proposed LFN is very robust, especially on high-error and noisy datasets, and our LCSF also outperforms others on both synthetic and real datasets. The execution time of LFN and LCSF are $O(mn)$ the same as DTW and LCSS if w and c are fixed.

For future works, we would like to propose a DTW-based LCSF with smaller number of parameters. In partial contour matching algorithm [15], a dynamic programming algorithm similar to LCSS but in 2D using triangle similarity properties, could be speeded up by utilizing LSF idea to obtain potential baseline candidates.

References

1. Sakoe, H., Chiba, S.: Dynamic programming algorithm optimization for spoken word recognition. IEEE Trans. Acoust. Speech Signal Process. **26**, 43–49 (1978)
2. Vlachos, M., Hadjieleftheriou, M., Gunopulos, D., Keogh, E.: Indexing multi-dimensional time-series with support for multiple distance measures. In: Proceedings of the 9th

ACM SIGKDD International Conference on Knowledge Discovery and Data Mining, pp. 216–225. ACM, New York, USA (2003)
3. Das, G., Gunopulos, D., Mannila, H.: Finding similar time series. In: Principles of Data Mining and Knowledge Discovery, pp. 88–100. Springer, Berlin, Heidelberg (1997)
4. Rakthanmanon, T., Campana, B., Mueen, A., Batista, G., Westover, B., Zhu, Q., Zakaria, J., Keogh, E.: Searching and mining trillions of time series subsequences under dynamic time warping. In: Proceedings of the 18th ACM SIGKDD International Conference on Knowledge Discovery and Data Mining, pp. 262–270. ACM, New York, USA (2012)
5. Gavrila, D.M., Davis, L.S.: 3-D model-based tracking of human upper body movement: a multi-view approach. In: Proceedings of International Symposium on Computer Vision—ISCV, pp. 253–258 (1995)
6. Crouch, D., Huang, H.: Simple EMG-driven musculoskeletal model enables consistent control performance during path tracing tasks. In: 2016 38th Annual International Conference of the IEEE Engineering in Medicine and Biology Society (EMBC), pp. 1–4 (2016)
7. Yun, Y., Dancausse, S., Esmatloo, P., Serrato, A., Merring, C.A., Agarwal, P., Deshpande, A. D.: Maestro: an EMG-driven assistive hand exoskeleton for spinal cord injury patients. In: 2017 IEEE International Conference on Robotics and Automation (ICRA), pp. 2904–2910 (2017)
8. Chen, X., Zhang, X., Zhao, Z.Y., Yang, J.H., Lantz, V., Wang, K.Q.: Hand gesture recognition research based on surface EMG sensors and 2D-accelerometers. In: 2007 11th IEEE International Symposium on Wearable Computers, pp. 11–14 (2007)
9. Fredman, M.L.: On computing the length of longest increasing subsequences. Discret. Math. **11**, 29–35 (1975)
10. Peterson, L.E.: K-nearest neighbor. Scholarpedia **4**, 1883 (2009)
11. Chen, Y., Keogh, E., Hu, B., Begum, N., Bagnall, A., Mueen, A., Batista, G.: The UCR Time Series Classification Archive. http://www.cs.ucr.edu/~eamonn/time_series_data/
12. Keogh, E.J., Pazzani, M.J.: Derivative dynamic time warping. In: First SIAM International Conference on Data Mining (SDM'2001 (2001)
13. Lucey, S.: Patient education flyers and videos about orthopedic conditions and treatments. http://www.smjrortho.com/education.php
14. Friedrich, M., Cermak, T., Maderbacher, P.: The effect of brochure use versus therapist teaching on patients performing therapeutic exercise and on changes in impairment status. Phys. Ther. **76**, 1082–1088 (1996)
15. Chonbodeechalermroong, A., Chalidabhongse, T.H.: Dynamic contour matching for hand gesture recognition from monocular image. In: 2015 12th International Joint Conference on Computer Science and Software Engineering (JCSSE), pp. 47–51 (2015)

Perceiving Attributes of Game AI Using Fuzzy Logic

Saadman Shahid Chowdhury, Ruhul Mashbu, Ariq Ahnaf Shaan,
Kazi Al Ashfaq, Fazal Mahmud Niloy and Rashedur M. Rahman

Abstract This paper introduces the concept of the SEA Model: "Skill, Experience, and Aggression"—a novel way of representing the behavior of Video Game AI in a linguistic or graphical manner. The SEA Model takes inspiration from the "Big Five Factors of Psychology"—an apparatus used to describe the personalities of different human beings. In the SEA Model, the ideas from the Big Five Factors are modified and applied to Video Game AI, processed through the "Fuzzy Inference System", and presented in a manner that is easily perceivable by humans.

Keywords Fuzzy logic · Video Game AI · OCEAN Model
Big Five Factors of Psychology · SEA Model · Perception of personality
AI behavior · Fuzzy Inference System

S. S. Chowdhury · R. Mashbu · A. A. Shaan · K. Al Ashfaq · F. M. Niloy
R. M. Rahman (✉)
Department of Electrical and Computer Engineering, North South University, Plot-15,
Block-B, Bashundhara Residential Area, Dhaka, Bangladesh
e-mail: rashedur.rahman@northsouth.edu

S. S. Chowdhury
e-mail: saadman.shahid@gmail.com

R. Mashbu
e-mail: mashbu111@gmail.com

A. A. Shaan
e-mail: ariq.new91@gmail.com

K. Al Ashfaq
e-mail: kazialashfaq@gmail.com

F. M. Niloy
e-mail: niloooy@gmail.com

© Springer International Publishing AG, part of Springer Nature 2018   161
A. Sieminski et al. (eds.), *Modern Approaches for Intelligent Information
and Database Systems*, Studies in Computational Intelligence 769,
https://doi.org/10.1007/978-3-319-76081-0_14

1 Introduction

The mark of a good video game is transparency: designers need to provide as much information as they can to the players—about the environments and adversaries present in the game world. This is especially true for sports games, competitive games, and strategy games—where players require very detailed information about every aspect of the game.

Unfortunately, describing each of these aspects is not an easy task, as there may be over a thousand of them (depending on the complexity of the game). The sheer number of variables that influence the actions of video game AI would be overwhelming to human players. With that in mind, the primary goal of this paper is to introduce a process that provides designers the ability to convey information about the game AI to the human players in a concise, understandable, and useful way. We have named this process the "SEA Model".

We have developed a game of air hockey to illustrate the use of the SEA model. This game involves two players with "paddles", trying to hit a "puck" into each other's goals. This game comes with different versions of difficulty levels. We have created these difficulty levels using the SEA model. A player who is less skilled, less experienced and less aggressive will come across as easy opponent to a person in comparison to a player who has high skill, experience and aggression. The SEA model provides the "fuzzy" representation of these two AI players' skills.

The purpose of fuzzy logic is to generalize the crisp sets and control the degree of membership of an element of that set, e.g. how strongly does the element belong in that set? The membership of the elements of a set is described by a membership function. Values of the membership function are fuzzy, meaning they can have any value between 0 and 1. 0 being—does not belong at all, and 1 being—fully belongs in the set.

This paper is structured so that readers would be systematically introduced to the concepts that shape the SEA Model. In (Sect. 2) Background, we define the Big Five Factors of Psychology and its fuzzy interpretation: OCEAN Model. In (Sect. 3) Methodologies describe the SEA Model, and our application of it in the air-hockey game—this section contains the bulk of our work. The (Sect. 4) Findings section contains the results of a survey we have conducted to evaluate the effectiveness of the SEA Model. We concluded the paper and provided an outline for future works on this topic in Sect. 5.

2 Background

Human beings, like Game AI, have innumerable variables contributing to their personality. In 1961, Tupes and Christal [1], noticed similarities in the behavior of different people and identified "five key factors" that could be used to describe an individual's personality [2]—these became known as the Big Five Factors of

Psychology, and each of the 5 factors describes a separate aspect of a character's personality [3]. The factors being: Openness (Creativity), Conscientiousness (Dedication), Extraversion (Sociability), Agreeableness (Niceness), Neuroticism (Emotional stability) [3]. These factors are arranged in the acronym—OCEAN. Each of the 5 factors in the OCEAN Model is a spectrum from "very low" to "very high" (consider as 0 to 1 in fuzzy terms); For example, an extroverted person would have "very high extraversion", whereas an introvert would have "very low extraversion".

Oren and Aghaee describe in [4] that each of the 5 "Factors" in the OCEAN model can be broken down into multiple sub-factors, called "Facets"—these sub-factors are also in a spectrum from "very low" to "very high"—and the combinations of the sub-factors give us a more meaningful representation of the parent factors [4, 5]. The SEA Model relies on the concept of using multiple sub-factors to define a few main-factors.

3 Methodology

Let us begin by creating a distinction between the "Player Agents" and the "AI Agents". In video games, the "Player Agent" is any entity that represents the human player in the game world, e.g. in role-playing-games: the game's protagonist is the player agent—as the human player has full control of the protagonist's actions and motives. The human player basically puts himself/herself in the shoes of the Player Agent.

In contrast, the "AI Agent" are entities that seem intelligent from a human player's perspective, but are not under the control of the human player, e.g. in role-playing-games: individual villagers, animals, soldiers, villains, etc. are AI Agents. The human player can interact with them, talk to them, and at best influence their actions, but ultimately, the AI Agent's actions are controlled by the algorithm in the game. Note: It is important that the human player "perceives" the AI Agent as intelligent—this improves the game world's realism.

For the purpose of this paper, the Player Agent in the air-hockey game is the "bottom paddle" controlled by the player, while the AI Agent is the "top paddle" controlled by the Artificial Intelligence of the game; and the main focus of this paper is on the AI Agent—specifically on how the human player perceives the "personality" of the AI Agent.

3.1 Reasoning—Summary

The AI Agent—being only a paddle—cannot talk, cannot make facial expressions, and cannot display emotions. Attempting to linguistically describe such an agent's personality using the OCEAN model would be futile. However, we can identify other aspects of the AI Agent: the way it moves, the speed at which it reacts, the type of shots it plays, etc. We will call these observable aspects the "Facets" of the AI Agent. They are: "Accuracy", "Reaction", "Position", "Cunning", "Speed", and "Power".

These Facets are also the core functionality of the air-hockey game's AI Agent: they have values, are programmable and are chosen by the game's developers. This paper has 6 facets for the AI Agent in an air-hockey game (this game is simple—more complex games would require more facets). All of these facets are observable by the human player through the actions of the AI Agent.

Furthermore, these 6 Facets can be grouped into 3 clusters—we will call these three groups the "Factors". They are: "Skill", "Experience", and "Aggression". Inspired from the OCEAN model, we call this the "SEA model". The 3 Factors are fuzzy functions of the values of the 6 Facets, such that:

$$
\begin{aligned}
\text{Skill} &\leftarrow \text{S(Accuracy, Reaction)} \\
\text{Experience} &\leftarrow \text{E(Positioning, Cunning)} \\
\text{Aggression} &\leftarrow \text{A(Speed, Power)}
\end{aligned}
$$

Just like the OCEAN model's factors, the SEA Model Factors are abstract in nature—the values they hold do not make sense to the game's programming. However, they serve an important role in conveying a concise "message" to the human players: How "skilled" is the AI Agent? How "experienced" is the AI agent? And how "aggressive" is the AI Agent?

Here in, lies the fundamental concept of this paper: human beings easily understand behavior of agents in terms of few, linguistic, abstract terms (the Factors); and are confused by multiple, specific, values (the Facets). The SEA model provides a way to convert the developer's input values that the game utilizes, into linguistic terms that the game's consumer can easily understand.

3.2   Inputs—Explanation

When developing the air-hockey game, we included the 6 Facets in the game's programing. Each Facet has a specific value and can control/limit/influence the AI Agent's behavior. They are: "Accuracy", "Reaction", "Position", "Cunning", "Speed", and "Power".

Accuracy: The AI Agent is programmed to target the opponent's goal. So by default, the AI Agent always hits the puck in such a way that it is shot at the center of the opponent's goal. The Accuracy modifier is a probability that the default action is performed; i.e., if set to 100% Accuracy (MAX), the AI Agent performs the default action 100% of the time; and when set to 50% Accuracy (MIN), the AI Agent performs the default operation 50% of the time and intentionally misses (it targets the bars of the goal) 50% of the time. An AI agent with "perfect aim" has Accuracy set to 100%, and one with "terrible aim" Accuracy set to 50%.

Reaction: To make the game easier to play, the AI Agent has a time delay (delay in reaction) which it must suffer before reacting to a shot from the Player Agent. Reaction is a value from −1.5 to 0 s: describing how long the AI Agent waits before making a move against the Player Agent's attack. An AI Agent with "lazy reaction" has the variable set to −1.5 s (MIN), and one with "instant reaction" has the variable set to 0 s.

Positioning: Is a discreet input for (1–6) that defines the AI Agent's knowledge of positioning in the playing field. An agent with Positioning set to 1 (MIN) is thought to be "reactionary": it has no understanding of positioning and only runs after the puck (i.e. reacting to the opponents shots). An agent with positioning set to 2–5 shows varying levels of position adjustments, e.g., an AI Agent with rating of 3 "mimics" the Player Agent after hitting the puck. An agent with positioning set to 6 (MAX) is said to be "predictive", based on the player's movements, it calculates where the Player Agent is going to shoot the puck and readies itself in that position.

Cunning: There are 3 trick-shots in air-hockey [6], and the discreet input, Cunning, unlocks these. A rating of 1 (MIN) mean the AI Agent cannot perform any trick-shots and must resort to making only "direct" shots towards the goal. A rating of 2, means the AI can perform angled shots, a rating of 3 "tricky" unlocks the ability to perform bank shots, and a rating of 4 (MAX) "deceiving" unlocks the ability to perform double-bank shots.

Speed: The game has a top speed, T, at which any object in the game is allowed to move at. AI Agents always move at the highest speed they are allowed to. The Speed modifier is a percentage input that acts as a limit to that top speed. An input of 100% (MAX) means the AI Agent can move at T * 100% speed—the agent is "explosive" in nature; whereas, an agent with 30% (MIN) is able to move at T * 30%—making it "sluggish" in nature.

Power: The Power modifier is a percentage that describes the strength with which the AI Agent hits the puck. We have developed the game to calculate momentum when making shots. To make the game playable we have made the AI Agent "defy the laws of physics" by reducing momentum in an instant. Consider the transfer of momentum to the puck from the AI Agent to be, M. With 100% (MAX), the puck retains M * 100% of its momentum; and with 30% (MIN), the puck retains M * 30% of its momentum.

3.3   Reasoning—Explanation

Skill: From the perspective of a human being, Skill is a trait that describes someone's latent abilities. Linguistically speaking, a "talented" player has: "quick"

Reaction and "perfect" accuracy; where as an "inept" player has: "lazy" Reaction and "terrible" accuracy. For our AI Agent, Skill can be defined as a function of Accuracy and Reaction.

$$\text{Skill} \quad \leftarrow \quad S(\text{Accuracy}, \text{Reaction})$$

Furthermore, the relationship between the Facets: ("Accuracy", "Reaction") and the Factor: ("Skill") can be represented in a table.

However, the problem in this table is that it shows relationship between "linguistic" variables—whereas, the inputs we have used in the game are both floating numbers.

The "linguistic values" of Accuracy are: Terrible, Precise, and Perfect. The "linguistic values" of Reaction: Sluggish, Attentive, and Instant. Whereas, in the game, the input range for Accuracy: [50, 100] %, and the input range for Reaction: [−1.5, 0] s (Fig. 1).

In order to represent the relationship between Accuracy and Reaction as Skill, we need to find a way of representing the floating number inputs in the form of linguistic variables. This is where the Fuzzy Inference System (FIS) plays an important role. The FIS allows us to represent the "linguistic values" of Accuracy and Reaction in the form of membership functions (Figs. 2 and 3).

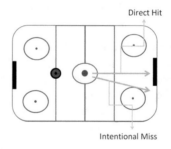

Fig. 1 Air hockey game

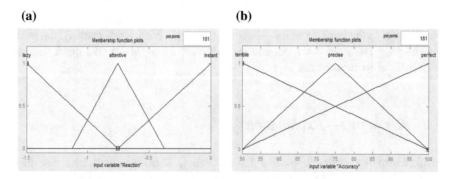

Fig. 2 **a** Reaction, **b** Accuracy

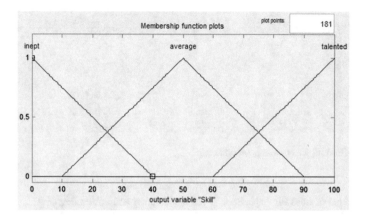

Fig. 3 Skill membership

	Reaction	Lazy	Attentive	Instant
Accuracy	**Skill**			
Terrible		Inept	Inept	Average
Precise		Inept	Average	Talented
Perfect		Average	Talented	Talented

Table 1 Table of rules for Skill, given the Accuracy and Reaction of the AI

The input values for Reaction and Accuracy provide a fuzzy membership value (between 0 and 1) for each linguistic value (depending on the membership function of each linguistic value). The membership values are compared with the Relationship Rules in the FIS (which is the Table 1). Finally, depending on the membership functions of the linguistic values of Skill, the results is a value between [1, 100] for Skill, This is simply a generalization of how the FIS works. A full explanation of the membership functions and FIS is beyond the scope of this paper.

Experience: From the perspective of a human being, experience can only be gained over time and through practice—individuals may learn from through trial and error on where to "position" themselves given the situation, and by observing more experienced players, individuals may pick up knowledge on how to play "cunning" shots. For this reason, the AI Agent's experience factor can be represented in the form of a function of Positioning and Cunning (Fig. 4).

$$\text{Experience} \leftarrow \text{E(Positioning, Cunning)}$$

The relationship can be represented in terms of linguistic values in the form of a Table 2 (Figs. 5, 6 and 7).

Aggression: Unlike experience and skill, aggression is dependent on a person's mentality. An aggressive individual moves around the board quickly and plays

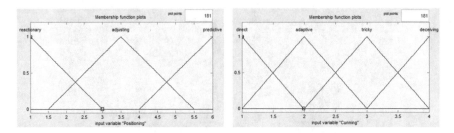

Fig. 4 Positioning and cunning membership

Table 2 Table of rules for Experience, given the Cunning and Positioning of AI

	Positioning	Reactionary	Adjusting	Predictive
Cunning	**Experience**			
Direct		newbie	amateur	Moderate
Adaptive		amateur	moderate	Moderate
Tricky		moderate	Professional	Professional
Deceiving		moderate	legendary	Legendary

Fig. 5 Speed membership function

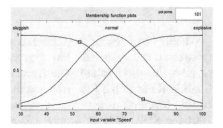

Fig. 6 Power membership function

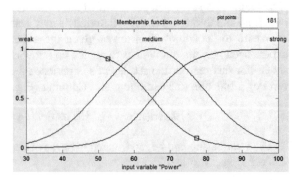

Fig. 7 Aggresion
membership function output

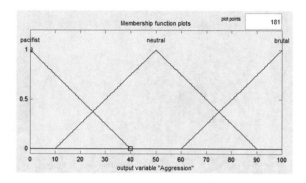

Table 3 Table of rules for
Aggression, given the Speed
and Power of the AI

	Power	Weak	Medium	Strong
Speed	**Aggression**			
Sluggish		Pacifist	Pacifist	Neutral
Normal		Neutral	Neutral	Brutal
Explosive		Neutral	Neutral	Brutal

powerful shots as a display of intimidation. For this reason, the Aggression factor of our AI Agent is a function of Speed and Power.

$$\text{Aggression} \leftarrow A(\text{Speed}, \text{Power})$$

Similarly, the relationship can be represented in terms of linguistic values in the form of a Table 3.

By itself the values are between [0, 100] for the Factors: Skill, Experience, and Aggression is enough to give human players a good understanding of what the "play-style" and in-turn the "personality" of different AI Agents are like. However, we can make it even easier for the human players by displaying the Factors in the form of a 3-armed Kiviat Chart (or RADAR Chart).

4 Findings

Given predefined Air-hockey AIs, we wanted to know how closely the SEA Model describes them compared to how humans describe them. The input to the SEA Model FIS are the Facets of the AI; and the input to the humans are their individual experiences of playing against the AIs. After that, both the SEA Model and the humans would rate the AIs in terms of Factors: "Aggression" and "Skill".

We conducted the survey on 20 people picked randomly from a pool of North South University students who volunteered to play the game. There were 12 male

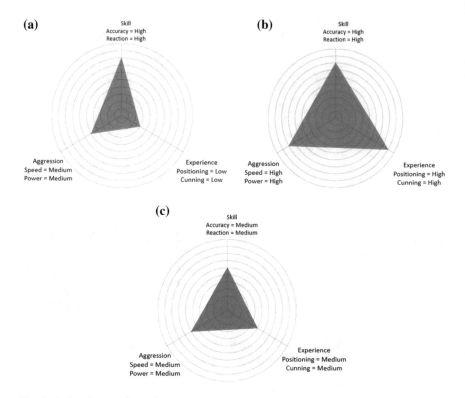

Fig. 8 Radar charts with varying parameters

and 8 female respondents, and they had mixed gamer and non-gamer backgrounds. They played against 2 AI opponents named, "Opponent 1"—an easy AI, and "Opponent 2"—a difficult AI. Each volunteer played 10 min games against both AIs.

Skill and Aggression of Opponent 1 (easy AI)

80% of the respondents felt that the AI was of average or below average skill, with 50% claiming that the AI was "kind of unskilled". And 100% of the respondents felt that the AI was moderately or less than moderately aggressive, with 50% claiming the AI as being "not very aggressive". Interestingly, this is reflected in the FIS's output (Fig. 8c), where the FIS claims that the AI is "Medium" in both Aggression and Skill (Fig. 9).

Skill and Aggression of Opponent 2 (difficult AI)

In contrast, every respondent rated Opponent 2 as being above average in both Skill and Aggressiveness! This is also reflected in the FIS's output (Fig. 8b), where the FIS rated that the AI as "High" in both Aggression and Skill.

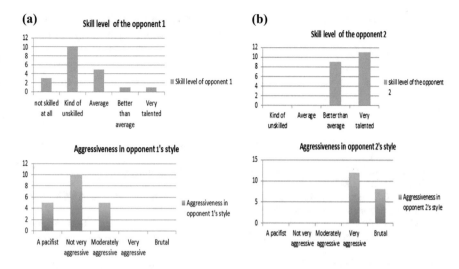

Fig. 9 **a** Skill and aggression of Opponent 1, **b** Skill and aggression of Opponent 2

Reasoning

From this survey, it can be assumed that the SEA Model is close to human perception. It rated Opponent 2 just as humans did. As for Opponent 1, while a rating of "Low" would have been better, a rating of "Medium" is also acceptable.

5 Conclusion and Future Work

This paper introduces the concept of the "SEA Model", for describing the behavior of a computer controlled agents in a concise and simple manner. To demonstrate the model, we have developed an Air-Hockey video game with 2 AI-Agents, extracted the SEA Model's interpretations of their behavior, and displayed them in web-charts. We conducted a survey where we asked respondents to play against the 2 AI-Agents and rate their behavior—the survey resulted in majority of the respondents agreeing with the output from the SEA Model. This paper fulfills its purpose of introducing this new concept of "perceiving game AI using Fuzzy Logic". However, at its infancy, the SEA Model is still not a completely ground breaking tool. Therefore we have outlined our next path:

The application of the SEA Model on a game as simple as Air-Hockey is a bit of an exaggeration. Here, the model converts 6 facets to 3 factors—a ratio of 2. We expect the SEA Model would shine in more complex games, where it might be necessary to convert over a thousand facets to a handful of factors—having a conversion ratio of hundreds.

As such, our next path of research would be to refine the processes in the SEA Model and apply it to a more complex video game AI—and perform further testing on this concept.

References

1. Tupes, E.C., Christal, R.E.: Recurrent personality factors based on trait ratings. J. Pers. **60**(2), 225–251 (1992)
2. McCrae, R.R., John, O.P.: An introduction to the five-factor model and its applications. J. Pers. **60**(2), 175–215 (1992)
3. Howard, P.J., Howard, J.M.: The Big Five Quickstart: An Introduction to the Five-Factor Model of Personality for Human Resource Professionals (1995)
4. Oren, T.I., Ghasem-Aghaee, N.: Personality representation processable in fuzzy logic for human behavior simulation. In: Summer Computer Simulation Conference, pp. 11–18. Society for Computer Simulation International (2003)
5. Ghasem-Aghaee, N., Oren, T.I.: Towards fuzzy agents with dynamic personality for human behavior simulation. In: Summer Computer Simulation Conference, pp. 3–10. Society for Computer Simulation International (2003)
6. Air Hockey Trick Shots—Deceptive Shooting Techniques. Bubble & Air Hockey. www.bubbleairhockey.com/trick-shots.html

Approaches to Building a Detection Model for Water Quality: A Case Study

Fitore Muharemi, Doina Logofătu, Christina Andersson and Florin Leon

Abstract Predicting failure or success of an event or value is a problem that has recently been addressed using data mining techniques. By using the information we have from the past and the information of the present, we can increase the chance to take the best decision on a future event. In this paper, we evaluate some popular classification algorithms to model a water quality detection system. The experiment is carried out using data gathered from Thüringer Fernwasserversorgung water company. We briefly introduce baseline steps we followed in order to achieve a descent model for this binary classification problem. We describe the algorithms we have used, and the purpose of using each algorithm, and in the end we come up with a final best model. Representative models are compared using the F1 score, as a performance measurement. Finding the best model allows for early recognition of undesirable changes in the drinking water quality and enables the water supply companies to counteract in time.

Keywords Classification · Watter quality · Performance metrics

1 Introduction

Water covers 71% of Earth's surface and is vital for all known forms of life. The purity of drinking water is an essential task for water supply companies all over the world. By using different sensors we try to monitor relevant water and environmental

F. Muharemi (✉) · D. Logofătu (✉) · C. Andersson
Frankfurt University of Applied Sciences, Frankfurt Am Main, Germany
e-mail: muharemi@stud.fra-uas.de

D. Logofătu
e-mail: logofatu@fb2.fra-uas.de

C. Andersson
e-mail: andersso@fb2.fra-uas.de

F. Leon
Technical University of Iaşi, Iaşi, Romania
e-mail: florin.leon@tuiasi.ro

© Springer International Publishing AG, part of Springer Nature 2018
A. Sieminski et al. (eds.), *Modern Approaches for Intelligent Information and Database Systems*, Studies in Computational Intelligence 769,
https://doi.org/10.1007/978-3-319-76081-0_15

173

data at several measuring points on a regular basis. Despite these measurements, we also need a detection system which accurately notifies for water quality changes based on measured values.

The Goal of the GECCO 2017 Industrial Challenge[1] was to develop a change detection system to accurately predict any kind of changes in time series of drinking water composition data. An adequate and accurate alarm system that allows for early recognition of all kinds of changes is a basic requirement for the provision of clean and safe drinking water.

In this paper, we present the solution offered for this system, by analyzing and comparing some classification algorithms on how they perform and how well they can predict and detect changes in our data. There is a large number of classification algorithms today and each one has its benefits, but it is important to know which one to use for the particular task [3].

Here we explain the steps we took in order to achieve the winning model of Gecco International Challenge 2017—Water Monitoring System competition.[2] The methodology of solving the problem includes four classification algorithms on the same classification task. When sufficiently representative training data were used, most algorithms perform reasonably well, but in our experiment even when we had a large dataset, not every algorithm gave much promising results. The goal of this paper is to find the most suitable algorithm for the problem under investigation. There is a number of dimensions we can look at to give a sense of what will be a reasonable algorithm: number of training data, number of features, the dependency of features, etc. We have also to consider that features selection is a very important stage to decide on the performance of the algorithms.

There are many methods to help us decide to go with a tested model or not. In this research, we use the F1 score, considered as one of the best performance metrics for classification algorithms. So far, the results obtained provide a satisfactory justification for the choice of the Logistic Regression algorithm, by using F1 score.

The rest of the paper is organized as follows. Section 2 outlines related works on the analysis made on water quality. Section 3 briefly describes the dataset. Sections 4 and 5 explain preprocessing and feature selection on our dataset, respectively. Sections 6, 7 and 8 respectively presents the tools, experimental evaluation and results. Conclusions and future work are included in the last section.

2 Related Work

Water as a very important factor of life makes crucial devising new methodologies for analyzing water quality and forecast future water quality trends. Many researchers have analyzed water quality problem using machine learning techniques. In [12] the problem with water quality is investigated, by using three different algorithms, SVM,

---

[1] http://gecco-2017.sigevo.org/index.html/HomePage.

[2] http://www.spotseven.de/gecco/gecco-challenge/gecco-challenge-2017/.

and two methods of artificial neural networks. The performance is compared using R^2, RMSE, MAE. On the results they achieved, the SVM algorithm is competitive with neural networks. Also in [10] a model for water quality prediction was developed. The best result was achieved using Artificial Neural Network with Nonlinear Autoregressive terms. In 2009, Xiang and Jiang applied least squares support vector machine (LS-SVM) with particle swarm optimization methods to predict the water quality and defeat the weaknesses of customary back propagation algorithms as being moderate to meet and simple to achieve the extreme minimum value. They discovered that through simulation testing, the model shows high proficiency in estimating the water quality of the Liuxi River [17].

3 Dataset

This experiment extracts data from the public water company Thüringer Fernwasserversorgung, located at the heart of Germany. The Thüringer Fernwasserversorgung performs measurements at significant points throughout the whole water distribution system, in particular at the outflow of the waterworks and the in- and outflow of the water towers. For this purpose, a part of the water is bypassed through a sensor system where the most important water quality indicators are measured. The data that is supplied for this challenge has been measured at different stations near the outflow of a waterworks. The analysis done here can serve in general for every water company to monitor water quality. A total of 122334 samples and 10 variables were used. The variables of the data are: Time, Tp, Cl, pH, Redox, Leit, Trueb, Cl_2, Fm, Fm_2, Event, see Table 1. The EVENT is the target variable, which

Table 1 Description of the given time series data

Variable name	Description
Time	Time of measurement, given in following format: yyyy-mm-dd HH:MM:SS
Tp	The temperature of the water, given in C
Cl	Amount of chlorine dioxide in the water, given in mg/L (MS1)
pH	PH value of the water
Redox	Redox potential, given in mV
Leit	Electric conductivity of the water, given in μS/cm
Trueb	Turbidity of the water, given in NT
Cl_2	Amount of chlorine dioxide in the water, given in mg/L (MS2)
Fm	Flow rate at water line 1, given in m^3/h
Fm_2	Flow rate at water line 2, given in m^3/h
EVENT	True if there is any Event, False otherwise

one we want to predict with a high accuracy. The possible values of EVENT are true or false, so the problem is a classification problem. The dataset contains one time series denoting water quality data and operative data on a minutely basis.

4 Preprocessing

Unfortunately, real-world databases are highly influenced by negative factors such the presence of noise, inconsistent and superfluous data and huge sizes in both dimensions, examples, and features. Data preprocessing is an essential step for data mining processes. If data are not prepared, the results offered will not make sense, or often they are not accurate as we expect. Some important steps on data preparation are: data cleaning, data transformation, data integration, data normalization, noise identification [6].

In case we have missing values we can choose to either omit these elements from the dataset or to impute values. In our experiment, we tried both approaches to handle the problem and based on the given values by the F1 score, the second approach proved to be more appropriate. In our data, there were 11519 missing values, and if we remove them we lose too much information. There are many methods on filling in the missing values, imputation of the mean, fill with zero, kNN-imputation, etc. [4, 16]. Specifically, in our case study, we tried different approaches: filling with mean value, filling with knn- algorithm, and filling with zero. Surprisingly, the best model comparing with F1 score, was the one we filled the missing values with zero. So in our case study we continued with this approach.

5 Features Selection

We know that some machine learning algorithms can have poor performance if there are highly correlated data. In our data the predictors are highly correlated with each other, we can see this on Table 2.

As a consequence, if we use all variables, our models will be unstable, and by choosing the appropriate method we must ensure the stability of the model we choose. Usually, it is suggested that pairs of columns with correlation coefficient higher than a threshold are reduced to only one of these variables.

A very important step when we have more than 5 predictors and when there exists a correlation between them is to do feature selection. Today we can find many algorithms on feature selection which can help on deciding on the importance of our predictor variables. In our research experiment, we have tried two algorithms for feature selection, Boruta with Random Forest and Recursive Feature Selection with Random Forest.

Random forests is a very good ensemble-based classification method, while the Boruta is developed as an improvement of Random Forest. We tested both

Table 2 Correlation table

	Tp	Cl	pH	Redox	Leit	Trueb	Cl_2	Fm	Fm_2
Tp	1.000000	0.948834	0.955430	0.944517	0.898868	0.528958	0.846250	0.938023	0.923177
Cl	0.948834	1.000000	0.979351	0.9973239	0.948435	0.516091	0.846250	0.938023	0.923177
pH	0.955430	0.979351	1.000000	0.996609	0.968112	0.517524	0.934405	0.949046	0.940577
Redox	0.944517	0.9973239	0.996609	1.000000	0.962424	0.507129	0.941441	0.943918	0.936754
Leit	0.898868	0.948435	0.968112	0.962424	1.000000	0.4991666	0.940569	0.905860	0.902853
Trueb	0.528958	0.516091	0.517524	0.507129	0.4991666	1.000000	0.426762	0.516625	0.478793
Cl_2	0.846250	0.846250	0.934405	0.941441	0.940569	0.426762	1.000000	0.883319	0.877863
Fm	0.938023	0.942389	0.949046	0.943918	0.905860	0.516625	0.883319	1.000000	0.919937
Fm_2	0.923177	0.923177	0.940577	0.936754	0.902853	0.478793	0.877863	0.919937	1.000000

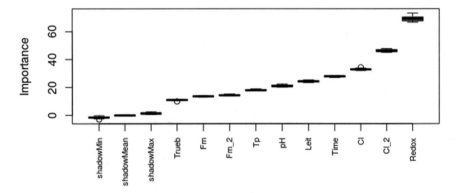

Fig. 1 Boruta result plot for our data. Boxplot of each feature shows that the mean on each feature is higher than shadowMax (maximum Z score of a shadow attribute). None of the attributes is rejected

algorithms and based on the results we got, we tested different models using the obtained importance of features. Finding the most important predictor variables that explain the major part of the variance of the response variable is the key to identify and build high-performing models. Boruta is a feature selection algorithm, which one works as a wrapper algorithm around Random Forest. While fitting a random forest model on the dataset, it can get rid of attributes in each iteration, so it does not perform well in the process. On the other hand, Boruta can find all features which are either strongly or weakly relevant to the decision variable.

To use Boruta algorithm for feature selection we have used the Boruta package in R. By running the Boruta algorithm in our dataset we got that all variables are important, and is suggested to use all of them in the model. Figure 1 shows the importance of the features for our dataset, using Boruta package. The features will be classified as confirmed or rejected by comparing the median Z score of the attributes with the median Z score of the best shadow attribute. So from the results we got from 'Boruta', no attributes deemed unimportant, and all 10 attributes were confirmed as important and by using the results we must consider all features on the models we create.

Even there are correlated predictors, Boruta algorithm can come up with all predictors as important, as the way how it works is that it trains a Random Forest on the set of original and randomized feature. This Random Forest during training, as every Random Forest only sees a subset of all features at every node. Hence sometimes it will not have choice between X_1 and X_2 when picking the variable for the current node and it can not prefer one of the two variables X_1 and X_2 over the other [11].

Not every feature selection algorithm end up with the same set of important features. Using Recursive Feature Selection we got only five features as important (Redox, Cl, Trueb, Leit and pH) [7]. Figure 2 shows the variable importance chart selected with random forest algorithm.

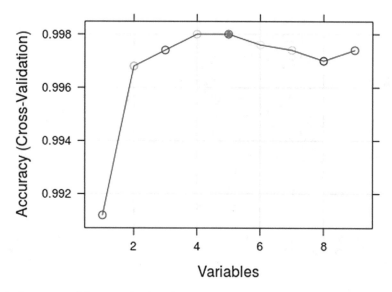

Fig. 2 Importance of features—Random Forest

6 Modeling Tools

We have trained a set of models using classification algorithms: logistic regression, SVM, Linear Discriminant Analysis, and Neural Networks algorithm. We have tested 9 different models with different parameters and compared them.

Logistic Regression one of the most popular machine learning algorithms. It is a simple algorithm that performs very well on a wide range of problems. In our modeling we tested this algorithm as it corresponds with the data we want to predict, pre dict TRUE or FALSE events. We have considered three different models using the logistic regression. In two models we have used attributes chosen as important from the two feature selection algorithms, and one model where we have used interaction terms [9].

Linear Discriminant Analysis can be used to find a linear combination or features characterized by two or more events. LDA works well when measurements of the independent variables are continuous quantities. In our problem it is not preferred much because of the assumptions about data, it assumes that data distribution is Gaussian. We tested two variants of LDA [9].

Support Vector Machine SVM is a supervised machine learning algorithm which can be used for both classification and regression problems. The SVM algorithm, in the past, had promised to be very efficient for binary classification, so it was expected to be good for our experiment. We tested two variants using SVM, by changing the parameters (kernel, cost). After parameters tuning, we used kernel = "linear", cost = 100 [5].

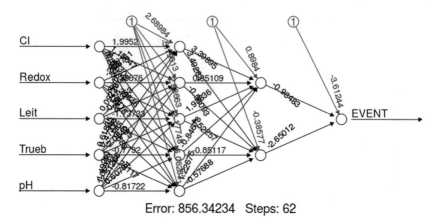

Error: 856.34234 Steps: 62

Fig. 3 Neural network plot from a sample of our dataset. Input layer—five predictor variables (left side). Two hidden layers, with five and two neurons respectively (middle). Output Layer (Response Variable)

Artificial Neural Network ANN is known to be a good methodology for classification of complex datasets. The structure of the algorithms tends to simulate the structure of the human brain. We have a large training set and running the algorithm takes to much time, so we were not able to test different variants by changing parameters. We experimented with only one simple variant using neural network with five hidden layers and five neurons on each layer. Figure 3 visualize the plot of this simple neural network model. We used only five variables, as extracted by using Random Forest algorithm, since ANN does not perform well when there are highly correlated data [1, 8].

7   Evaluation of Models

When we want to compare models we have to decide first what metrics to use. Commonly-accepted performance evaluation measures are Accuracy, Precision, Recall, F-score [14]. In our research experiment, we have to deal with a binary classification problem and is often suggested to use F-score for evaluation of classification problems.

7.1   Problems with Accuracy

Usually, this approach has many problems and is not preferred to be used alone. Not always 99% accuracy of the model indicates that it is a good model. It can be excellent or very poor based on the problem or training set. In all models we modeled for our water monitoring system data, we investigated that the accuracy is 98–99%, which at first sight looks very promising, but the truth is that it is useless since if we would use a model that only predict FALSE(no event) we would get the same accuracy (98–99%). We excluded Accuracy as a performance measurements for our models.

7.2 Precision, Recall and F1

The number of correctly recognized class examples (true positives), the number of correctly recognized examples that do not belong to the class (true negatives), and examples that either were incorrectly assigned to the class (false positives) or that were not recognized as class examples (false negatives), constitute a confusion matrix. Most often used measures for binary classification are based on the values of the confusion matrix, precision, recall and F-measure [2, 14].

Precision: the number of correctly classified positive examples divided by the number of examples labeled by the system as positive

$$Precision = \frac{TP}{TP + FP} = \frac{positive\ predicted\ correctly}{all\ positive\ predictions} \qquad (1)$$

Recall: the number of correctly classified positive examples divided by the number of positive examples in the data

$$Recall = TPR = \frac{TP}{TP + FN} = \frac{TP}{P} = \frac{predicted\ positive}{all\ positive\ obs} \qquad (2)$$

F1 score: the harmonic mean of precision and recall.

$$F1 = 2 \times \frac{Precision \times Recall}{Precision + Recall} \qquad (3)$$

In this paper we were concerned to have a better F1 score, since it was used for the selection of the best model by GECCO Industrial Challenge. Table 3 visualize the F1 score of four different models, modeled with four different algorithms.

8 Experimental Results

This section presents the results obtained by running experiments to trained models for each algorithm. Using the methodology we described before, we obtained 9 variants of event detection in water quality. Here we are trying to check best results

Table 3 Results of four of the best models

Algorithm	True Pos	False Pos	True negative	False negative	F1
Log regression	**795**	**187**	**120407**	945	**0.5841293**
LDA	603	17427	103167	1137	0.06100152
SVM	487	9707	111599	**541**	0.08679380
Neural networks	242	24392	96914	786	0.01886057

obtained using F1 score as one of the best performance metrics for classification problems. Table 3 shows scores of the best four models according to the F1 score. The best parameters are highlighted, but the most important value is F1 = 0.58412932.

We have to consider interaction terms, based on the domain knowledge. A decrease of one pH unit is accompanied by an increase in Redox Potential, and the pH of pure water decreases as the temperature increases [13]. From prior researches in water quality we see how water factors impact each other, and we can use them to build the model. We used this prior knowledge to add interaction terms in the logistic regression algorithm.

Interaction terms played a very important role in achieving the best F1 score. When the interaction terms have been added to logistic regression we obtained a much more promising result than when we removed correlated features. This may not be quite an optimal model for interpreting the behavior of attributes in our dataset, but it enhanced our model in an obvious way [15].

9　Conclusion and Future Work

Nowadays we can follow different ways of modeling a forecasting system, none of these methods are proven to be 100% as the best method. By using this methodology we created a model which resulted as the winner model in GECCO International Challenge. The bad results we got from SVM and Neural Network compared with Logistic Regression were expected as we tested these two algorithm only for two different parameters, and not relevant one. The dataset was large enough so using different parameters needs more time. We were focused to experiment the logistic regression and interaction terms. In past experiments both mentioned algorithms SVM and Neural Network proved to be very useful in ecologically-related problems, bur here the merits for the results of Logistic Regression goes to interaction terms. We found that interaction terms can have a huge impact when we have collinearity between many variables, and we have some prior knowledge on the domain.

As future work, we have to try to optimize our model, by decreasing False Negative and False Positive prediction. Variable selection remains an important part of ANN model development, due to the negative impact that poor selection can have on the performance of ANNs during training and deployment post-development. We will experiment more with Artificial Neural Network and Support Vector Machine for further research in Water Quality detection system, as both are very promising.

References

1. Angelov, P., Manolopoulos, Y., Iliadis, L., Roy, A., Vellasco, M.: In: Advances in Big Data: Proceedings of the 2nd INNS Conference on Big Data, 23–25 Oct 2016, Thessaloniki, Greece, vol. 529. Springer (2016)
2. Bottenberg, R.A., Ward, J.H.: Applied multiple linear regression. Technical report. Personnel Research Lab Lackland AFB TEX (1963)
3. Chandrasekaran, S., Freise, M., Stork, J., Rebolledo, M., Bartz-Beielstein, T.: GECCO 2017 Industrial Challenge: Monitoring of Drinking-Water Quality (2017)
4. Darlington, R.B., Hayes, A.F.: Regression Analysis and Linear Models: Concepts, Applications, and Implementation. Guilford Publications (2016)
5. Demšar, J.: Statistical comparisons of classifiers over multiple data sets. J. Mach. Learn. Res. 7(Jan), 1–30 (2006)
6. García, S., Luengo, J., Herrera, F.: Data Preprocessing in Data Mining. Springer (2015)
7. Hartshorn, S.: Machine Learning with Random Forests and Decision Trees (2016)
8. Hassoun, M.H.: Fundamentals of Artificial Neural Networks. MIT press (1995)
9. James, G., Witten, D., Hastie, T., Tibshirani, R.: An Introduction to Statistical Learning, vol. 112. Springer (2013)
10. Kang, G.K., Gao, J.Z., Xie, G.: Data-driven Water Quality Analysis and Prediction: A survey
11. Kursa, M.B., Rudnicki, W.R., et al.: Feature selection with the boruta package. J. Stat. Softw. 36(11), 1–13 (2010)
12. Mohammadpour, R., Shaharuddin, S., Chang, C.K., Zakaria, N.A., Ab Ghani, A., Chan, N.W.: Prediction of water quality index in constructed wetlands using support vector machine. Environ. Sci. Pollut. Res. 22(8), 6208–6219 (2015)
13. Rodkey, F.L.: The Effect of Temperature on the Oxidation-reduction Potential of the Diphosphopyridine Nucleotide System
14. Sokolova, M., Lapalme, G.: A systematic analysis of performance measures for classification tasks. Inf. Process. Manag. 45(4), 427–437 (2009)
15. Vapnik, V.: The Nature of Statistical Learning Theory. Springer Science & Business Media (2013)
16. Wong, J.: Imputation: imputation. R Package Version 2.0, 1 (2013)
17. Xiang, Y., Jiang, L.: Water quality prediction using LS-SVM and particle swarm optimization. In: Second International Workshop on Knowledge Discovery and Data Mining, 2009. WKDD 2009, pp. 900–904. IEEE (2009)

A Deep Learning Approach to Case Based Reasoning to the Evaluation and Diagnosis of Cervical Carcinoma

José Neves, Henrique Vicente, Filipa Ferraz,
Ana Catarina Leite, Ana Rita Rodrigues, Manuela Cruz,
Joana Machado, João Neves and Luzia Sampaio

Abstract Deep Learning (DL) is a new area of Machine Learning research introduced with the objective of moving Machine Learning closer to one of its original goals, i.e., Artificial Intelligence (AI). DL breaks down tasks in ways that makes all kinds of machine assists seem possible, even likely. Better preventive healthcare, even better recommendations, are all here today or on the horizon.

J. Neves (✉) · H. Vicente · F. Ferraz
Centro Algoritmi, Universidade do Minho, Braga, Portugal
e-mail: jneves@di.uminho.pt

H. Vicente
e-mail: hvicente@uevora.pt

F. Ferraz
e-mail: filipatferraz@gmail.com

H. Vicente
Departamento de Química, Escola de Ciências e Tecnologia,
Universidade de Évora, Évora, Portugal

A. C. Leite · A. R. Rodrigues · M. Cruz
Departamento de Informática, Universidade do Minho, Braga, Portugal
e-mail: anacleite@gmail.com

A. R. Rodrigues
e-mail: anaritavvr@gmail.com

M. Cruz
e-mail: manuelavalecruz@gmail.com

J. Machado
Farmácia de Lamaçães, Braga, Portugal
e-mail: joana.mmachado@gmail.com

J. Neves
Mediclinic Arabian Ranches, PO Box 282602, Dubai, UAE
e-mail: joaocpneves@gmail.com

L. Sampaio
Dubai Healthcare City, PO Box 118855, Dubai, UAE
e-mail: luzia.sampaio@dbaj.ae

© Springer International Publishing AG, part of Springer Nature 2018
A. Sieminski et al. (eds.), *Modern Approaches for Intelligent Information and Database Systems*, Studies in Computational Intelligence 769,
https://doi.org/10.1007/978-3-319-76081-0_16

However, keeping up the pace of progress will require confronting currently AI's serious limitations. The last but not the least, Cervical Carcinoma is actuality a critical public health problem. Although patients have a longer survival rate due to early diagnosis and more effective treatment, this disease is still the leading cause of cancer death among women. Therefore, the main objective of this article is to present a DL approach to Case Based Reasoning in order to evaluate and diagnose Cervical Carcinoma using Magnetic Resonance Imaging. It will be grounded on a dynamic virtual world of complex and interactive entities that compete against one another in which its aptitude is judged by a single criterion, the Quality of Information they carry and the system's Degree of Confidence on such a measure, under a fixed symbolic structure.

Keywords Artificial Intelligence · Deep Learning · Machine Learning
Cervical Carcinoma · Magnetic Resonance Imaging · Logic Programming
Knowledge Representation and Reasoning · Case Based Reasoning

1 Introduction

Improving *Patient Care* with *Artificial Intelligence* (*AI*) is transforming the world of *Medicine*. *AI* can help doctors make faster, more accurate diagnosis or predict the risk of a disease in time to prevent it. Indeed, although *AI* has been around for decades, new advances have ignited a boom in the *AI* field, namely due to the advent of *Deep Learning* (*DL*) [1, 2]. This *AI* technique has been powering self-driving cars, image recognition, and even life-saving advances in *Medicine*, just to name a few. Undeniably, *DL* helps researchers analyze medical data to treat diseases. It is advancing the future of *Personalized Medicine*. But the peculiar thing with the present approach to *DL* is just how old its ideas are. Especially, we wonder if we are at the beginning of a revolution; if a real intelligence breaks when slightly someone changes the problem or if we are we endorsing the case where *AI* is riding a one-trick pony? [3]. On the one hand, cervical cancer still is portrayed as the second most common one occurring in females [4], with 70% of cases up in developed countries [5] (e.g., in 2013, 11 955 women in the United States were diagnosed with cervical cancer, of which 4 217 died [6]). This type of cancer arises in the cervix, resulting in an abnormal growth of cells with the ability to invade or spread to other parts of the body. It comes into two major sub-types, namely *Cervical Squamous Cell Carcinoma* (*CSCC*) and *Endocervical Adenocarcinoma* (*EA*) [7]. On the other hand, *Magnetic Resonance Imaging* (*MRI*) is used in medical imaging field to visualize in detail internal structures of the body. Compared with other medical imaging techniques such as *Computer Tomography* (*CT*) and *X-rays*, it provides good contrast between the soft tissues of the body [8]. In this study *MRI* is used to evaluate the stage of the *CSCC* of a given patient. Features of resonance images, such as *Tumor Volume*, were extracted using a dataset of 54 patient images that stand for previously experienced and studied concrete situations (cases).

Last but not least, cleanliness of the data will factor heavily here. One's view of *DL* begins when a data item is analyzed to break it down into its component parts. It relies on Logic Programming to create such combinations and score them, i.e., one may say that a data item will be dissected in a way analogous to the mode used to analyze the inner part of an atom. *DL* is more an evolution than a machine process. It brings *Machine Learning* closer to *AI* through *enhanced cognition* or the functional equivalent *augmented intelligence*. It is under this setting that we intend to answer the questions referred to above. In terms of the computational equivalent it will be used *Case Based Reasoning* (*CBR*) once one of its fundamental tenets is on the fact that reasoning from prior experiences improves performance [9].

The article is subdivided into five sections, being the former one an opening where the problem to be addressed is made known, followed by a background's one where issues related to it are object of attention. The third section introduces the time-line of the problem solving process. The next one addresses the way one comes to a solution to the problem using *CBR*. Finally, a conclusion is presented and directions for future work are outlined.

2 Background

2.1 Cervical Squamous Cell Carcinoma

Cervical Cancer (*CC*) is the second most common cancer in women, being only surpassed by *Breast Cancer* [10]. *CC* is located in the lower part of the uterus and, like other cancers, is related to the abnormal growth of cells that have the capacity to invade other parts of the body [7]. *CC* presents two different parts and is lined by two types of cells, namely glandular cells in the area closest to the body of the uterus and termed the endocervical, and squamous ones lining the part closest to the vaginae (exocervical) [7]. Depending on the cells lining the cervical, there are also different types of cancer, i.e., there are squamous cell carcinoma with cancer cells, adenocarcinoma developed from glandular cells and, less frequently, adenosquamous or mixed carcinomas because they have characteristics of carcinomas of squamous cells and adenocarcinomas [7]. Here, *CSCC* was the only one studied. Some of the major risk factors for the development of *CSCC* are related to numerous factors, ranging from *Human Papillomavirus* (*HPV*) infection, smoking, multiple sexual partners, a weakened immune system, *HIV* infections or organ transplantation [11]. Once the carcinoma has been identified, it is evaluated the way it has spread, thus determining the staging [12].

2.2 Deep Learning Versus Knowledge Representation and Reasoning

On the one hand, different approaches to integrate *Deep Learning* (*DL*) with *Knowledge Representation* and *Reasoning* (*KRR*) are based on the fact that one must give up on having a *fixed symbolic structure* [1]. However, when working on such a *relaxation* process, in situations where the *KRR* systems are *induced* by learning algorithms, the process turns out to be mostly opaque to the programmers [1, 2]. Putting the things in this form, the distinctiveness about *DP* is just how old their ideas are. This stands for the key distinction between the former approaches (in which it is asserted that the work done is symbolic logic in vector spaces, remaining *discrete* the essential features, and nothing is gained), to the one that will be presented here, although having a symbolic logic in vector spaces, the elements or attributes of the logical functions there described go from discrete to continuous, allowing for the representation and handling of *unknown, incomplete, forbidden* and even *self-contradictory information* or *knowledge*. On the other hand, multiple approaches for *KRR* have been proposed using the *Logic Programming* (*LP*) paradigm, namely in the area of *Model Theory* [13, 14] and *Proof Theory* [15, 16]. In this work it is followed the proof theoretical one in terms of an extension to the *LP* language [13, 14, 16], that embed our view of *DL* and how it influences one's *KRR* paradigm, i.e., advances in observational data have transformed the *KRR* paradigm from a largely speculative science into a predictive one with precise agreement between theory and observation. A data item is itemized as being fixed to a given interval fitting its domain, plus the *Quality-of-Information* (*QoI*) and *Degree-of-Confidence* (*DoC*) values [17–19], viz.

- If the program clauses have discrete values as attributes, either in terms of *positive* or *negative* information, the *QoI* of each clause is set to 1;
- If the information is *unknown*, the *QoI* = 0;
- If the extension of a given predicate is given as a disjoint set of clauses, $QoI_i = 1/Card$, where *Card* denotes the clauses' set cardinality;
- If the clauses set is not disjoint, but the clauses have discrete values as attributes, the set cardinality is given by $C_1^{Card} + \cdots + C_{Card}^{Card}$, and the *QoI*'s are given by $QoI_{i_{1 \leq i \leq Card}} = 1/C_1^{Card}, \ldots, 1/C_{Card}^{Card}$, where C_{Card}^{Card} is a card-combination subset, with *Card* elements (Fig. 1); and
- If the clause set is not disjoint and the clauses do not have only discrete values as attributes, the *QoI*'s are depicted as shown in Fig. 2. $K = C_1^{Card} + C_2^{Card} + \cdots + C_{Card}^{Card}$, and $\sum_{i=1}^{n}(QoI_i \times p_i)/n$ denotes the attributes QoI's average for each clause; p_i stands for the relative weight of attribute *i* with respect to its homologous, being $\sum_{i=1}^{n} p_i = 1$.

Now, one is in position to consider a new evaluation factor, denoted as *DoC*, which stands for the system confidence that, having into consideration the clauses

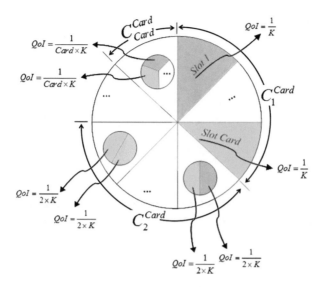

Fig. 1 *QoI's* values for the case where the clause set is not disjoint, but the clauses have discrete values as attributes

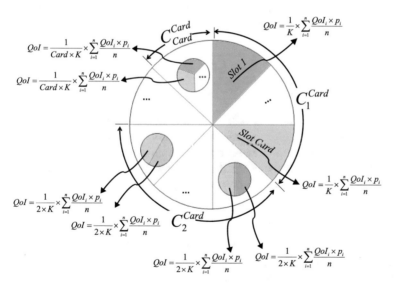

Fig. 2 *QoI's* values for the case where the defined clauses are not only disjoint, but the clauses do not have only discrete values as attributes

domains, the clauses attribute values fit into a given interval [19]. The *DoC* for each clause attribute is evaluated as it is illustrated in Fig. 3, i.e., $DoC = \sqrt{1 - \Delta l^2}$, where Δl stands for the arguments interval length.

Fig. 3 *DoC's* evaluation

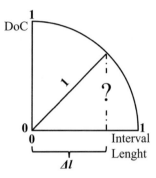

It is now possible to engender the universe of discourse according to the information presented in the extensions of a set of n ($n \geq 0$) *predicates* according to productions of the type, that engender one's view to *DL*.

$$predicate_{1 \leq i \leq n} - \bigcap_{1 \leq j \leq m} clause_j(([A_{x_1}, B_{x_1}](QoI_{x_1}, DoC_{x_1})), \ldots$$
$$\ldots, ([A_{x_m}, B_{x_m}](QoI_{x_m}, DoC_{x_m})))::QoI_j::DoC_j$$

where n, $\cap$, m and A_{xj}, B_{xj} stand for, respectively, predicates set cardinality, conjunction, *predicate* extension cardinality, and the interval ends where the predicates attributes values may be located.

2.3 Case Based Reasoning

To make a change, a different *CB* cycle was induced (Fig. 4). Although the *CB* cycle is self-explanatory, the optimization process will be explained in detail. Such a operation is applied to the case set retrieved from the *Case Base* whenever they do not lead to a solution. It uses *Genetic Algorithms* [15], and the optimized case set must be in conformity with the invariant, viz.

$$\bigcap_{i=1}^{n} (B_i, E_i) \neq \emptyset$$

i.e., the intersection of the attribute's values ranges for the cases in the case set retrieved from the *Case Base* (i.e., B_i, being n its cardinality), and those that resulted from the optimization process (i.e., E_i), can not be empty.

2.4 Pre-processing and Segmentation

Pre-processing aims to improve the visual appearance of images and improve the manipulation of datasets. To conduct this process the *SimpliFilters* module of *3D*

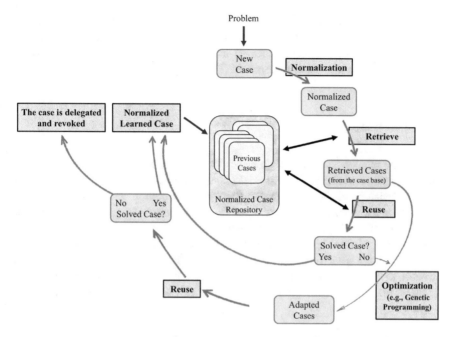

Fig. 4 The updated view of the *CBR* cycle proposed by Neves et al. [20, 21]

Slicer was used, which provides a simple interface for hundreds of basic and advanced *ITK* filters. In this module, a large number of filters were found, and in this study *AddImageFilter* was used, which allowed the addition of pixels of two images, being the most frequently used the *BinaryCoutourImageFilter*, which marks the pixels at the edge of objects [22]. The algorithms provided included binary morphology, grayscale morphology, weighting, threshold, manipulation of image intensity, growing region, fast Fourier transform, just to name a few. Image segmentation is the process of dividing an image into sub regions using discontinuity and similarity properties, such as the gray level and the precise definition of its spatial extent, to which there exists three targeting techniques, i.e., the region-based, contour and texture-based segmentation [23–25].

2.5 Feature Extraction

At this stage and in terms of the feature set the following issues were considered:

- *Tumor Volume*; and
- *Tumor Diameter*.

Once having accomplished the segmentation step, the *3D Slicer Ruler* was used to measure the diameter as well as the *LabelStatistics* module that allowed for the estimate of the tumor's volume. These extracted characteristics will produce a feature vector that will represent the dataset. Other characteristics, such as *Age*, *Weight*, or *Risk Factors* were extracted in the form of a text file from the image repository *The Cancer Imaging Archive (TCIA)* [26].

3 Case Study

Chosen the patients sample and described the procedures that served as a basis for pre-processing and extraction of the *MRI* features, such data is given below in terms of the extensions of the relations the *Patient's Information*, *Risk Factors*, and the *Data Extracted from Uterine Resonance Magnetic Images* (Fig. 5). The extensions of these relations become the knowledge base of a set of complex and virtual entities (i.e., software agents) that will make one's computational system and will lead to the objective function that maps the system behavior. Thus, it is now possible to set the objective function with respect to the problem under analysis in terms of the extension of predicate *diagnosis of Cervical Squamous Cell Carcinoma* ($diag_{CSCC}$), viz [19].

$$
\begin{aligned}
diag_{CSCC} : Age, &\, W_{eight}, H_{eight}, M_{enopause}S_{tatus}, N_{umber}T_{otal of}P_{regnancies}, \\
&\, L_{ive}B_{irth}P_{regnancies}, T_{umor}G_{rade}, T_{umor}S_{tatus}, P_{athologic}S_{tage}, \\
&\, V_{ital}S_{tatus}, T_{umor}D_{iameter}, T_{umor}V_{olume}, T_{obacco}S_{moking}H_{istory}, \\
&\, S_{moking}Y_{ears}, O_{ral}C_{ontraceptives} \rightarrow \{0,1\}
\end{aligned}
$$

where 0 and 1 stand for *true* and *false*.

Face to a new case with feature vector *Age = 52; Weight = 66; Height = 156; Menopause Status = [0, 1], Number Total of Pregnancies = 2, Live Birth Pregnancies = 1, Tumor Grade = ⊥, Tumor Status = [1, 2], Pathologic Stage = [6, 7], Vital Status = [0, 1]; Tumor Diameter = 0.4670; Tumor Volume = 0.0650; Tobacco Smoking History = 1; Smoking Years = ⊥; Oral Contraceptives = [0, 1]*, one may procedure in order to rewrite the feature vector as a $diag_{CSCC}$ clause, viz.

$$
diag_{CSCC_{New Case}} \frac{(((0.49, 0.49)(1,1)), \ldots, ((0, 0.5)(1, 0.87)))}{\substack{attribute's\ values\ ranges\ once\ normalized \\ and\ respective\ QoI\ and\ DoC\ values}} :: 1 :: 0.84
$$

Now this clause is compared with each case retrieved clause from the *Case Base* or its optimize counterpart using as similarity function the mean of the module of the arithmetic difference between the arguments of each selected case and the arguments of the new one, viz.

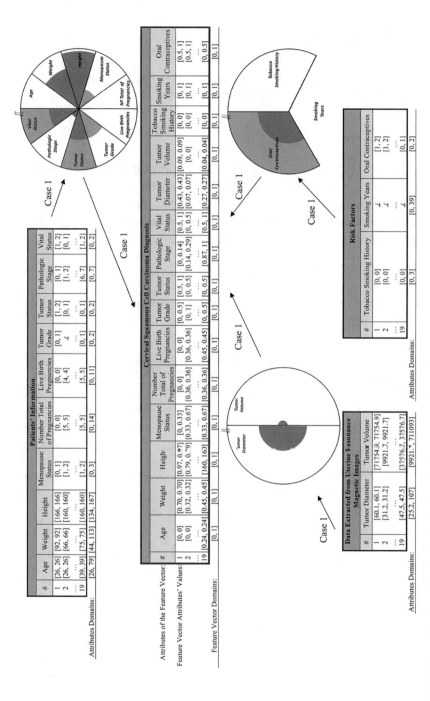

Fig. 5 Knowledge base for the diagnosis of cervical squamous cell carcinoma

$$\frac{\substack{retrieved_{case_1}(((0.58,0.58)(1,1)),\,\ldots,((0,0.5)(1,0.87))):\,:1::0.96 \\ \vdots \\ retrieved_{case_n}(((0.47,0.47)(1,1)),\,\ldots,((0,1)(1,0))):\,:1::0.84}}{\textit{normalized cases that make the retrieved cluster}}$$

Assuming that each attribute has equal weight, the *DoC's* similarity between the new case and *retrievedcase*$_1$ is evaluated in the form, viz.

$$sim^{DoC}_{New\,Case \to 1} = 1 - \frac{|1-1| + \cdots + |0.87-1|}{15} = 1 - 0.022 = 0.978$$

where $sim^{DoC}_{new\,case \to 1}$ denotes the *similarity* in terms of *DoC*, between the *new case* and the retrieved ones (in this example the *retrieved case*$_1$).

An analogous process was followed in order to compute the *similarity*, in terms of *QoI*, between the *new case* and *retrieved case*$_1$, returning $sim^{QoI}_{new\,case \to 1} = 1$. The *overall similarity*, $sim^{QoI,DoC}_{new\,case \to 1}$, is the product of the metrics computed above, i.e.:

$$sim^{QoI,DoC}_{New\,Case \to 1} = 1 \times 0.978 = 0.978$$

This procedure was extended to all retrieved cases leading to a set of solutions to the problem under study, which are presented to the physicians in charge.

Table 1 The coincidence matrix with respect to the proposed model

Output	Model output	
	True (1)	False (0)
True (1)	9	2
False (0)	1	7

Fig. 6 The ROC curve

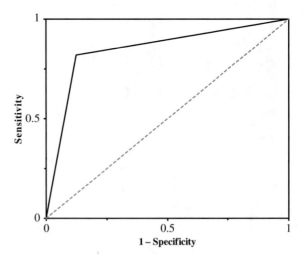

The coincidence matrix regarding the *CBR* model is presented in Table 1. The values displayed refer to the average of 25 (twenty five) experiments. A reading from Table 1 reveals that the *CBR* model classifies properly 16 of a total of 19 cases, being the model accuracy 84.2%. Regarding sensitivity and specificity the results obtained were 81.8%, and 87.5% respectively. The *ROC* curve is depicted in Fig. 6, being the area under the curve 0.85, i.e., the curve follows the left-hand border and then the top one of the ROC space, suggesting that the proposed model exhibits an acceptable performance in the diagnosis of *Cervical Squamous Cell Carcinoma*.

4 Conclusions

The purpose of this paper was to present a framework to construct a dynamic virtual world of complex and interactive entities that map real cases of *Cervical Squamous Cell Carcinoma* in order to develop a decision support system to help predict the different stages of this disease. Based on a new *DL* approach to *KRR*, it also proved that when you boil it down, *AI* may be close to *DL*, but *DL* is not only *Backprop*. Indeed, it was described the extraction of features that allowed the classification of stages of the *Cervical Squamous Cell Carcinoma* through a system of segmentation and classification based on a simple and fixed symbolic structure, that sets a new understanding to *DL*. This leads us to the construction of an inductive theory using case studies, specifying the research questions for closure, in the field of *Carcinoma* evaluation. The basic idea of the dynamic mechanism of evolutionary heuristics follows the basic rule of the *CBR* approach to evolution, where the *Quality-of-Information* (*QoI*) maintained in the *Case Base* is essential.

Acknowledgements This work has been supported by COMPETE: POCI-01-0145-FEDER-007043 and FCT—Fundação para a Ciência e Tecnologia within the Project Scope: UID/CEC/00319/2013.

References

1. Rumelhart, D.E., Hinton, G.E., Williams, R.J.: Learning internal representation by error propagation. In: Rumelhart, D.E., McClelland, J.L. (eds.) Parallel Distributed Processing: Explorations in the Microstructure of Cognition, vol. 1, pp. 318–362. MIT Press, Cambridge (1986)
2. Rumelhart, D.E., Hinton, G.E., Williams, R.J.: Learning representations by back-propagating errors. Nature **323**, 533–536 (1986)
3. Sommers, J.: Is AI Riding a One-Trick Pony? https://www.technologyreview.com/s/608911/is-ai-riding-a-one-trick-pony/
4. Sousa, A.: Cervical Cancer: Trends and Studies. Fernando Pessoa University Edition, Oporto (2011)

5. Deng, S., Zhu, L., Huang, D.: Predicting hub genes associated with cervical cancer through gene co-expression networks. IEEE/ACM Trans. Comput. Biol. Bioinf. **13**, 27–35 (2016)
6. U.S. Cancer Statistics Working Group: United States Cancer Statistics: 1999–2014 Incidence and Mortality Web-based Report. http://www.cdc.gov/uscs
7. American Cancer Society: What is cervical cancer?. https://www.cancer.org/cancer/cervical-cancer/about/what-is-cervical-cancer.html
8. Remya, V., Lekshmi-Priya V.L.: Simultaneous segmentation and tumor detection in MRI cervical cancer radiation therapy with hierarchical adaptive local affine registration. In: Proceedings of the 2014 International Conference on Computer Communication and Informatics, 6 pp. IEEE Edition (2014)
9. Leake, D.B.: Case-Based Reasoning: Experiences, Lessons and Future Directions. MIT Press Cambridge, Massachusetts (1996)
10. Mithlesh, A., Namita, M., Girdhari, S.: Cervical cancer detection using segmentation on Pap smear images. In: Proceedings of the 1st International Conference on Informatics and Analytics, Article 29, Association for Computing Machinery, New York (2016)
11. Vaccarella, S., Lortet-Tieulent, J., Plummer, M., Franceschi, S., Bray, F.: Worldwide trends in cervical cancer incidence: Impact of screening against changes in disease risk factors. Eur. J. Cancer **49**, 3262–3273 (2013)
12. Pecorelli, S.: Corrigendum to "Revised FIGO staging for carcinoma of the vulva, cervix, and endometrium". Int. J. Gynecol. Obstet. **108**, 176 (2010)
13. Kakas, A., Kowalski, R., Toni, F.: The role of abduction in logic programming. In: Gabbay, D., Hogger, C., Robinson, I. (eds.) Handbook of Logic in Artificial Intelligence and Logic Programming, vol. 5, pp. 235–324. Oxford University Press, Oxford (1998)
14. Pereira, L., Anh, H.: Evolution prospection. In: Nakamatsu, K. (ed.) Studies in Computational Intelligence, vol. 199, pp. 51–64. Springer, Berlin (2009)
15. Neves, J., Machado, J., Analide, C., Abelha, A., Brito, L.: The halt condition in genetic programming. In: Neves, J., Santos, M.F., Machado, J. (eds.) Progress in Artificial Intelligence. LNAI, vol. 4874, pp. 160–169. Springer, Berlin (2007)
16. Neves, J.: A logic interpreter to handle time and negation in logic databases. In: Muller, R., Pottmyer, J. (eds.) Proceedings of the 1984 annual conference of the ACM on the 5th Generation Challenge, pp. 50–54. Association for Computing Machinery, New York (1984)
17. Machado J., Abelha A., Novais P., Neves J., Neves J.: Quality of service in healthcare units. In Bertelle, C., Ayesh, A. (eds.) Proceedings of the ESM 2008, pp. 291–298. Eurosis – ETI Publication, Ghent (2008)
18. Fernandes, A., Vicente, H., Figueiredo, M., Neves, M., Neves, J.: An adaptive and evolutionary model to assess the organizational efficiency in training corporations. In: Dang, T.K., Wagner, R., Küng, J., Thoai, N., Takizawa, M., Neuhold, E. (eds.) Future Data and Security Engineering. Lecture Notes on Computer Science, vol. 10018, pp. 415–428. Springer International Publishing, Cham (2016)
19. Fernandes, F., Vicente, H., Abelha, A., Machado, J., Novais, P., Neves J.: Artificial Neural Networks in Diabetes Control. In Proceedings of the 2015 Science and Information Conference (SAI 2015), pp. 362–370. IEEE Edition (2015)
20. Quintas, A., Vicente, H., Novais, P., Abelha, A., Santos, M.F., Machado, J., Neves, J.: A case based approach to assess waiting time prediction at an intensive care unity. In: Arezes, P. (ed.) Advances in Safety Management and Human Factors. Advances in Intelligent Systems and Computing, vol. 491, pp. 29–39. Springer International Publishing, Cham (2016)
21. Silva, A., Vicente, H., Abelha, A., Santos, M.F., Machado, J., Neves, J., Neves, J.: Length of stay in intensive care units—a case base evaluation. In: Fujita, H., Papadopoulos, G.A. (eds.) New Trends in Software Methodologies, Tools and Techniques, Frontiers in Artificial Intelligence and Applications, vol. 286, pp. 191–202. IOS Press, Amsterdam (2016)
22. 3D Slicer. https://www.slicer.org/wiki/Documentation/Nightly/Modules/SimpleFilters
23. Perdigão, N., Tavares, J.M., Martins, J.A., Pires, E.B., Jorge, R. M.: Sobre a Geração de Malhas Tridimensionais para fins computacionais a partir de imagens médicas. In Aparicio, J.L.,

Ferran, A.R., Martins, J.A., Gallego, R., Sá, J.C. (eds.) Proceedings of the Congreso de Métodos Numéricos en Ingeniería, 16 pp. SEMNI, Barcelona (2005)

24. Hasan, D.I., Enaba, M.M., El-Rahman, H.M.A., El-Shazely, S.: Apparent diffusion coefficient value in evaluating types, stages and histologic grading of cancer cervix. Egypt. J. Radiol. Nuclear Med. **46**, 781–789 (2015)

25. Camisão, C., Brenna, C., Lombardelli, S., Djahjah, K., Zeferino, L.: Magnetic resonance imaging in the staging of cervical cancer. Radiologia Brasileira **40**, 207–215 (2007)

26. The Cancer Imaging Archive (TCIA). http://www.cancerimagingarchive.net/

A Fuzzy Approach for the Diagnosis of Depression

Abhijit Thakur, Md. Sakibul Alam, Md. Rashidul Hasan Abir,
Mahir Ashab Ahmed Kushal and Rashedur M. Rahman

Abstract The main objective of this study is to develop a software prototype for diagnosing the risk factors and grading depression in the developing region of South East Asia especially in Bangladesh. World Health Organization (WHO) identified depression to be the most prevalent psychological disorder and according to global burden of disease survey it will be the second leading cause of long term disability. For various social constructs depression or any kind of psychiatric disorder is considered a taboo subject. Hence it is very difficult to collect data on the context of a developing country like Bangladesh. We are using a hybrid model questionnaire based on Diagnostic and Statistical Manual of Mental Disorders (DSM-IV) and Patient Health Questionnaire (PHQ-9) by consulting psychiatric experts of Bangladesh. This study proposed a model based on Fuzzy Logic (FL). An experimental study of the system was conducted using 50 anonymous medical dataset of depression patients' cases obtained from two experts in psychiatry from National Mental Health Institute of Bangladesh and Sheikh Mujib Medical College Hospital.

Keywords Depression model · Risk diagnosis · Fuzzy logic
Neuro-fuzzy controller · Hybrid tool · Depression in Bangladesh

A. Thakur · Md. Sakibul Alam · Md. Rashidul Hasan Abir · M. A. A. Kushal
R. M. Rahman (✉)
Department of Electrical and Computer Engineering, North South University,
Plot-15, Block-B, Bashundhara, Dhaka 1229, Bangladesh
e-mail: rashedur.rahman@northsouth.edu

A. Thakur
e-mail: abhijit.thakur@northsouth.edu

Md. Sakibul Alam
e-mail: sakibul.alam@northsouth.edu

Md. Rashidul Hasan Abir
e-mail: rashidul.abir@northsouth.edu

M. A. A. Kushal
e-mail: mahir.ashab@northsouth.edu

1 Introduction

Depression is the silent phenomena of our mental health and one of the burning issues in our present world. It is a disease whose symptoms in primary care are controversial, vague, imprecise and ambiguous. The disease is a comorbid factor in many chronic health conditions such as Body Mass Index (BMI), Blood Pressure (BP), diabetes, cancer, alcohol abuse, cardiovascular disease. Several studies have shown that the disease is a major public health problem with a high prevalence amongst the adult population [1–14]. It is very sensitive to take any kind of medical decision. As depression depends on various types of issues so the experts have to take decisions through intricate process. To take any kind of authentic decision it is necessary to handle the sensitive data of the concern and make proper connection between the instinct and medical perceptions. The more the symptoms of the concern are matched with the recommended disease, the more the decisions of the expertise will be authentic. This research paper starts data collection but being in the list of a developing country we do not have the proper data storage management system in our medical sectors. Therefore, diagnosing depression can be a demanding task because of the non-availability of data. Several reasons can be seen as the primary cause hindrance for example, (i) Taboo factor with psychiatric disease, (ii) Conservative social management system, (iii) Tendency of our population to go to physicians first instead of considering psychiatrists for their problem, (iv) Non existing system of collecting data by the doctors and health care system alike, (v) Negligent health management system of Bangladesh's hospitals and clinics.

We have to face various issues to collect our desired information. We have gathered authentic information from indoor and outdoor of hospitals and mental clinics. In medical decision making, experts use their knowledge base and experience to arrive at a conclusion. Medical doctors put some values to symptoms and calculate the overall disease result. Doing these procedures manually is a complex task and often error prone. That is why the objective of this paper is to device a model which will compute the risk factors or depression and diagnose the severity of depression. This is not a fully expert tool, but it is an addition by merging the theory of fuzzy logic and fields of psychiatry.

The paper is organized as follows. In literature review we have discussed some of the existing studies. In methodology we discussed the symptom parameters with the help of psychiatric experts'. Then there is description of our process and data set. The 1st and 2nd phase of the study is designed using fuzzy inference system and the outcome has been explained with comparison to other studies. At the end of the study future work to enhance the process has been discussed.

2 Literature Review

With advances in artificial intelligence (AI), intelligent computing has accelerated new approaches that can enhance medical decision support services. The authors in [1–5] statistically modeled seven psychotic diseases using the Brief psychiatric rating scale version-F2 (BPRS-F2) and Plackett–Burman design (PBD) of experiments. Then using multiple regressions, the authors identified significant predictors behind each of these illnesses. It was evident that Fuzzy C-Means (FCM)-based Fuzzy Logic Controller (FLC) predicted psychoses more accurately compared to Entropy-based Fuzzy Clustering (EFC) and its extensions. To find the association of recurrent suicide attempts and overall self-harm of Taiwanese soldiers Artificial Neural Network (ANN) had been used in [6]. 10 factors were identified which could have impact on suicidal. Network training with these factors was conducted using the Radial basis function (RBF). The trained model showed about 82% sensitivity and 86% specificity. In a recent study, based on severity, a group of researchers used a BPNN in categorizing a set of real-world adult depression [7]. The authors at first collected fifteen symptoms of depression under four major constructs, which are 'motivational', 'emotional', 'cognitive', and 'vegetative and physical'. Information about symptoms was then captured by a questionnaire. Answer to each question was then quantified and the corresponding grade was assigned as 'mild'–'moderate'–'severe' under a three-point scale. The information was then fed to the network for training (incremental mode). Finally the study concluded that by the BPNN approach the network was able to classify depression cases with 89% average accuracy. Performance of a neuro-fuzzy model in predicting weight changes in chronic schizophrenic patients exposed to atypical or typical antipsychotics for more than a number of years had been examined. The authors used Fuzzy tech 5.54 software package to generate the rule base. The model was able to predict 93% of weight changes. To differentiate between epileptic and non-epileptic events various fuzzy arithmetic operations have been investigated in [8]. A total of 244 patients were studied using the NEFCLASS (Neuro Fuzzy Classification) architecture with the Artificial Neural Network and Back propagation algorithm (ANNB). The authors were able to detect the differences with a 85% sensitivity, while 95.65% was achieved by ANNB. 16 symptoms across psychological and physiological factors were identified after consultation with medical experts [9]. This soft computing approach consists of Data Acquisition Model (DAM), a knowledge based system (KBS), an Inference Engine (IE). Cognitive and Emotional filters were used to design a support engine. They defined the membership function of severity of depression into five categories. Later they analyzed their accuracy with 20 cases and found out 83.33% match for near absent and 100% match for mild, moderate and severe cases. A computer-aided CBT (CCBT) was described for delivering self-help for sufferers with phobic, panic, anxiety, obsessive-compulsive and depressive disorders [10]. Although the approach greatly reduced the demand on therapist's time, it lacked the patient–doctor feedback mechanism which is crucial for depression therapy. An ANN approach was described for diagnosing depression using radial basis function (RBF) and back propagation neural networks [14]. The approach showed great promise in accurately identifying the psychiatric problem among patients, but it lacked an explanation mechanism.

3 Methodology

The system is implemented and simulated using MATLAB fuzzy tool box. The result of the system is consistent with an expert specialist's opinion on evaluating the performance of the system. Basis on experts' suggestion we reduce the number of symptoms we consider first. We deal with seven major symptoms along with other parameter like age, Body Mass Index (BMI), blood pressure and diseases. We divide our work in two phases, in first phase we find out the possibility of a human being to get depressed and in second phase we measure the depression level of depressed people basis on seven parameters (Table 2) which are sorted from the fourteen symptoms [1] (Table 1) according to psychiatric experts' suggestions [15–17].

These seven symptoms are extracted based on the values of priority table of their weighted average significance.

3.1 Dataset

We collected 50 patient data from hospitals, mental institutions and treatment centers across Dhaka. They include Bangabandhu Medical College Hospital, Mental Health Institute of Bangladesh, Dhaka medical college, Shohorawardi

Table 1 Fourteen symptoms

Symptoms	Short form
Sadness	S
Loss of interest	LI
Self-criticism	SC
Change in sleep pattern	CS
Loss of energy	LE
Change appetite	CA
Concentration difficulties	CD
Impaired social function	IF
Indecision	I
Worthlessness	W
Guilty feelings	GF
Punishment feeling	PF
Suicidal thoughts	ST

Table 2 Sorted seven symptoms

Symptoms
Sadness
Loss of interest
Self-criticism
Loss of energy
Worthlessness
Guilty feelings
Suicidal thoughts

Medical Hospital, and 3 more private mental Health institute. Outdoor patients' data were collected through on spot questionnaire with the help of medical expert. All of the data were sorted and used on the basis of expert suggestion.

3.2 Designing Fuzzy Inference System (1st Phase)

We designed a fuzzy inference system with four input variables with multiple membership value and one output with four membership value [Not at all, Low, Moderate, and High] (Figs. 1 and 2).

1. Fuzzy Input (Age and BP)

We used Triangular membership function for both age and BP input values. The function is given below:

$$traiangle(x; a, b, c) = \begin{cases} 0 & x \leq a \\ \frac{x-a}{b-a} & a \leq x \leq b \\ \frac{c-x}{c-b} & b \leq x \leq c \\ 0 & c \leq x \end{cases} \tag{1}$$

We modified range for both age [Young, Adult, and Old] and BP [Low, Ideal, and High]. For age, Value between 0–20 is considered as Young, 20–35 is considered as Adult and 35–50 is ranged as old. For blood pressure we took diastolic

Fig. 1 Adaptive FIS

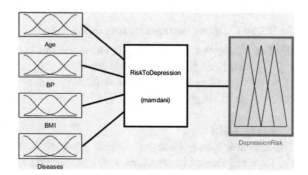

Fig. 2 Surface of FIS

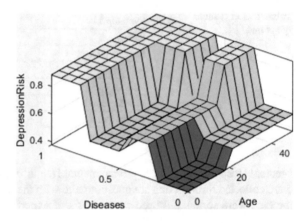

Fig. 3 Fuzzy input (age)

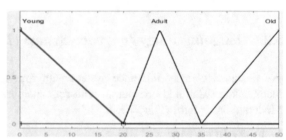

blood pressure and fixed the range from 0–60 as Low, 60–80 as Ideal and 80–100 as high blood pressure. Here one important thing to mention is that we have treated both age and BP as independent parameters discarding their possibilities of their co-dependence on one another (Fig. 3).

2. Fuzzy Input (BMI and Diseases)

Here we use Triangular membership function for BMI and Diseases inputs. We changed the range for both input BMI [Underweight (0–20), Normal (20–25), Overweight (25–30), Obesity (30–35)]. For Disease the representation is like this [None of them (0–0.25), one of them (0.25–0.50), several of them (0.50–0.75), all of them (0.75–1)].

3. Fuzzy Output (Depression Possibility Level)

We use Triangular membership function for our output. The function is same as Eq. (1). We have four values with modified ranges. [Not at all, Low, Moderate, High] Not at all = $0 \leq 25$, Low = $0.25 \leq 0.375 \leq 0.50$, Moderate = $0.50 \leq 0.625 \leq 0.75$, High = $0.75 \leq 0.875 \leq 1.00$ (Fig. 4).

4. Fuzzy Rules

The reason for using Fuzzy rule based approach is for the simplicity of it and also the for similarity of its structure with our motivation to diagnose depression.

Fig. 4 Fuzzy output
(depression level)

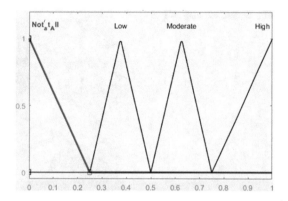

Among 110 rules we defined few of the rules which are used to control 1st system:

  i. If Age is (Young) and BP is (Ideal) and BMI is (Normal Weight) and Diseases are (None of them) Depression risk is (None)

  ii. If Age is (Adult) and BP is (Low) and BMI is (Normal Weight) and Diseases are (None of them) Depression risk is (Low)

 iii. If Age is (Young) and BP is (Low) and BMI is (Over Weight) and Diseases are (One of them) Depression risk is (Moderate)

 iv. If Age is (Young) and BP is (High) and BMI is (Obese) and Diseases are (All of them) Depression risk is (High)

  v. If Age is (Young) and BP is (Low) and BMI is (Over Weight) and Diseases are (All of them) Depression risk is (High).

3.3 Designing Fuzzy Inference System (2nd Phase)

We designed a fuzzy inference system with four input variables with multiple membership values and one output with three membership value [Not at all, Low, Moderate, and High] (Figs. 5 and 6).

Fig. 5 Adaptive FIS

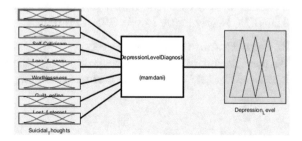

Fig. 6 Surface map of FIS

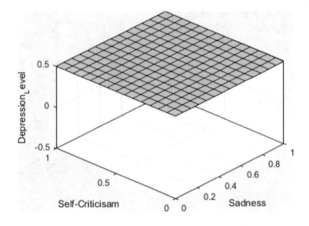

Fig. 7 Fuzzy inputs (all)

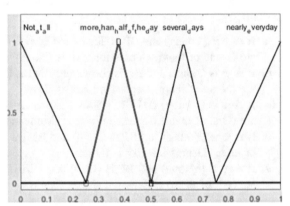

1. **Fuzzy Input (Sadness, Loss of interest, Self-Criticism, Loss of Energy, Worthlessness, Guilty feelings, Suicidal Thoughts)**

We used Triangular membership function for all the input values. The function is given below.

$$f(x; a, b, c) = \max\left(\min\left(\frac{x-a}{b-a}, \frac{c-x}{c-b}\right), 0\right) \tag{2}$$

We have fixed the following ranges according patients' everyday behavior.

Not at all = 0–0.25, More than half of the day = 0.25–0.50, Several Days = 0.50–0.75, Nearly Every Day = 0.75–1.00 (Fig. 7).

2. **Fuzzy Output (Depression Level)**

We use Triangular function for measuring depression level in output. The function is same as Eq. (2) We have fixed four ranges for output like this: Not at all = 0 ≤

Fig. 8 Fuzzy output
(depression level)

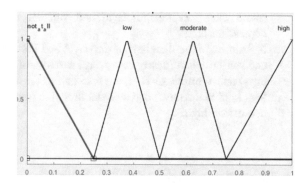

25, Low = 0.25 ≤ 0.375 ≤ 0.50, Moderate = 0.50 ≤ 0.625 ≤ 0.75, High =
0.75 ≤ 0.875 ≤ 1.00 (Fig. 8).

3. Fuzzy Rules

Based on expert's suggestion and the analysis on the patients' symptoms for
achieving better accuracy from our system we figure out more than 80 if-then rules.
Brief explanation of the parameters is given below:

Not at all = parameters are not present at all in a week's time.

Several days = parameters are present at least once in less than half of the days in a
week.

More than half of the days = parameters are present at least once in more than half
of the week.

Nearly every day = parameters are persistence nearly every day of the week.

Some of the rules are defined bellow:

  i. If Sadness (several days) and loss of interest (not at all) and self-criticism (not
at all) and loss of energy (not at all) and Worthlessness (not at all) and Guilty
feeling (not at all) and suicidal thoughts (not at all) then depression low.

  ii. If Sadness (not at all) and loss of interest (not at all) AND self-criticism (not at
all) and loss of energy (not at all) and Worthlessness (several days) and Guilty
feeling (not at all) and suicidal thoughts (not at all) then depression low.

  iii. If Sadness (not at all) and loss of interest (not at all) and self-criticism (several
days) and loss of energy (several days) and Worthlessness (several days) and
Guilty feeling (not at all) and suicidal thoughts (not at all) then depression
moderate.

  iv. If Sadness (not at all) and loss of interest (not at all) and self-criticism (not at
all) and loss of energy (several days) and Worthlessness (several days) and
Guilty feeling (several days) and suicidal thoughts (several days) then
depression moderate.

  v. If Sadness (several days) and loss of interest (more than half of the days) and
self-criticism (more than half of the days) and loss of energy (more than half of
the days) and Worthlessness (more than half of the days) and Guilty feeling

(more than half of the days) and suicidal thoughts (several days) then depression high.

vi. If Sadness (more than half of the days) and loss of interest (nearly every day) and self-criticism (nearly every day) and loss of energy (more than half of the days) and Worthlessness (more than half of the days) and Guilty feeling (more than half of the days) and suicidal thoughts (more than half of the days) then depression high.

4   Results and Analysis of Results

We have worked with 50 patients' data and 10 non-patient data. Our two systems work well on the basis of our IF-THEN rules. Using MSE (Mean Square Error) we defined the accuracy level of two systems. We calculated the error by taking the midpoint value of output's four membership functions as constant and calculated the deviation using midpoint and system's output base on theinputs. Table 3 and Fig. 9 depict the result for first phase and Table 4 and Fig. 10 depict it for second phase.

We have calculated the MSE for error measurement with the formula (3):

$$MSE = \frac{1}{N} \sum_{i=1}^{N} (f_i - y_i)^2 \qquad (3)$$

Here,

f_i midpoint value
y_i System's output

1. **1st system:**

See Table 3 and Fig. 9.

2. **2nd system:**

See Table 4.

From our findings we conclude that our proposed IF-THEN fuzzy rule based system is quite compatible in diagnosing patients. In the previous studies [1–14] the

Table 3   Error of each membership function

Four membership function	
Depression risk level	Error
Not at all	0.00297
Low	0.00231
Moderate	0.04914
High	0.05647

Fig. 9 Output error
measurement graph

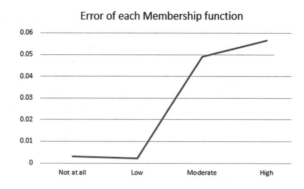

Table 4 Error of each
membership function

Four membership function	
Depression risk level	Error
Not at all	0.00202
Low	0.01562
Moderate	0.06250
High	0.08479

Fig. 10 Output error

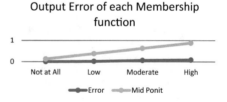

accuracy found is in the range of 90–95% on their dataset, however in our work, the accuracy surpasses all the accuracies reported earlier. As the authors did not disclose their dataset for analysis to others we could not directly compare our model with them.

5 Conclusion and Future Work

In this modern era of science and technology people are getting surrounded by devices. They are getting more and more detached from one another and thus the depression among them is increasing in an alarming rate. The main objective of this paper is to focus on predetermination and diagnosis of the depressed patient using FIS to get the correct level of severity of depression among the patients. In the end it is expected that the use of this model and soft computing using fuzzy system will help to diagnose faster and more accurately. It is also expected that the theory may be applied in other area of medical science. In future along with these two systems we will design a new system which will help to reduce suicide rate by predicting the possibility of suicide on the basis of depression level.

Acknowledgements The authors of this paper fully acknowledge the doctors, the psychiatrists and experts who have shared their valuable knowledge and help in collecting the required data, analyzing them, developing the correct system, and validating the results to make this research successful.

References

1. Chattopadhyay, S., Pratihar, D.K., De Sarkar, S.C.: Some studies on fuzzy clustering of psychosis data. Int. J. Bus. Intell. Data Min. **2**, 143–159 (2007)
2. Chattopadhyay, S., Ray, P., Lee, M.B., Chen, H.S.: Towards the design of an e-health system for suicide prevention. In: Proceedings of the Eleventh IASTED International Conference on Artificial Intelligence, Palma de Mallorca, Spain, 2010, pp. 191–196
3. Chattopadhyay, S., Pratihar, D.K., De Sarkar, S.C.: Developing fuzzy classifiers to predict the chance of occurrence of adult psychoses. Knowl. Based Syst. **20**, 479–497 (2008)
4. Chattopadhyay, S., Pratihar, D.K., De Sarkar, S.C.: Fuzzy logicbased screening and prediction of adult psychoses: a novel approach. IEEE Trans. Syst. Man Cybern. A Syst. Hum. **39**, 381–387 (2009)
5. Chattopadhyay, S., Pratihar, D.K., De Sarkar, S.C.: Statistical modelling of psychoses data, Comput. Methods Programs Biomed. **100**, 222–236 (2010)
6. Tai, Y.-M., Chiu, H.-W.: Artificial neural network analysis on suicide and self-harm history of Taiwanese soldiers. In: Proceedings of the Second International Conference on Innovative Computing, Information and Control (ICICIC), Kumamoto, Japan, p. 363 (2007)
7. Chattopadhyay, S., Kaur, P., Rabhi, F., Acharya, U.R.: Neural network approaches to grade adult depression. J. Med. Syst. **36**(5), 2803–2815 (2012)
8. de Carvalho, L.M.F., Nassar, S.M., de Azevedo, F.M., de Carvalho, H.J.T., Monteiro, L.L., Rech, C.M.Z.: A neurofuzzy system to support in the diagnostic of epileptic events and non-epileptic events using different arithmetical operations. Arq. Neuropsiquiatr. **66**(2a), 179–183 (2008)
9. Ekong, V.E., Inyang, U.G., Onibere, E.A.: Intelligent decision support system for depression diagnosis based on neuro-fuzzy-CBR hybrid. Modern Appl. Sci. **6**(7) (2012). https://doi.org/10.5539/mas.v6n7p79
10. Marks, I., Kenwright, M., McDonough, M., Whittaker, M., O'Brien, T., Mataix-Cols, D.: Saving clinicians' time by delegating routine aspects of therapy to a computer: a randomised controlled trial in Panic/Phobia disorder. Psychol. Med. **34**, 9–17 (2004). https://doi.org/10.1017/S003329170300878X
11. Ashish, K., Dasari, A., Chattopadhyay, S., Hui, N.B.: Genetic-neuro-fuzzy system for grading depression. Appl. Comput. Inform. (2017). https://doi.org/10.1016/j.aci.2017.05.005
12. Chattopadhyay, S.: Psyconsultant I: a DSM-IV-based screening tool for adult psychiatric disorders in Indian rural health center. Internet J. Med. Inform. **3** (Serial Online) (2006)
13. Chattopadhyay, S.: A computerized tool for screening of adult psychiatric illnesses: a third-world perspective. J. Clin. Inform. Telemed. **3**, 1–5 (2006)
14. Suhasini, A., Palanivel, S., Ramalingam, V.: Multi decision support model for psychiatry problem. Int. J. Comput. Appl. **1**, 61–69 (2010)

15. DSM-IV-TR. https://dsm.psychiatryonline.org/doi/abs/10.1176/appi.books.9780890420249.
    dsm-iv-tr (2000). Accessed 14 Dec 2017
16. Patient Health Questionnaire. http://www.apa.org/pi/about/publications/caregivers/practice-
    settings/assessment/tools/patient-health.aspx (n.d.). Accessed 14 Dec 2017
17. Kushal, D.: Expert opinion on parameter establishment. Unpublished raw data (2017)

A Weighted Approach for Class Association Rules

Loan T. T. Nguyen, Bay Vo, Thang Mai and Thanh-Long Nguyen

Abstract Class association rule mining is one of the most important studies supporting classification and prediction. Multiple researches recently focus on mining class association rules using support and confidence user-defined thresholds. However, in the real datasets, each attribute is associated with an indicator value. Based on the actual needs, in this paper, we propose a new approach which combines support, confidence and an interestingness measure (weight) to quickly improve the accuracy of class association rules.

Keywords Classification · Class association rules · Data mining

L. T. T. Nguyen
Division of Knowledge and System Engineering for ICT, Ton Duc Thang University,
Ho Chi Minh City, Vietnam
e-mail: nguyenthithuyloan@tdt.edu.vn

L. T. T. Nguyen
Faculty of Information Technology, Ton Duc Thang University,
Ho Chi Minh City, Vietnam

B. Vo (✉)
Faculty of Information Technology, Ho Chi Minh City University of Technology,
Ho Chi Minh City, Vietnam
e-mail: bayvodinh@gmail.com

T. Mai
Software Development Department, NashTech Global,
Ho Chi Minh City, Vietnam
e-mail: mhthang.it@gmail.com

T.-L. Nguyen
Center for Information Technology, Ho Chi Minh City University of Food Industry,
Ho Chi Minh City, Vietnam
e-mail: longthng@gmail.com

T.-L. Nguyen
VŠB-Technical University of Ostrava, Ostrava-Poruba, Czech Republic

© Springer International Publishing AG, part of Springer Nature 2018
A. Sieminski et al. (eds.), *Modern Approaches for Intelligent Information
and Database Systems*, Studies in Computational Intelligence 769,
https://doi.org/10.1007/978-3-319-76081-0_18

1   Introduction

Data mining is the process of extracting knowledge, which is transformed into an understandable structure, from enormous amounts of data involving methods at the intersection of artificial intelligence, machine learning, statistics, and database systems. Currently, data mining mainly focuses on three problems, namely mining association rules, classification, and clustering. Association rule mining is about retrieving the relationships between items in the transaction database. In the association rules, the consequence of a rule is an itemset. A class association rule is an association rule which has consequence as a class label.

Mining association rules requires finding the relationship between items. Unlike classification rules, the right-hand side of an association rule can contain one or more values. Therefore, the complexity of association rule mining is larger than that of classification rule mining.

Classification is one of the most major issues in data mining and knowledge discovery techniques. The goal of classification rule mining (CRM) is to mine a small set of rules in the dataset that forms an accurate classifier. Association rule mining (ARM) requires finding all rules in the dataset that satisfy the minimum support and minimum confidence thresholds in terms of frequency. The target is not pre-determined for ARM, while there is a pre-determined target, i.e., the class attribute, for CRM.

Associative classification (AC), which integrates association mining and classification [6, 12], is an efficient classification approach. A particular subset of association rules whose their right-hand side is restricted to the class attribute is mined. This subset of rules is denoted as CARs. There have been some methods include classification based on predictive association rules [16], classification based on multiple association rules [5], classification based on associations [4, 5–9] multi-class, multi-label association classification [14], multi-class classification based on association rules [13].

Address the problem that in real datasets, each attribute is associated with an indicator value, mining class association rules based on support, confidence and this indicator should provide a set of rules having more accuracy as well as high confidence. We then provide an approach to generate class association rules from weighted datasets.

The rest of the paper is organized as follows: In Sect. 2, some concepts of CARs are briefly given. Section 3 presents problem related to class association rules. Section 4 describes the proposed algorithm mining class association rules based on weighted approach. Section 5 shows experimental results on the standard database. Finally, conclusions and future works are described in Sect. 6.

2 Preliminary Concepts

Let $D = \{((A_1, a_{i_1}), (A_2, a_{i_2}), \ldots, (A_n, a_{i_n}), (C, c_i)) : i = 1, \ldots, |D|\}$ be a training dataset, where $A_1, A_2, \ldots, A_n$ are attributes and $a_{i_1}, a_{i_2}, \ldots, a_{i_n}$ are their values. Attribute C is the class attribute for which different values $c_1, c_2, \ldots, c_k$ $(k \leq |D|)$ represent classes in D. Without inconsistency, we can use the symbol C to denote the set of classes, that is we can write $C = \{c_1, c_2, \ldots, c_k\}$. One of the attributes among $A_1, A_2, \ldots, A_n$ is used to identify the records in D, often denoted by the symbol OID. Let $A = \{A_1, A_2, \ldots, A_n\} \backslash \{OID\}$. Attributes from A are often called *description attributes*.

For example, let D be the training dataset shown in Table 1, which contains 5 records ($|D| = 5$), where $A = \{A1, A2\}$, *class* is the decision attribute, and $C = \{yes, no\}$, i.e., there are 2 classes.

Definition 1 An itemset is a set of pairs (a, v) where a is an attribute from A, and v is its value. Any attribute from A can appear at most once in the itemset.

From Table 1, we have many itemsets, such as $\{(A2, b3)\}$, $\{(A1, a1), (A2, b1)\}$, etc.

Definition 2 A class association rule r is in the form of $X \rightarrow c$, where X is an itemset and c is a class.

Definition 3 The support of an itemset X, denoted by $Supp(X)$, is the number of records in D containing X.

Definition 4 The support of a class association rule $r = X \rightarrow c_i$, denoted by $Supp(r)$, is the number of records in D containing X and (C, c_i).

Definition 5 The confidence of a class association rule $r = X \rightarrow c_i$, denoted by $Conf(r)$, is defined as:

$$Conf(r) = \frac{Supp(r)}{Supp(X)}.$$

For example: consider rule r: $(A1, a1) \rightarrow y$ with $X = (A1, a1)$ and $c_i = y$. We have $Supp(X) = 3$, $Supp(r) = 2$, and $Conf(r) = \frac{Supp(r)}{Supp(X)} = \frac{2}{3}$.

Table 1 An example of training dataset

OID	A1	A2	class
1	a1	b1	yes
2	a1	b3	yes
3	a2	b3	no
4	a2	b1	yes
5	a1	b2	no

3 Related Work

Liu et al., 1998 introduced an integration of two mining techniques, association rule mining and classification rule mining [6]. This pioneered approach led to the CBA-RG algorithm (a rule generator of Classification Based on Association Rules (CBA)), an Apriori based algorithm, and focused on mining class association rules (CARs). This algorithm is based on heuristics for selecting the strongest rules to form a classifier. Hu et al., 1999 also investigated on mining CARs, however the approach was based on lattice framework [4].

Li et al., 2001 proposed CMAR (Classification based on Multiple Association Rules) [5] for mining CARs. CMAR uses the FP-tree to mine CARs and CR-tree to store the set of rules. The prediction of CMAR is based on multiple rules. In order to predict a record with an unlabeled, CMAR obtains the set of rules (R) that satisfies that record and divides them into l groups corresponding to l existing classes in R. A weighted $\chi2$ is calculated for each group and the class with the highest weighted $\chi2$ is selected and assigned to this record. CMAR algorithm was evaluated having best performance as well as providing more accuracy CARs.

Thabtah et al., 2004 proposed a new approach to find association rules, called multi-class, multi-label associative classification (MMAC) [14]. This algorithm firstly discovers the frequent itemset in datasets, then produces rules set which satisfies minimum confidence user-defined threshold (*MinConf*). The generated rules will then be ranked. The algorithm will have a recursive learning process before finishing and merging all rule sets to produce a multi-label classifier for classification. Thabtah et al., 2005 then also proposed another classification method called multi-class classification based on association rules (MCAR) [13] which has an effective technique to discover frequent itemsets and provide a rule evaluation approach to ensure the rules to be generated have high confidence and accuracy.

ECR-CARM was proposed by Vo and Le 2008 [15] having these main contributions:

- They proposed ECR-tree (Equivalence Class Rule-tree) structure.
- They proposed a class association rule mining algorithm based on ECR-tree, named ECR-CARM (Equivalence Class Rules-tree Classification based on Association Rule Mining).
- They developed an algorithm for pruning redundant rules with a confidence of 100%. The algorithm for pruning redundant rules is called pCARM (pruning Classification based on Association Rule Mining).

Unlike CBA, CMAR, and ECR-CARM, MAC is based on the Apriori approach and uses the intersection between TIDs (record IDs, the same Obidsets as ECR-CARM). This approach has two main steps: rule discovery and classifier building.

Nguyen e al., 2013 proposed a CAR-Miner [9] which was inherited and supported for multiple later on researches, such as Efficient strategies for parallel mining class association rules by Nguyen et al., 2014 [10] MapReduce solution for associative classification of big data by Bechini et al., 2016 [3]. A new strategy for case-based reasoning retrieval using classification based on association by Aljuboori et al., 2016 [1] An Effective Mining of Exception Class Association Rules from Medical Datasets by AI-Tapan et al., 2017 [2], etc. This CAR-Miner algorithm was then improved by CAR-Miner-Diff-Sort [8] which as provided by Nguyen et al., 2015. CAR-Miner-Diff-Sort help to overcome the CAR-Miner's performance issue such that CAR-Miner builds a MECR-Tree (a modified data structure of ECR-Trree), then use the sub-trees of MECR-Tree for each item is characterized different paths. According to the specified paths, CAR-Miner is composed and sorted frequent itemsets based on their corresponding extensions. Despite CAR-Miner computes all intersections between Obidsets (sets of object identifiers that contain itemsets). CAR-Miner-Diff-Sort measures only the difference between two Obidsets (d2O) to reduce the memory is required for storing and time required for calculating the intersection of two Obidsets.

Nguyen and Nguyen (2015) presented a method for mining CARs from incremental datasets [8] which used the modified equivalence class rules tree (MECR-tree) to generate set of rules quickly. The algorithm, Modified-CAR-Miner use the CAR-Miner algorithm to build the MECR tree in the original datasets with adding few modifications. The author also delivered a theorem, "given two nodes l_1 and l_2 in the MECR-tree, if l_1 is parent node of l_2 and $Sup(l_1.itemset \rightarrow Cl_1.pos) < minSup$, then $Sup(l_2.itemset \rightarrow Cl_2.pos) < minSup$", which help to prune infrequent nodes in MECR-tree then improve the process of updating tree.

Song and Lee (2017) proposed predictability-based collective class association rules (PCAR) [11] which applied Eclat algorithm structure and rules ranking to prune set of infrequent itemset uses cross-validation. The algorithm offers a better and more accurate prediction because of generating class association rules based on final classifier.

All of the above methods used support and confidence threshold to eliminate the weakness rules, so the rules set are not higher in accuracy.

4 Mining Class Association Rules with Weighted

This section briefly describes the works, we assume that, besides support and confidence threshold we use the interestingness measurement, a kind of indicator measurement (example: *weight*) to improve the accuracy and use diffset strategy [8].

Based on the above needs, we proposed a weighted approach for mining CARs on dataset, in which, each attribute is associated with a weighted value. The main advantage of the approach algorithm is that it uses the interestingness measure combines with support and confidence threshold instead of support and confidence to improve the accuracy on the rules set.

Table 2 An example of training dataset

OID	A	B	C	class
1	a1	b1	c1	1
2	a1	b2	c1	2
3	a2	b2	c1	2
4	a3	b3	c1	1
5	a3	b1	c2	2
6	a3	b3	c1	1
7	a1	b3	c2	1
8	a2	b2	c2	2

Table 3 An example of weighted indicator for each attribute

OID	Attribute	Weight
1	A	1
2	B	1, 5
3	C	2

Algorithm: Weighted-CARs

Input: Dataset D, *minSup*, *minConf*, and *weight values of attributes*

Output: all Rs that satisfy *minSup*, *minConf*

Step 1: 1-itemsets were generated (itemsets that their weighted supports satisfy *minSup*) and stored on the first level of tree.

Step 2: The rules whose satisfy *minConf* with 1-itemset were generated.

Step 3: Generate the candidates with frequent k-itemsets from $(k - 1)$-itemsets and compute all the information of k-itemsets. Find all k-itemsets that their weighted supports satisfy *minSup*.

Step 4: Generate all rules with k-itemsets and these have confidence satisfy *minConf*.

Step 5: Repeat step 3 until no new 1-itemset are generated.

For illustrative purpose, we use the training dataset example in Table 2 and the list of weighted values for each attribute in Table 3. We executed the algorithm with *minSup* = 20%, *minConf* = 60%. Figure 1 shows the results of this process.

Table 3 shows that the attribute and the weight of each attribute. These values can be determined by users, experts, etc.

Generate rules:

$$Weighted(r) = f(r) * Supp(r)$$

where f is a generic function to evaluate the weights of the left-hand side of r. f can be an average, max, or min, etc. function (In Table 4, we use the average function).

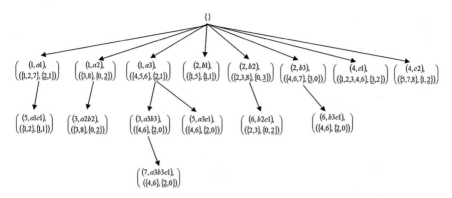

Fig. 1 The result of tree for the dataset in Table 2

Table 4 The satisfied class association rules from training dataset in Table 2

No.	Rule	Weighted support (r)	Conf(r)
1	a1 → 1	2	0.67
2	a2 → 2	2	1
3	a3 → 1	2	0.67
4	b2 → 2	4.5	1
5	b3 → 1	4.5	1
6	c1 → 1	6	0.6
7	c2 → 2	4	0.67
8	a2b2 → 2	2.5	1
9	a3b3 → 1	2.5	1
10	a3c1 → 1	3	1
11	b2c1 → 2	3.5	1
12	b3c1 → 1	3.5	1
13	a3b3c1 → 1	3	1

5 Experimental Results

In this section, we executed the Weight-CARs algorithm with different parameter sets. The experiments were carried out on a system with the following configuration: Intel Core I7-7500U 2.5 GHz (4 CPUs) processor, 8 GB of RAM, and running 64-bit Windows 10. The proposed algorithm was implemented in C# using Visual Studio 2015 Community, .NET framework 4.5. The datasets that were used in the tests having characteristics described in Table 5.

We executed the CAR-Miner and Weighted-CARs with the same set of parameters (*minSup*, *minConf*). In our experiments, we kept *minConf* = 60% and assigned different values to *minSup*. Our testing approach helped evaluate the process of mining class association rules with different size of itemset. Besides that, this approach also delivered the accuracy comparison between CAR-Miner

Table 5 Characteristics of testing datasets

Dataset	Number of records	Number of attributes	Number of classes
Breast cancer	699	11	2
Glass	214	10	6
Iris	150	5	3
Tictac	958	10	2

Table 6 Accuracy comparison between CAR-Miner and Weighted-CAR-Miner

Dataset	minSup (%)	Accuracy (%)	
		CAR-Miner	Weighted-CARs
Breast cancer	10	78.26	78.26
	5	86.23	86.38
	3	87.39	87.54
Glass	10	59.52	65.24
	5	65.24	67.62
	3	67.62	71.43
Iris	10	81.33	82
	5	74.67	74
	3	74	74
Tic-tac	10	65.16	70.32
	5	66.11	70.95
	3	71.68	71.26

algorithm and Weighted-CARs algorithm. The details of this comparison are listed in Table 6.

From the above experimental results, the Weighted-CARs helped gaining better accuracy than the CAR-Miner. For example, we had the accuracy as 86.23% in mining class association rules with $minSup = 5\%$ and $minConf = 60\%$, however, with the same parameters, the Weighted-CARs returned 86.38%. Similarly, with $minSup = 3\%$ and $minConf = 60\%$, Weighted-CARs also had 0.15% accuracy higher than CAR-Miner had. In Breast Cancer treatment as well as other medical treatment, the accuracy is the most important value supporting doctors to select correct solutions.

We also tested on the datasets which has high number of classes, such as Glass. The results also indicated that Weighted-CARs gained more accuracy than CAR-Miner. In another testing approach, we tried our approach with the dataset having high number of records, Tictac, Weighted-CARs algorithm also returned good results as expected. For example, with $minConf = 60\%$, we adjusted $minSup$ from 3 to 10%, with $minSup = 3\%$, the accuracy of class association rules from Weighted-CARs was not higher than that of CAR-Miner, however, with $minSup = 5\%$ or $minSup = 10\%$, the accuracy of class association rules from Weighted-CARs was much higher than that of CAR-Miner (greater than 4.8%). We also had a good accuracy comparison in testing with the Iris dataset.

6 Conclusion and Future Works

In this work, we investigated the actual datasets in the real world, in which, each attribute is assigned with a weighted value. We then proposed a new approach to mine class association rules based on weighted datasets. We also carried out the experiments with 4 different datasets having different characteristics in size, attributes and number of classes. Overall, our proposed algorithm, the Weighted-CARs algorithm improves the accuracy of class association rules. In the future works, we intent to investigate more solutions to provide higher accuracy class association rules from weighted datasets. Furthermore, improving the performance of the mining process is also one of our concerns.

Acknowledgements This research is funded by Vietnam National Foundation for Science and Technology Development (NAFOSTED) under grant number 102.05-2015.10.

References

1. Aljuboori, A., Meziane, F., Parsons, D.: A new strategy for case-based reasoning retrieval using classification based on association. In: Machine Learning and Data Mining in Pattern Recognition, pp. 326–340 (2016)
2. Al-Tapan, A.A., Al-Maqaleh, B.M.: An effective mining of exception class association rules from medical datasets. Int. J. Comput. Sci. Eng. **7**, 191–198 (2017)
3. Bechini, A., Marcelloni, F., Segatori, A.: A MapReduce solution for associative classification of big data. Inf. Sci. **322**, 33–55 (2016)
4. Hu, K., Lu, Y., Zhou, L., Shi, C.: Integrating classification and association rule mining: a concept lattice framework. In: Proceedings of the International Workshop on New Directions in Rough Sets, Data Mining, and Granular-Soft Computing, pp. 443–447 (1999)
5. Li, W., Han, J., Pei, J.: CMAR: accurate and efficient classification based on multiple class-association rules. In: Proceedings of the 1st IEEE International Conference on Data Mining, San Jose, California, USA, pp. 369–376 (2001)
6. Liu, B., Hsu, W., Ma, Y.: Integrating classification and association rule mining. In: Proceedings of the 4th International Conference on Knowledge Discovery and Data Mining, New York, USA, pp. 80–86 (1998)
7. Liu, B., Ma, Y., Wong, C.K.: Improving an association rule based classifier. In: Proceedings of the 4th European Conference on Principles of Data Mining and Knowledge Discovery, Lyon, France, pp. 80–86 (2000)
8. Nguyen, L.T.T., Nguyen, N.T.: An improved algorithm for mining class association rules using the difference of Obidsets. Expert Syst. Appl. **42**(9), 4361–4369 (2015)
9. Nguyen, L.T.T., Vo, B., Hong, T.P., Thanh, H.C.: CAR-Miner: an efficient algorithm for mining class-association rules. Expert Syst. Appl. **40**(6), 2305–2311 (2013)
10. Nguyen, D., Vo, B., Le, B.: Efficient strategies for parallel mining class association rules. Expert Syst. Appl. **41**(10), 4716–4729 (2014)
11. Song, K., Lee, K.: Predictability-based collective class association rule mining. Expert Syst. Appl. **79**, 1–7 (2017)
12. Thabtah, F.A., Cowling, P.I.: A greedy classification algorithm based on association rule. Appl. Soft Comput. **7**(3), 1102–1111 (2007)
13. Thabtah, F., Cowling, P., Peng, Y.: MCAR: multi-class classification based on association rule. In: Proceedings of the 3rd ACS/IEEE International Conference on Computer Systems and Applications, Tunis, Tunisia, pp. 33–39 (2005)

14. Thabtah, F., Cowling, P., Peng, Y.: MMAC: a new multi-class, multi-label associative classification approach. In: Proceedings of the 4th IEEE International Conference on Data Mining, Brighton, UK, pp. 217–224 (2004)
15. Vo, B., Le, B.: A novel classification algorithm based on association rule mining. In: Proceedings of the 2008 Pacific Rim Knowledge Acquisition Workshop (Held with PRICAI'08), pp. 61–75 (2008)
16. Yin, X., Han, J.: CPAR: classification based on predictive association rules. In: Proceedings of SIAM International Conference on Data Mining (SDM'03), San Francisco, CA, USA, pp. 331–335 (2003)

Fast and Memory Efficient Mining of Periodic Frequent Patterns

Vincent Mwintieru Nofong

Abstract Periodic frequent pattern mining, the process of finding frequent patterns which occur periodically in databases, is an important data mining task for various decision making. Though several algorithms have been proposed for their discovery, most employ a two stage process to evaluate the periodicity of patterns. That is, by firstly deriving the set of periods of a pattern from its coverset, and subsequently evaluating the periodicity from the derived set of periods. This two step process thus make algorithms for discovering periodic frequent patterns both time and memory inefficient in the discovery process. In this paper, we present solutions to reduce both runtime and memory consumption in periodic frequent pattern mining. We achieve this by evaluating the periodicity of patterns without deriving the set of periods from their coversets. Our experimental results show that our proposed solutions are efficient both in reducing the runtime and memory consumption in the discovery of periodic frequent patterns.

Keywords Frequent patterns · Periodic frequent patterns · Efficient measures

1 Introduction

Frequent pattern mining (the process of identifying patterns which occur frequently in databases) over the past years has been widely studied for knowledge discovery in databases by works such as [1, 4, 17]. Though algorithms for mining frequent patterns are helpful in revealing frequently occurring patterns in databases (for decision making), they often fail in revealing the occurrence shapes of patterns. For instance, in crime data analysis, frequent pattern mining algorithms will fail to report the periodic occurrence shapes of crime patterns, which are vital in decision making to curb crime. This limitation in frequent pattern mining, and the importance of patterns'

V. M. Nofong (✉)
Univeristy of Mines and Technology, P. O. Box 237, Tarkwa, Ghana
e-mail: vnofong@umat.edu.gh

© Springer International Publishing AG, part of Springer Nature 2018   223
A. Sieminski et al. (eds.), *Modern Approaches for Intelligent Information
and Database Systems*, Studies in Computational Intelligence 769,
https://doi.org/10.1007/978-3-319-76081-0_19

shapes in decision making resulted in the start of research on periodic frequent pattern mining.

Mining periodic frequent patterns from transactional databases has been widely researched on, in works such as [3, 5, 7, 11, 14, 15]. Though many algorithms have been proposed for mining periodic frequent patterns in transactional databases, they have the following limitations:

- Periodic frequent pattern mining algorithms such as [5–7, 14] which mine periodic frequent patterns (PFPs) based on the maximum periodicity threshold (proposed in [15]) will often miss some relevant periodic patterns if such patterns have just one periodic interval being greater than the maximum periodicity threshold.
- Periodic frequent pattern mining algorithms such as [9, 12, 13] which mine periodic frequent patterns based on the maximum variance threshold (proposed in [13]) will often report a set of PFPs with distinct periods for decision making.
- As mentioned previously, to test a pattern as either periodic or non-periodic, existing algorithms firstly derive the set of periods from its coverset, and subsequently evaluate its periodicity from the derived set of periods. This two stage process thus make existing algorithms inefficient in both runtime and memory usage in discovering PFPs.

Though some of these challenges have been addressed, the case of time and memory inefficiency in the discovery of periodic frequent patterns to the best of our knowledge is yet to be addressed. In this paper, we address this issue by proposing to evaluate the periodicity of patterns without deriving the set of periods. This in turn reduces both runtime and memory consumption in the discovery of periodic frequent patterns.

This paper makes the following contributions in the discovery of periodic frequent patterns:

- Effective and efficient techniques are introduced for evaluating the periodicity of patterns without deriving their set of periods from their coversets.
- The proposed techniques are incorporated on existing periodic frequent pattern mining frameworks which showed a drastic reduction in both runtime and memory usage in the discovery of PFPs.

2   Related Work

The associated notations for periodic frequent pattern mining in transactional databases can be given as follows.

Let $I = \langle i_1, i_2, \ldots, i_n \rangle$ be a set of literals, called items. Then, a transaction is a nonempty set of items. A pattern S is a set of items satisfying some conditions of measures like frequency. A pattern is of length-k if it has k items, for instance, $S = \{a, b, c\}$ is a length-3 pattern.

Given a transactional database of n transactions, $\mathbf{D} = <T_1, T_2, T_3, \ldots, T_n>$, where each T_m in $\mathbf{D}$ is identified by m called TID, the *cover* of a pattern S in $\mathbf{D}$, $cov_{\mathbf{D}}(S)$, is the set of TIDs of transactions that contain S. That is,

$$cov_{\mathbf{D}}(S) = \{m : T_m \in \mathbf{D} \wedge S \subseteq T_m\} \tag{1}$$

The *support* of a pattern S in $\mathbf{D}$, $sup_{\mathbf{D}}(S)$, is defined as,

$$sup_{\mathbf{D}}(S) = \frac{|cov_{\mathbf{D}}(S)|}{|\mathbf{D}|} \tag{2}$$

where $|cov_{\mathbf{D}}(S)|$ is called the *support count* of S in $\mathbf{D}$.

A pattern S in $\mathbf{D}$ is said to be frequent if its support in $\mathbf{D}$ is larger than or equal to a user specified minimum support (ε).

Let S be a pattern in a transactional database $\mathbf{D}$ and $cov_{\mathbf{D}}(S)$ be its coverset in $\mathbf{D}$. The notation $e.cov_{\mathbf{D}}(S)$ is used to indicate the extension of $cov_{\mathbf{D}}(S)$ by inserting a starting time 0 and the last time n to $cov_{\mathbf{D}}(S)$. That is,

$$e.cov_{\mathbf{D}}(S) = \{0 \cup cov_{\mathbf{D}}(S) \cup n\} \tag{3}$$

where $n = |\mathbf{D}|$. The last time, n, will be duplicated if it is already in $cov_{\mathbf{D}}(S)$. For instance, given $|\mathbf{D}| = 6$ and $cov_{\mathbf{D}}(S) = \{1, 4, 6\}$, then, $e.cov_{\mathbf{D}}(S) = \{0\} \cup \{1, 4, 6\} \cup \{6\} = \{0, 1, 4, 6, 6\}$.

Let $(m_j, m_{j+1}) \in e.cov_{\mathbf{D}}(S)$ be two consecutive occurrence times (TIDs) of S in $\mathbf{D}$, then $p_j^S = m_{j+1} - m_j$ is the jth period of S in $\mathbf{D}$. The set of all periods of S, P^S, obtained from its extended cover is denoted as:

$$P^S = \{p_1^S, \ldots, p_r^S\} \tag{4}$$

where $r = |e.cov_{\mathbf{D}}(S)| - 1$.

For instance, given $e.cov_{\mathbf{D}}(S) = \{0, 1, 4, 6, 6\}$, then $p_1^S = (1 - 0) = 1, p_2^S = (4 - 1) = 3$, $p_3^S = (6 - 4) = 2, p_4^S = (6 - 6) = 0$, and, $P^S = \{1, 3, 2, 0\}$. It can thus be derived that for any pattern S, $|P^S| = |cov_{\mathbf{D}}(S)| + 1$.

To mine the set of patterns with periodic occurrence shapes in transactional datasets for decision making, Tanbeer et al. in [15] proposed a periodicity measure on patterns as follows.

Definition 1 [15] Given a database $\mathbf{D}$, a pattern S and its set of periods P^S in $\mathbf{D}$, the periodicity of S, $Per(S)$, is defined as, $Per(S) = \max\{p | p \in P^S\}$.

Based on the proposed periodicity measure in Definition 1, Tanbeer et al. [15] subsequently defined a periodic frequent pattern as frequent pattern whose periodicity is not greater than a user defined maximum periodicity threshold, *maxPer*.

For a pattern S and its set of periods P^S, the proposition in Tanbeer et al. [15] returns S as periodic if the maximal occurring period (maximal time interval between any consecutive occurrence times) of S is not greater than the maximum periodicity threshold, *maxPer*. This idea of mining periodic frequent patterns based on the maximal occuring period proposed in [15] have been used in mining periodic frequent patterns in transaction-like datasets in works such as [6–8, 10, 14].

Rashid et al. in [13] however argued that mining periodic frequent patterns based on the proposed periodicity measure in [15] is inappropriate since it returns the maximum period for which a pattern does not appear in a database as its periodicity. To avoid reporting the maximal occuring period as a patterns' periodicity, Rashid et al. [13] defined the periodicity of a pattern under the name *regularity* as follows.

Definition 2 [13] Given a database **D**, a pattern S and its set of periods P^S in **D**, the regularity of S, $Reg(S)$, is defined as $Reg(S) = var(P^S)$, where $var(P^S)$ is the variance of P^S.

With the regularity (periodicity) measure in Definition 2, Rashid et al. [13] defined a regular (periodic) frequent pattern as a frequent pattern whose variance among its set of periods is not greater than a user desired maximum regularity threshold, *maxReg*. The concept of mining regular (periodic) frequent patterns proposed in [13] has also been used in mining regular (periodic) frequent patterns in works such as [9, 12].

Nofong in [11] however argued that though the concept proposed [13] will not discard interesting periodic frequent patterns as in [15], PFP mining algorithms based on the propositions in both [13, 15] will always report PFPs with totally distinct periods. To ensure only PFPs with similar periods are reported, Nofong [11] defined a periodic frequent pattern as follows.

Definition 3 [11] Given a database **D**, minimum support threshold ε, periodicity threshold p, difference factor p_1, a pattern S and P^S, S is a periodic frequent pattern if $sup_{\mathbf{D}}(S) \geq \varepsilon$, $(p - p_1) \leq Prd(S) - std(P^S)$ and $Prd(S) + std(P^S) \leq (p + p_1)$.

where, $Prd(S)$ (the mean of P^S, that is, $\bar{x}(P^S)$) is the periodicity of S and $std(P^S)$ the standard deviation in P^S.

Nofong [11] observed that with Definition 3, though PFPs with similar periods will be reported, some may be periodic due to random chance without inherent item relationship. To address this issue, the productiveness measure (proposed in [16]) was incorporated in defining the productive periodic frequent patterns as the set of PFPs with inherent item relationship.

Philippe et al. in [3] introduced PFPM, an efficient algorithm with novel pruning techniques for mining periodic frequent patterns. Unlike the methods proposed in [11, 13, 15], Philippe et al. [3] introduced three periodicity measures (the *minimum*, *maximum* and *average* periodicity measures) for mining user desired periodic frequent patterns. Employing the three measures proposed in [3] gives users the advantages of more flexibility in PFP mining.

As mentioned previously, the propositions in [3, 11, 13, 15] and works based on these propositions ([6–9, 12, 14]) are faced with the challenges of time and memory inefficiency, and, difficulty in finding early termination mechanisms in the PFP mining process.

3 Proposed Periodicity Evaluation Measures

We adopt the PFP definition (Definition 3) proposed in [11]. To enable us mine PFPs based on Definition 3 while addressing the time and memory inefficiencies in PFP discovery, we show that the periodicity of a pattern can be evaluated from its support and the size of the database as follows:

Lemma 1 *Given a database* $\mathbf{D} = <T_1, T_2, T_3, \ldots, T_n>$ *from which a pattern* S *is mined, the periodicity of* S *in* $\mathbf{D}$ *can be expressed as* $Prd_D(S) = |D| / \left(sup_D(S) \times |D|\right) + 1.$

Proof Let $cov_D(S) = (T_1, T_2, T_3, \ldots, T_{m-1}, T_m)$, then P^S can be expressed as, $P^S = \left((T_1 - 0), (T_2 - T_1), (T_3 - T_2), \ldots, (T_m - T_{m-1}), (|\mathbf{D}| - T_m)\right)$. Then, $Prd_D(S)$ (which is the mean of P^S) can thus be expressed as $Prd_D(S) = \frac{\sum_{i=1}^{|P^S|} P_i^S}{|P^S|}$. Hence, $Prd_D(S) = \frac{(T_1 - 0) + (T_2 - T_1) + (T_3 - T_2) + \cdots + (T_m - T_{m-1}) + (|\mathbf{D}| - T_m)}{|P^S|}$. This simplifies to $Prd_D(S) = \frac{|D|}{|P^S|}$. However, since $|P^S| = |cov_D(S)| + 1$, the periodicity of S, $Prd_D(S)$ can thus be expressed as $Prd_D(S) = \frac{|D|}{(sup_D(S) \times |D|) + 1}$. □

For instance, given $|\mathbf{D}| = 6$ and $cov_{\mathbf{D}}(S) = \{1, 4, 6\}$ (i.e. $sup_D(S) = 0.5$), then, based on Lemma 1, $Prd_D(S) = \frac{6}{(0.5 \times 6) + 1} = 1.5$. Though $Prd_D(S)$ can be evaluated without Lemma 1, it will require more time and memory compared to evaluating $Prd_D(S)$ with Lemma 1 explained below.

Let $\mathbf{D}$ be a dataset and n the set of frequent patterns in $\mathbf{D}$ whose periodicities are to be evaluated. The functions for evaluating the periodicities based on Lemma 1 (Function 1) and without Lemma 1 (Function 2) are as shown below.

Function 1
1 **for** $i = 1 \rightarrow n$ **do**
2 $\quad\lfloor\ Prd_D(i) = \frac{|D|}{(sup_D(i) \times |D|) + 1}$
3 **return** $Prd_D(i)$

Function 2
1 **for** $i = 1 \rightarrow n$ **do**
2 $\quad$ Get $|cov_D(i)|$
3 $\quad$ Let $m = |cov_D(i)|$
4 $\quad$ Create $|P^i|$
5 $\quad$ **for** $k = 1 \rightarrow m$ **do**
6 $\quad\quad\lfloor$ Obtain elements of P^i
7 $\quad$ $Prd_D(i) = \frac{\sum P^i}{|P^i|}$
8 **return** $Prd_D(i)$

With the Big-O notation, the runtime complexity of Function 1 (based Lemma 1) turns out to be $O(n)$ while that of Function 2 is $O(n^2)$. This shows that employing Lemma 1 in evaluating the periodicity of patterns will result in a significant reduction in runtime for discovering PFPs.

Though Lemma 1 will evaluate the periodicity of a pattern, it will not be able to detect the standard deviation (which will be required to identify PFPs with similar periodicities—see Definition 3) among the set of periods of patterns. In exiting works on periodic frequent pattern mining, the set of periods for each pattern have to be obtained (from their coversets) to derive the standard deviation among the set of periods. Obtaining the sets of periods and subsequently evaluating the periodicity is both time and memory expensive.

To avoid this situation, we show that the standard deviation among the set of periods can be directly evaluated from the coverset without necessarily obtaining the set of periods as follows.

Lemma 2 *Given* $cov_D(S) = \{n_1, n_2, n_3, \ldots, n_{m-1}, n_m\}$, *the standard deviation among the set of periods of S can be evaluated as* $std(P^S) = \sqrt{\frac{X_S + Y_S + Z_S}{|cov_D(S)| + 1}}$ *where:*

$$X_S = n_1^2 + Prd(S)^2 \tag{5}$$

$$Y_S = n_m^2 + |D|^2 + Prd(S)^2 - 2|D|(n_m + Prd(S)) \tag{6}$$

$$Z_S = \sum_{j=2}^{m-1}(n_j^2 + n_{j-1}^2 + Prd(S)^2 - 2n_j n_{j-1}) \tag{7}$$

Proof Let $\bar{x} = Prd(S)$. If $cov_D(S) = \{n_1, n_2, n_3, \ldots, n_{m-1}, n_m\}$, then set of periods of S will be $P^S = \{n_1 - 0, n_2 - n_1, n_3 - n_2, \ldots, n_m - n_{m-1}, |D| - n_m\}$. Hence, the variance among the periods of S will be $Var(P^S) = \sum_{i=1}^{|P^S|}(\frac{(P_i^S - \bar{x})^2}{|P^S|})$. This expands to $Var(P^S) = \frac{((n_1 - 0) - \bar{x})^2 + ((n_2 - n_1) - \bar{x})^2 + \cdots + ((n_m - n_{m-1}) - \bar{x})^2 + ((|D| - n_m) - \bar{x})^2}{|cov_D(S)| + 1}$.

Expanding the expressions in the numerator, we get:

$((n_1 - 0) - \bar{x})^2 = n_1^2 + \bar{x}^2 - 2n_1\bar{x}$.

$((n_2 - n_1) - \bar{x})^2 = n_1^2 + n_2^2 + \bar{x}^2 - 2n_1 n_2 - 2n_2\bar{x} + 2n_1\bar{x}$.

$((n_3 - n_2) - \bar{x})^2 = n_2^2 + n_3^2 + \bar{x}^2 - 2n_2 n_3 - 2n_3\bar{x} + 2n_2\bar{x}$.

$\vdots$

$((n_m - n_{m-1}) - \bar{x})^2 = n_{m-1}^2 + n_m^2 + \bar{x}^2 - 2n_{m-1}n_m - 2n_m\bar{x} + 2n_{m-1}\bar{x}$.

$((|D| - n_m) - \bar{x})^2 = n_m^2 + |D|^2 + \bar{x}^2 - 2n_m|D| - 2|D|\bar{x} + 2n_m\bar{x}$.

Since expansions are being summed, we get the following:

$$X_S = n_1^2 + Prd(S)^2 \text{ (for the first expansion)} \ldots ①$$

where X_S is the variance value for the first period in P^S.

$$Y_S = n_m^2 + |D|^2 + Prd(S)^2 - 2n_m|D| - 2|D|Prd(S) \text{ (for the last expansion)} \ldots ②$$

where, Y_S is the variance value for the last period in P^S.

$$Z_S = \sum_{j=2}^{m-1}(n_j^2 + n_{j-1}^2 + Prd(S)^2 - 2n_j n_{j-1} \text{ (for any other period in } P^S) \ldots ③$$

where, Z_S is the variance value for any other period in P^S which is not the first or last period.

As such with only the coverset of a pattern, the variance and standard deviations among its periods can be obtained respectively as: $Var(P^S) = \frac{X_S + Y_S + Z_S}{|cov_D(S)| + 1}$, and $std(P^S) = \sqrt{\frac{X_S + Y_S + Z_S}{|cov_D(S)| + 1}}$. $\qquad\square$

Our proposed measures for evaluating the periodicity, standard deviation and variance among the set of periods unlike the existing measures have the advantage of reducing both runtime and memory consumption in PFP discovery.

We incorporate our proposed periodicity measures on existing PFP discovery algorithms to experimentally check their effectiveness.

4 Experimental Results

The following implementations were used in our experimental analysis:

- PFP*: This is an implementation of the approach for detecting all periodic frequent patterns. PFP detects and reports the set of all periodic frequent patterns.
- PFP+: This is our improved implementation of PFP based on our proposed periodicity detection measures.
- PPFP: This is an implementation of the approach proposed in [11]. PPFP detects and reports the set of productive periodic frequent patterns for decision making.
- PPFP+: This is our improved implementation of PPFP based on our proposed periodicity detection measures.

We conduct experimental analysis with regards to (i) execution time and (ii) memory usage. Table 1 describes the datasets used in our experimental analysis.

All compared approaches are implemented in Java and experiments carried on a 64-bit windows 10 PC (Intel Core i7, CPU 2.10 GHz, 12 GB). The outcome of the analysis are discussed below.

Table 1 Datasets

Dataset	Origin	Characteristics
Kosarak10K	SPMF [2]	10,000 Transactions—partly dense
Accident	FIMI repository	7593 Transactions—very dense
Tafeng Nov. 2000	AIIA lab	31,807 Transactions—very sparse

4.1 Execution Time Comparison

For time performance and scalability, we compare the runtimes of the above mentioned implementations on the datasets shown in Table 1.

Figures 1, 2 and 3 show the runtime comparison of the above mentioned implementations on the Kosarak10K, Accident and Tafeng datasets respectively. As can be seen in all three Figures, employing our proposed techniques significantly reduces the runtime required in discovering PFPs. For instance, in Fig. 1, PFP+, an implementation based on our proposed techniques is almost twice as efficient as PFP* in periodic frequent pattern discovery. Also, as can be seen in Figs. 2 and 3, PFP+ and PPFP+ (based on our proposed techniques) are also slightly more efficient compared to than PFP* and PPFP in periodic frequent pattern discovery.

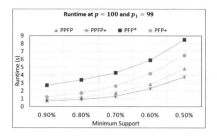

Fig. 1 PFP discovery: runtime in Kosarak10K dataset

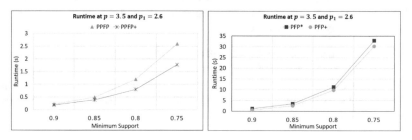

Fig. 2 PFP discovery: runtime in Accident dataset

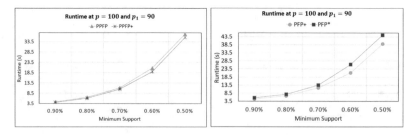

Fig. 3 PFP discovery: runtime in Tafeng dataset

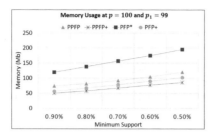

Fig. 4 PFP discovery: memory usage Kosarak10K dataset

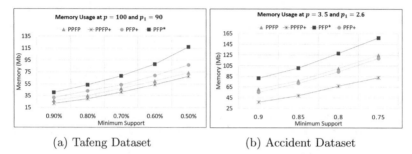

Fig. 5 PFP discovery: memory usage

4.2 Memory Usage Comparison

For memory usage, we compare the memory used by the above mentioned implementations on the datasets shown in Table 1 on PFP discovery.

Figures 4 and 5a, b show the memory usage comparison of the above mentioned implementations on the Kosarak10K, Tafeng and Accident datasets respectively. As can be seen in all three Figures, employing our proposed techniques significantly reduces the memory used in discovering PFPs. In Fig. 5b for instance, both PFP+ and PPFP+ (implementations based on our proposed techniques) are almost twice as efficient as PFP* and PPFP with regards to memory usage in periodic frequent pattern discovery.

5 Conclusion

Existing PFP mining algorithms often employ a two stage process for discovering PFPs hence making them both time and memory inefficient in the discovery process. In this paper, we propose techniques to reduce both runtime and memory usage in PFP discovery. We have incorporated these techniques on existing PFP mining algorithms and show experimentally that our proposed techniques are efficient in reducing both the runtime and memory used in discovering periodic frequent patterns.

References

1. Agrawal, R., Imieliński, T., Swami, A.: Mining association rules between sets of items in large databases. ACM SIGMOD Rec. **22**(2), 207–216 (1993)
2. Fournier-Viger, P., et al.: The SPMF open-source data mining library version 2. In: Proceedings of European Conference on Machine Learning and Knowledge Discovery in Databases 2016. Springer International Publishing (2016)
3. Fournier-Viger, P., Lin, C.W., Duong, Q.H., Dam, T.L., Ševčík, L., Uhrin, D., Voznak, M.: PFPM: discovering periodic frequent patterns with novel periodicity measures. In: Proceedings of the 2nd Czech-China Scientific Conference 2017. InTech (2017)
4. Jian, P., Jiawei, H., Hongjun, L., Shojiro, N., Shiwei, T., Dongqing, Y.: H-Mine: fast and space-preserving frequent pattern mining in large databases. IIE Trans. **39**(6), 593–605 (2007)
5. Kiran, R.U., Kitsuregawa, M.: Discovering quasi-periodic-frequent patterns in transactional databases. In: Bhatnagar, V., Srinivasa, S. (eds.) BDA 2013. LNCS, vol. 8302, pp. 97–115. Springer International Publishing (2013)
6. Kiran, R.U., Kitsuregawa, M.: Novel techniques to reduce search space in periodic-frequent pattern mining. In: Bhowmick, S.S., Dyreson, C.E., Jensen, C.S., Lee, M.L., Muliantara, A., Thalheim, B. (eds.) DASFAA 2014. LNCS, vol. 8422, pp. 377–391. Springer International Publishing (2014)
7. Kiran, R.U., Reddy, P.K.: Towards efficient mining of periodic-frequent patterns in transactional databases. In: Bringas, P.G., Hameurlain, A., Quirchmayr, G. (eds.) DASFAA 2010. LNCS, vol. 6262, pp. 194–208. Springer, Heidelberg (2010)
8. Kiran, R.U., Reddy, P.K.: An alternative interestingness measure for mining periodic-frequent patterns. In: Yu, J.X., Kim, M.H., Unland, R. (eds.) DASFAA 2011. LNCS, vol. 6587, pp. 183–192. Springer, Heidelberg (2011)
9. Kumar, V., Valli Kumari, V.: Incremental mining for regular frequent patterns in vertical format. Int. J. Eng. Tech. **5**(2), 1506–1511 (2013)
10. Lin, J.C.W., Zhang, J., Fournier-Viger, P., Hong, T.P., Zhang, J.: A two-phase approach to mine short-period high-utility itemsets in transactional databases. Adv. Eng. Inf. **33**, 29–43 (2017)
11. Nofong, V.M.: Discovering productive periodic frequent patterns in transactional databases. Ann. Data Sci. **3**(3), 235–249 (2016)
12. Rashid, M.M., Gondal, I., Kamruzzaman, J.: Regularly frequent patterns mining from sensor data stream. In: Lee, M., Hirose, A., Hou, Z.G., Kil, R. (eds.) NIP 2013. LNCS, vol. 8227, pp. 417–424. Springer, Heidelberg (2013)
13. Rashid, M.M., Karim, M.R., Jeong, B.S., Choi, H.J.: Efficient mining regularly frequent patterns in transactional databases. In: Lee, S., Peng, Z., Zhou, X., Moon, Y., Unland, R., Yoo, J. (eds.) DASFAA 2012. LNCS, vol. 7238, pp. 258–271. Springer, Heidelberg (2012)
14. Surana, A., Kiran, R.U., Reddy, P.K.: An efficient approach to mine periodic-frequent patterns in transactional databases. In: Cao, L., Huang, J.Z., Bailey, J., Koh, Y.S., Luo, J. (eds.) PAKDD 2011 Workshops. LNAI, vol. 7104, pp. 254–266. Springer, Heidelberg (2012)
15. Tanbeer, S.K., Ahmed, C.F., Jeong, B.S., Lee, Y.K.: Discovering periodic-frequent patterns in transactional databases. In: Theeramunkong, T., Kijsirikul, B., Cercone, N., Ho, T. (eds.) PAKDD 2009. LNAI, vol. 5476, pp. 242–253. Springer, Heidelberg (2009)
16. Webb, G.I.: Self-sufficient itemsets: an approach to screening potentially interesting associations between items. ACM Trans. Knowl. Discov. Data **4**(1), 3:1–3:20 (2010)
17. Zaki, M.J., Gouda, K.: Fast vertical mining using diffsets. In: Proceedings of the 9th ACM SIGKDD International Conference on Knowledge Discovery and Data Mining, pp. 326–335 (2003)

A Coupling Support Vector Machines with the Feature Learning of Deep Convolutional Neural Networks for Classifying Microarray Gene Expression Data

Phuoc-Hai Huynh, Van-Hoa Nguyen and Thanh-Nghi Do

Abstract Support vector machines (SVM) and deep convolutional neural networks (DCNNs) are state-of-the-art classification techniques in many real-world applications. Our investigation aims at proposing a hybrid model combining DCNNs and SVM (called DCNN-SVM) to effectively predict very-high-dimensional gene expression data. The DCNN-SVM trains the DCNNs model to automatically extract features from microarray gene expression data and followed which the DCNN-SVM learns a non-linear SVM model to classify gene expression data. Numerical test results on 15 microarray datasets from Array Expression and Medical Database (Kent Ridge) show that our proposed DCNN-SVM is more accurate than the classical DCNNs algorithm, SVM, random forests.

Keywords Microarray gene expression · Convolutional neural networks
Support vector machines

1 Introduction

Nowadays, the development of high-throughput technologies such as DNA microarray has led to incremental growth in the public databases such as the ArrayExpress [1] and NCBI Gene Expression Omnibus [2]. Microarray is technology which enables researchers to investigate and address issues which is once thought to be non traceable by facilitating the simultaneous measurement of the expression levels of thousands of genes in a single experiment [3]. A characteristic of microarray gene

P.-H. Huynh (✉) · V.-H. Nguyen (✉)
An Giang University, An Giang, Vietnam
e-mail: hphai@agu.edu.vn

V.-H. Nguyen
e-mail: nvhoa@agu.edu.vn

T.-N. Do (✉)
Can Tho University, Can Tho, Vietnam
e-mail: dtnghi@cit.ctu.edu.vn

© Springer International Publishing AG, part of Springer Nature 2018
A. Sieminski et al. (eds.), *Modern Approaches for Intelligent Information and Database Systems*, Studies in Computational Intelligence 769,
https://doi.org/10.1007/978-3-319-76081-0_20

expression data is that the number of variables (genes) m far exceeds the number of samples n, commonly known as curse of dimensionality problem. The vast amount of gene expression data leads to statistical and analytical challenges and conventional statistical methods give improper result due to high dimension of microarray data with a limited number of patterns [4]. It is not feasible when build machine learning model due to the extremely large features set with millions of features and high computing cost.

With the wealth of gene expression data from microarrays being produced, more and more new prediction, classification, and clustering techniques are being used for the analysis of the data. Many methods have been used for microarray gene expression data classification, and typical methods are support vector machines (SVM) [5–8], k-nearest neighbor classifier [9], C4.5 decision tree [10–12] and ensemble methods, such as random forests [13], random forests of oblique decision trees [14], bagging and boosting [15, 16].

In recent years, convolutional neural networks (CNNs) have achieved remarkable results in computer vision [17], text classification [18]. In addition, CNNs is also used for omics, biomedical imaging and biomedical signal processing [19]. Most data in bioinformatics are raw data such as gene sequences, proteins, microarray, medical image. Conventional machine learning algorithms have limitations in processing the raw form of data, so hybrid models often are used to combine the advantage of features extraction from the raw data of CNNs and performance classification of SVM or random forests (RF). The hybrid model neural network and SVM was initially proposed in [20]. In [21], model is later proposes in for handwritten digit recognition. More relevant previous work include [22], where a hybrid model approach is presented: the CNNs has trained using the back-propagation algorithm and the SVM is trained using a non-linear regression approach. It is noticeable that error classification rate gained by the hybrid model has achieved better results. In [23], the hybrid model uses for recognition for mobile swarm robotic systems. In addition, CNNs and RF are also combined to build hybrid model for electron microscopy images segmentation [24].

In this paper, we propose a hybrid model combining DCNNs and SVM (called DCNN-SVM) to effectively classify very-high-dimensional gene expression data. The main idea of our approach is to train a specialized DCNNs to extract robust hierarchical features from microarray gene expression data (MGE data) and provide them to SVM classifier using radial basis function kernel (RBF). Our approach differs from these previous ones as we build a single model instead of using disjoint classifiers trained separately. In relevant previous work, the CNNs is trained using the back-propagation algorithm and the SVM is trained using a non-linear regression approach, linear kernel function and random forest. The data in the relevant previous work was image such as: handwritten digit, medical image and video.

We have used 15 datasets of ArrayExpress [1] and Biomedical repository [25] to evaluate our model and also to compare to traditional classification methods such as DCNNs, support vector machines [26] and random forests [27]. The results showed that DCNN-SVM extract robust hierarchical features and improves classification accuracy. Our method shows an excellent performance in general with support vector machines classifier using radial basis function kernel.

The paper is organized as follows. Section 2 presents our approach, a hybrid model combining DCNNs and SVM. Section 3 shows the experimental results. We then conclude in Sect. 4.

2 Methods

2.1 Deep Convolutional Neural Networks

DCNNs are designed to process multiple data types, especially two-dimensional images, and are directly inspired by the visual cortex of the brain. In the visual cortex, there is a hierarchy of two basic cell types: simple cells and complex cells [28]. Simple cells react to primitive patterns in sub-regions of visual stimuli, and complex cells synthesize the information from simple cells to identify more intricate forms. Since the visual cortex is such a powerful and natural visual processing system, DCNNs are applied to imitate three key ideas: local connectivity, invariance to location, and invariance to local transition [29]. There are three main types of layers used to build DCNNs architectures: convolutional layer, pooling layer, and fully connected layer. Normally, a full DCNNs architecture is obtained by stacking several of these layers. In a DCNNs, the key computation is the convolution of a feature detector with an input signal. Convolutional layer computes the output of neurons connected to local regions in the input, each one computing a dot product between their weights and the region they are connected to in the input volume. The set of weights which is convolved with the input is called filter or kernel. Every filter is small spatially (width and height), but extends through the full depth of the input volume. For inputs such as images typical filters are small areas and each neuron is connected only to this area in the previous layer. The weights are shared across neurons, leading the filters to learn frequent patterns that occur in any part of the image. The distance between the applications of filters is called stride. Whether stride hyper parameter is smaller than the filter size the convolution is applied in overlapping windows.

2.2 Support Vector Machines

Support vector machines (SVMs) proposed by Vapnik [26] are systematic and properly motivated by statistical learning theory. SVMs are the most well known as class of learning algorithms using the idea of kernel substitution. SVM and kernel-based methods have shown practical relevance for classification, regression [30]. The SVM algorithm is to find the best separating plane furthest from the different classes. In order to achieve this purpose, a SVM algorithm tries to simultaneously maximize

Fig. 1 Linear separation of
the datapoints into two
classes

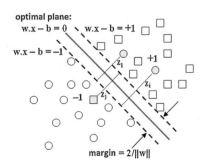

the margin (the distance between the supporting planes for each class) and minimize the error (any point falling on the wrong side of its supporting plane is considered to be an error). For binary classification problem (see Fig. 1), samples of one class are located on one side of the hyper-plane while samples of the other class are located on the other side of the hyper-plane.

For multiclass, one-versus-all [26], one-versus-one [31] are the most popular methods due to their simplicity. Let us consider k classes ($k > 2$). The one-versus-all strategy builds k different classiers where the ith classier separates the ith class from the rest. The one-versus-one strategy constructs $k(k − 1)/2$ classiers, using all the binary pairwise combinations of the k classes. The class is then predicted with a majority vote.

SVM can use some other classification functions, for example a polynomial function of degree d, a radial basis function (RBF) or a sigmoid function. More details about SVM and other kernel-based learning methods can be found in [32].

2.3 Support Vector Machines Using the Feature Extraction from Deep Convolutional Neural Networks

DCNNs are efficient at learning invariant features from data, but do not always produce optimal classification results. Conversely, a non-linear SVM cannot learn complex invariances, but produce good decision surfaces by maximizing margins using soft-margin approaches [33].

Our investigation is to propose a hybrid model architecture: A coupling SVM with the feature learning of DCNNs (denoted by DCNN-SVM) for classifying microarray gene expression data. The training task of DCNN-SVM consists of two main steps. First, the algorithm learns DCNNs to deeply extract functional features from high dimensional gene expression profiles. Next, it trains non-linear SVM models to perform the classification of the data representation extracted by the previous one.

The network architecture is shown in Fig. 2. Firstly, the first layer uses gene expression data. Secondly, the second and fourth layers of the network are convolution layers alternator with sub-sampling layers, which take the pooled maps as input.

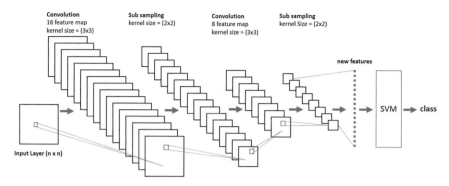

Fig. 2 The DCNN-SVM architecture

Consequently, they are able to extract features that are more and more invariant to local transformations of the input layer. The sixth layer is fully connected layer. The final layer is substituted by SVM with the RBF kernel for classification. The outputs from the hidden units are taken by the SVM as a feature vector for the training process. After that, the training stage continues till realizing good trained. Finally, classification on the test set is performed by the SVM classifier with such automatically extracted features.

3 Evaluation

We implement DCNN-SVM, SVM and random forests in python, using library SVM, LibSVM [34], tensorflow [35] and scikit library [36]. All tests were run under Linux Mint on a single 2.4 GHz Core I3 PC with 8 GB RAM.

3.1 Experiments Setup

In our experiments, we use datasets provided by ArrayExpress database [1] and the Medical Database (Kent Ridge) [25]. ArrayExpress archive of Functional Genomics Data stores data from high-throughput functional genomics experiments. We downloaded MGE datasets from the ArrayExpress. The criteria for selecting the datasets were that the experiments had been conducted in humans and in the field of cancer. Datasets published or updated after 2012 and provided processed data. To reduce the source of variability of classification model performances because of the array used in the experiments, we retained studies conducted with Affymetrix array. The datasets and their characteristics are summarized in Table 1.

The test protocols are presented in the column 5 of Table 1. Some datasets are already divided in training set (Trn) and testing set (Tst). For these datasets, we used

Table 1 Description of microarray gene expression datasets

ID	Name	Individuals	Attributes	Classes	Protocols	References
1	Lung cancer	181	12533	2	Trn-Tst	[37]
2	Prostate cancer	136	12600	2	Trn-Tst	[38] ID:68907
3	Astra Zeneca	627	54756	22	k-fold	[1] ID:57083
4	L. Leukaemia	575	22500	3	k-fold	[1] ID:33315
5	CCLE	917	19044	21	kfold	[1] ID:36133
6	GTex	837	22801	13	k-fold	[1] ID:45787
7	Breast cancer	327	54627	6	k-fold	[1] ID:20685
8	Breast cancer	97	24481	2	Trn-Tst	[39]
9	Leukemia	40	54675	2	loo	[1] ID:14858
10	Cancer cell. Project	950	54627	50	k-fold	[1] ID:MT-AB37
11	T-ALL & T-LL	29	15435	3	loo	[1] ID:1577
12	Sarcoma	105	22283	10	loo	[1] ID:6481
13	Lung cancer	203	12600	5	loo	[40]
14	Breast cancer	286	22283	3	loo	[1] ID:2034
15	Miscellaneous	50	10100	5	loo	[41]

the training data to build the our model. Then, we classified the testing set using the resulted model. With a datasets having less than 300 data points, the test protocol is leave-one-out cross-validation (loo). For the others, we used 10-fold cross-validation protocols remains the most widely to evaluate the performance [42]. Our evaluation used on the classification accuracy.

The DCNN-SVM architecture is shown in Table 2. It consist of 2 convolutional layers with 32 and 16 feature maps of (3×3) kernel, and each convolutional layer has a (2×2) average pooling layer followed. The features are taken from the last fully connected layer. SVM takes these outputs from the fully connected for classification. The one-versus-all method is utilized for the multi-class SVM that is possibly to be viewed as a trainable feature extractor. We have also tried other configurations of CNN, whereas this one gives the best performance. Input data are transformed the following way: we use microarray expression feature to represent each sample patient, which transform into a feature matrix. For deep convolutional neural networks configurations, we use ADAM method [43] for optimization, cross-entropy for loss function. The batch size is set to 16 and 50 epochs are used. We also tried to tune activation function with ReLU, Tanh and Sigmoid. The Tanh activation works better than other activation functions for microarray gene expression data.

We propose to use RBF kernel type in SVM models because it is general and efficient [44]. We also tried to tune parameters γ of RBF kernel and the cost C (a trade-off between the margin size and the errors) to obtain a good accuracy. These parameters are presented in Table 2.

Table 2 Hyper-parameters of SVM used in DCNN-SVM

ID	1	2	3	4	5	6	7	8	9	10	11	12	13	14	15
γ	0.01	0.01	3.85e−05	0.01	0.001	0.01	0.01	0.01	0.01	3.85e−05	0.01	0.01	0.01	0.01	0.001
C	10^5	10^5	10^5	10^5	10^6	10^5	10^5	10^5	10^5	10^5	10^6	10^5	10^5	10^5	10^6

In order to evaluate the effectiveness of our approach, we used two different experiments to classify microarray samples. First, we compare DCNN-SVM with SVM, random forests (RF) and traditional DCNNs. In this experiments, RF algorithms build 200 decision trees and we use linear kernel type in SVM models ($C = 10^5$, $\gamma = 0.01$). Second, we compare different kernel functions in the SVM classifier: a linear kernel (DCNN-SVM linear) and a radial basis function (DCNN-SVM) with best parameter in Table 2. In addition, we also compared DCNN-SVM with DCNNs using random forest (DCNN-RF) classifier.

3.2   Experiments Results

Numerical test results on 15 microarray datasets are shown in Table 3. Results on 15 datasets showed that DCNN-SVM is more accuracy than the classical DCNNs algorithm, SVM, random forests. DCNN-SVM has the best accuracy of 11 out of 15 datasets. SVM and RF have the best only 1 out of 15 datasets. Table 3 and Fig. 3 showed that DCNN-SVM uses the RBF kernel to achieve the best accuracy result of 11 over 15 datasets. The DCNN-SVM uses linear kernel to achieve the best accuracy of 6 out of 15 datasets and DCNN-RF uses RF classifier has the best accuracy of 5

Table 3 Classification results in terms of accuracy (%)

ID	SVM	RF	DCNNs	DCNN-RF	DCNN-SVM	
					Linear kernel	RBF kernel
1	99.33	97.99	100.0	100.0	100.0	100.0
2	91.18	94.12	97.00	97.00	97.00	97.00
3	80.54	71.45	89.30	88.51	88.96	89.32
4	83.30	80.35	87.14	86.62	87.13	87.13
5	90.73	87.24	91.18	91.40	91.17	91.40
6	97.73	97.37	97.86	97.74	97.86	97.86
7	88.69	84.40	88.80	88.30	88.80	89.20
8	68.42	68.42	75.00	75.00	75.00	75.00
9	87.50	85.00	77.50	85.00	100.0	100.0
10	61.41	56.42	94.8	95.00	95.4	95.4
11	89.66	96.55	86.21	96.55	93.10	93.10
12	71.43	63.81	65.71	62.86	69.52	69.52
13	93.60	92.12	93.60	94.09	93.60	93.60
14	87.41	86.01	87.06	87.76	88.11	88.11
15	68.00	64.65	56.00	62.00	58.00	96.00

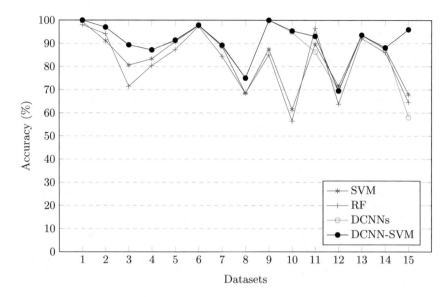

Fig. 3 Comparison of accuracy (%)

out of 15 datasets. DCNNs has the best accuracy of 4 out of 15 datasets. This superiority of DCNN-SVM (RBF) on CNNs, DCNN-SVM (RF) and DCNN-SVM (linear) showed in table results: 5 wins of DCNN-SVM (RBF) on DCNN-SVM (linear), 10 wins of DCNN-SVM (RBF) on DCNN-SVM (RF) and DCNNs on 15 datasets.

4   Conclusion and Future Works

We have presented a hybrid model combining DCNNs and SVM to classify very-high-dimension microarray gene expression data. The features are learned through a convolution process and then sent as input to a SVM classifier using RBF kernel to the objective of interest. After modifications through specified hyper parameters, the model performs quite comparatively well on the task tested on 15 different datasets from ArrayExpression and Medical Database. The numerical test results show that our proposal is more accurate than the classical DCNNs algorithm, support vector machines, random forests for classifying.

In the near future, we intend to provide more empirical test on large datasets of microarray gene expression and comparisons with other algorithms. Our proposal can be effectively parallelized. A parallel implementation that exploits the multicore processors can greatly speed up the learning and predicting tasks.

References

1. Brazma, A., et al.: ArrayExpress a public repository for microarray gene expression data at the EBI. Nucleic Acids Res. **31**(1), 68–71 (2003)
2. Edgar, R., Domrachev, M., Lash, A.E.: Gene expression omnibus: NCBI gene expression and hybridization array data repository. Nucleic Acids Res. **30**(1), 207–210 (2002)
3. Schena, M., et al.: Quantitative monitoring of gene expression patterns with a complementary DNA microarray. Science (New York then Washington) 467–470 (1995)
4. Pinkel, D., et al.: High resolution analysis of DNA copy number variation using comparative genomic hybridization to microarrays. Nat. Genet. **20**(2) (1998)
5. Brown, M.P.S., et al.: Support vector machine classification of microarray gene expression data. University of California, Santa Cruz, Technical Report UCSC-CRL-99-09 (1999)
6. Furey, T.S., Cristianini, N., Duffy, N., Bednarski, D.W., Schummer, M., Haussler, D.: Support vector machine classification and validation of cancer tissue samples using microarray expression data. Bioinformatics **16**(10), 906–914 (2000)
7. Guyon, I., Weston, J., Barnhill, S., Vapnik, V.: Gene selection for cancer classification using support vector machines. Mach. Learn. **46**(1), 389–422 (2002)
8. Hasri, N.N.M., et al.: Improved support vector machine using multiple SVM-RFE for cancer classification. Int. J. Adv. Sci. Eng. Inf. Technol. **7**(4–2), 1589–1594 (2017)
9. Yeang, C.H., Ramaswamy, S., Tamayo, P., Mukherjee, S., Rifkin, R.M., Angelo, M., Reich, M., Lander, E., Mesirov, J., Golub, T.: Molecular classification of multiple tumor types. Bioinformatics **17**(suppl-1), S316–S322 (2001)
10. Li, J., Liu, H.: Ensembles of cascading trees. In: 2003 Third IEEE International Conference on Data Mining, ICDM 2003, pp. 585–588. IEEE (2003)
11. Li, J., Liu, H., Ng, S.K., Wong, L.: Discovery of significant rules for classifying cancer diagnosis data. Bioinformatics **19**(suppl-2), ii93–ii102 (2003)
12. Tsai, M.H., et al.: A decision tree based classifier to analyze human ovarian cancer cDNA microarray datasets. J. Med. Syst. **40**(1), 21 (2016)
13. Díaz-Uriarte, R., De Andres, S.A.: Gene selection and classification of microarray data using random forest. BMC Bioinf. **7**(1), 3 (2006)
14. Do, T.N., Lenca, P., Lallich, S., Pham, N.K.: Classifying very-high-dimensional data with random forests of oblique decision trees. In: Advances in Knowledge Discovery and Management, pp. 39–55. Springer (2010)
15. Tan, A.C., Gilbert, D.: Ensemble machine learning on gene expression data for cancer classification. Bioinformatics (2003)
16. Dettling, M.: Bagboosting for tumor classification with gene expression data. Bioinformatics **20**(18), 3583–3593 (2004)
17. Krizhevsky, A., et al.: ImageNet classification with deep convolutional neural networks. In: Advances in Neural Information Processing Systems, pp. 1097–1105 (2012)
18. Lai, S., Xu, L., Liu, K., Zhao, J.: Recurrent convolutional neural networks for text classification. AAAI **333**, 2267–2273 (2015)
19. Min, S., Lee, B., Yoon, S.: Deep learning in bioinformatics. Brief. Bioinf. (2016). https://doi.org/10.1093/bib/bbw068
20. Suykens, J.A., Vandewalle, J.: Training multilayer perceptron classifiers based on a modified support vector method. IEEE Trans. Neural Netw. **10**(4), 907–911 (1999)
21. Bellili, A., Gilloux, M., Gallinari, P.: An hybrid MLP-SVM handwritten digit recognizer. In: Proceedings of the Sixth International Conference on Document Analysis and Recognition 2001, pp. 28–32. IEEE (2001)
22. Niu, X.X., Suen, C.Y.: A novel hybrid CNN-SVM classifier for recognizing handwritten digits. Pattern Recognit. **45**(4), 1318–1325 (2012)
23. Nagi, J., et al.: Convolutional neural support vector machines: hybrid visual pattern classifiers for multi-robot systems. In: 2012 11th International Conference on Machine Learning and Applications (ICMLA), vol. 1, pp. 27–32. IEEE (2012)

24. Cao, G., Wang, S., Wei, B., Yin, Y., Yang, G.: A hybrid CNN-RF method for electron microscopy images segmentation. Tissue Eng. J. Biomim. Biomater. Tissue Eng. **18**, 2 (2013)
25. Jinyan, L., Huiqing, L.: Kent ridge bio-medical data set repository (2002)
26. Vapnik, V.: Statistical Learning Theory, vol. 1. Wiley, New York (1998)
27. Breiman, L.: Random forests. Mach. Learn. **45**(1), 5–32 (2001)
28. Hubel, D., Wiesel, T.: Shape and arrangement of columns in cat's striate cortex. J. Physiol. **165**(3), 559–568 (1963)
29. LeCun, Y., Bengio, Y., Hinton, G.: Deep learning. Nature **521**(7553), 436–444 (2015)
30. Burges, C.J.: A tutorial on support vector machines for pattern recognition. Data Min. Knowl. Discov. **2**(2), 121–167 (1998)
31. Kreßel, U.H.G.: Pairwise classification and support vector machines. In: Advances in Kernel Methods, pp. 255–268. MIT press (1999)
32. Cristianini, N., Shawe Taylor, J.: An introduction to support vector machines and other kernel-based learning methods. Cambridge university press (2000)
33. Huang, F., LeCun, Y.: Large-scale learning with SVM and convolutional nets for generic object recognition. In: 2006 IEEE Computer Society Conference on Computer Vision and Pattern Recognition (2006)
34. Chang, C.C., Lin, C.J.: LIBSVM: a library for support vector machines. ACM Trans. Intell. Syst. Technol. **2**, 27:1–27:27 (2011). Software available at http://www.csie.ntu.edu.tw/~cjlin/libsvm
35. Abadi, M., et al.: TensorFlow: large-scale machine learning on heterogeneous systems (2015). Software available from http://www.tensorflow.org
36. Pedregosa, F., et al.: Scikit-learn: machine learning in python. J. Mach. Learn. Res. **12**, 2825–2830 (2011)
37. Gordon, G.J., et al.: Translation of microarray data into clinically relevant cancer diagnostic tests using gene expression ratios in lung cancer and mesothelioma. Cancer Res. **62**(17), 4963–4967 (2002)
38. Singh, D., et al.: Gene expression correlates of clinical prostate cancer behavior. Cancer Cell **1**(2), 203–209 (2002)
39. Veer, V., et al.: Gene expression profiling predicts clinical outcome of breast cancer. Nature **415**(6871), 530–536 (2002)
40. Bhattacharjee, A., et al.: Classification of human lung carcinomas by mRNA expression profiling reveals distinct adenocarcinoma subclasses. Proc. Natl. Acad. Sci. **98**(24), 13790–13795 (2001)
41. Subramanian, A., et al.: Gene set enrichment analysis: a knowledge-based approach for interpreting genome-wide expression profiles. Proc. Natl. Acad. Sci. U. S. A. **102**(43), 15545–15550 (2005)
42. Wong, T.T.: Performance evaluation of classification algorithms by k-fold and leave-one-out cross validation. Pattern Recognit. **48**(9), 2839–2846 (2015)
43. Diederik, P., Kingma, J.B.: Adam: a method for stochastic optimization. In: Proceedings of the 3rd International Conference on Learning Representations (ICLR) (2014)
44. Hsu, C.W., et al.: A practical guide to support vector classification (2003)

Development of Seawater Temperature Announcement System for Quick and Accurate Red Tide Estimation

Yu Agusa, Takuya Fujihashi, Keiichi Endo, Hisayasu Kuroda
and Shinya Kobayashi

Abstract Fisheries researchers who predict the occurrence of red tides want to be able to visualize the seawater temperature information measured using the marine buoy in a form that can be handled easily for researchers in order to improve the prediction accuracy. In this research, we develop a system to visualize measured seawater temperature information. In the system, it is possible to display latest seawater temperature information and time change of seawater temperature in the form of tables and graphs by using a web application. In addition, it is possible to download the sea water temperature information collected in the past into the csv format file. With this system, researchers can predict the occurrence of red tide at an early stage because seawater temperature information becomes easier to handle.

1 Introduction

The population of the world will exceed 7 billion people in 2011 and is expected to reach 9.6 billion people in 2050. Food difficulties are faced by the increase in the world population. Aquaculture fishery is an important food supply source under such circumstances. In Japan, although the catch amount of the fish is sluggish due to sluggish fish prices, soaring food prices, aging of aquaculture workers, etc., it is expected that importance of aquaculture production will increase in the future considering the world food situation.

However, the phenomenon called red tide occurs suddenly and it is the biggest factor impeding stabilization of cultured production in the aquaculture industry. Red tide is a phenomenon in which the color of water in the sea area changes due to the massive generation of plankton. The large amount of fish will die due to the drop in dissolved oxygen concentration in the sea area and asphyxiation due to clogging of plankton in fish gills when this phenomenon occurs.

Y. Agusa (✉) · T. Fujihashi · K. Endo · H. Kuroda · S. Kobayashi
Graduate School of Science and Engineering, Ehime University, Matsuyama, Japan
e-mail: agusa@koblab.cs.ehime-u.ac.jp

© Springer International Publishing AG, part of Springer Nature 2018
A. Sieminski et al. (eds.), *Modern Approaches for Intelligent Information and Database Systems*, Studies in Computational Intelligence 769,
https://doi.org/10.1007/978-3-319-76081-0_21

245

To give an example that the occurrence of red tide hinders the stabilization of aquaculture production, the red tide occurred in 2015 in Ehime Prefecture Uwakai caused damage of about 372 million yen [1].

Correspondence to red tides is an important issue in order to build a sustainable and stable aquaculture production system. The damage has already expanded already and the amount of damage can not be suppressed so much in case of responding after the red tide occurred, so it is strongly required that countermeasures against red tide be carried out in the phase before occurrence or early stage of occurrence.

It is necessary to prepare a system that can predict the occurrence of red tide beforehand in order to be able to respond to the red tide before the occurrence or at the initial stage of occurrence. Ehime prefecture is currently trying to predict the occurrence of red tide by two methods. The first method is predicting the occurrence from a biological point of view based on sea area information such as sea area abnormality and seawater concentration in seawater, which is conducted at South Ehime Fisheries Research Center. The second method is predicting the occurrence from an oceanophysical point of view by understanding the flow of the tide such as low temperature tide inflow (bottom lake) and warm tide inflow (steep tide) from the seawater temperature variation at multiple water depths and multiple observation points, which is conducted at Center for Marine Environmental Studies, Ehime University.

In this paper, out of these two methods, we deal with the latter prediction of occurrence in ocean physics viewpoint.

As for the measurement of seawater temperature, it is done using equipment that measures seawater temperature, dissolved oxygen concentration, etc. called marine buoy, which is installed at 14 places in Ehime prefecture as of October 2017. The data measured by the marine buoy is sent as an e-mail and stored in the mail server. However, since this type of data is not accumulated in one file, it is very hard for researchers. In addition, since we can not visualize the latest measurement data of each marine buoy and visualize the seawater temperature variation over time, we can not see seawater temperatures variation at multiple depths and multiple measurement point.

Ehime University proposed a system with the structure as shown in Fig. 1 in cooperation with Ehime University Coastal Environment Science Research Center to solve this problem.

This system based on requests from fisheries researchers, and aims to accumulate and visualize seawater temperature information and to communicate notifications to users of this system. Fishery researchers can store seawater temperature information in a manageable format, and can easily and quickly see changes in multiple depths and seawater temperatures at multiple stations with this system.

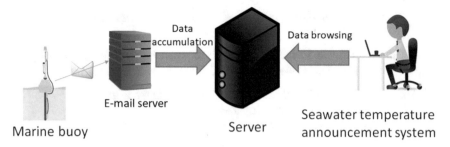

Fig. 1 System configuration

This system consists of three components. The first is the seawater temperature information accumulation system for accumulating seawater temperature information, the second is the server for storing and managing seawater temperature information, and the third is the display system for visualizing and announcing accumulated seawater temperature information. In this research, we aim to develop the display system among these three components.

2 Seawater Temperature Announcement System

2.1 Required Items

Based on a hearing from fishery researchers, our announcement system needs to satisfy the following requirements:

1. The display system should visualize seawater temperature information considering spatial spread of marine area
2. The display system should visualize the seawater temperature variation over time
3. The display system should store the past seawater temperature information

In addition to the above requirements, they desire that data can be narrowed down by a simple operation, the visualized data is easy to see for them, and the display can be completed in a short time, to enable them to predict the occurrence of red tide quickly and accurately.

To satisfy all the above-mentioned requirements, we realize the announcement system as an Web application because it can provide seawater temperature information regardless of platforms.

2.2 Implemented Functions and Web Page Structure

We list the implemented functions by each requirement item in constructing this system. For request item 1, we implemented functions as follows:

(a) Displaying the location of the observation point on the map
(b) Displaying measurement data in tabular form
(c) Displaying the latest measurement data as a graph

For request item 2, we implemented function as follows:

(d) Displaying seawater temperature variation over time

For request item 3, we implemented function as follows:

(e) Saving measured data as a file in csv format to the user's terminal

In addition, in order to improve the convenience of this system, we implemented function as follows:

(f) Saving graph display settings

Regarding page structure, we created public page besides page for researchers, because fishermen and others also emphasize seawater temperature information. In the page for researchers, it is possible to use all the functions except function (a). In the page for public release, all the functions except for the function (e) can be used (however, in the public page, the display of measurement data is restricted). The users of this system can switch display items between "current status" (Display of measurement data at each measurement point, and at the latest measurement date and time) and "past status" (Display of past measurement data and seawater temperature variation on time of each measurement point), bidirectionally. Functions (a), (b), and (f) are functions common to the respective displays, the function (c) includes functions included only in the "current status", the function (d) includes functions included only in the "past status".

2.3 Function Details

In this section, we respectively described the functions corresponding to each request item and the functions implemented for improving the convenience of the system.

Location	Fukuura	Shiokojima	Shimonada	Kitanada	Hiburijima	Shimonami	Miura	Komobuchi	Yusu	Uwajima	Yoshida	Akehama	Mikame	Yawatahama	Saijo
Update date※	2018/02/16 09:00:00	2018/02/16 09:00:00	2018/02/16 09:00:00	2018/02/16 09:32:51	2018/02/16 09:00:00	2018/02/16 09:00:00	2018/02/16 09:03:25	2018/02/16 09:00:00	2018/02/16 09:00:00	2018/02/16 09:15:53	2018/02/16 09:24:38	2018/02/16 09:00:00	2018/02/16 09:00:00	2018/02/16 09:00:00	2018/02/16 09:00:33
Depth =1m	15.9°C	15.3°C	NA	NA	12.9°C	NA	NA	13.7°C	12.3°C	NA	NA	12.1°C	11.5°C	NA	8.5°C
=3.5m	NA	NA	NA	NA	NA	NA	NA	NA	NA	NA	NA	NA	NA	NA	9.9°C
=5m	16.0°C	15.4°C	14.6°C	13.1°C	13.0°C	13.7°C	12.1°C	13.6°C	12.3°C	11.0°C	11.9°C	12.2°C	11.5°C	11.1°C	NA
=10m	15.9°C	15.4°C	14.7°C	13.1°C	12.9°C	NA	12.1°C	13.6°C	12.3°C	11.2°C	12.2°C	12.1°C	11.5°C	NA	NA
=20m	15.9°C	15.3°C	14.7°C	13.2°C	12.9°C	NA	12.1°C	13.6°C	12.3°C	11.9°C	11.9°C	12.2°C	11.4°C	NA	NA
=30m	15.9°C	15.2°C	NA	NA	13.0°C	NA	NA	13.5°C	12.4°C	NA	NA	12.2°C	11.3°C	NA	NA
=40m	15.9°C	15.1°C	NA	NA	12.8°C	NA	NA	13.3°C	12.2°C	NA	NA	12.2°C	11.2°C	NA	NA
=50m	15.8°C	15.0°C	NA	NA	12.8°C	NA	NA	NA	NA	NA	NA	NA	NA	NA	NA
=60m	15.7°C	15.0°C	NA	NA	12.9°C	NA	NA	NA	NA	NA	NA	NA	NA	NA	NA

(a) "current status"

Location	Fukuura	Shiokojima	Shimonada	Kitanada	Hiburijima	Shimonami	Miura	Komobuchi	Yusu	Uwajima	Yoshida	Akehama	Mikame	Yawatahama	Saijo
Depth =1m	15.9°C	15.3°C	NA	NA	12.9°C	NA	NA	13.6°C	12.3°C	NA	NA	12.1°C	11.5°C	NA	8.5°C
=3.5m	NA	NA	NA	NA	NA	NA	NA	NA	NA	NA	NA	NA	NA	NA	9.9°C
=5m	16.0°C	15.3°C	14.6°C	12.9°C	13.0°C	13.7°C	12.1°C	13.5°C	12.3°C	11.0°C	12.0°C	12.2°C	11.5°C	11.0°C	NA
=10m	15.9°C	15.4°C	14.7°C	13.5°C	13.0°C	NA	12.1°C	13.6°C	12.3°C	11.2°C	12.2°C	12.1°C	11.5°C	NA	NA
=20m	15.9°C	15.3°C	14.7°C	13.2°C	12.9°C	NA	12.1°C	13.5°C	12.3°C	11.9°C	11.9°C	12.2°C	11.5°C	NA	NA
=30m	16.0°C	15.3°C	NA	NA	12.9°C	NA	NA	13.5°C	12.4°C	NA	NA	12.2°C	11.2°C	NA	NA
=40m	15.9°C	15.2°C	NA	NA	12.8°C	NA	NA	13.4°C	12.2°C	NA	NA	12.2°C	11.2°C	NA	NA
=50m	15.8°C	15.0°C	NA	NA	12.8°C	NA	NA	NA	NA	NA	NA	NA	NA	NA	NA
=60m	15.7°C	15.0°C	NA	NA	12.9°C	NA	NA	NA	NA	NA	NA	NA	NA	NA	NA

Measuring date : 2018/02/16 10:00:00

(b) "past status"

Fig. 2 Displaying measurement data in a tabular

2.3.1 Visualization of Seawater Temperature Information Considering Spatial Spread of Marine Area

This system can display the position of each measurement point as a point on a map in function (a). The users can grasp the positional relationship of each measurement point with this function.

In addition, in function (b), this system can display the measurement data of each measurement point in a table format as shown Fig. 2a in the "current status" or as shown Fig. 2b in the "past status". For the table showing measurement data, the horizontal items is the location (measurement point) and the vertical items is the water depth in order to be able to grasp seawater temperature variation due to differences in latitude of the measurement point. In "past status", it is possible to measurement data of arbitrarily specified date and time by a simple operation such as sliding a slider or pressing a button.

Furthermore, in function (c), as shown Fig. 3, the latest measurement data can also be displayed as a line graph in which lines are color-coded according to the water depth, with the seawater temperature as the vertical axis and the measurement point as the horizontal axis. The users can change the upper and lower limits of the seawater temperature to be displayed by sliding the slider, and the water depth to be

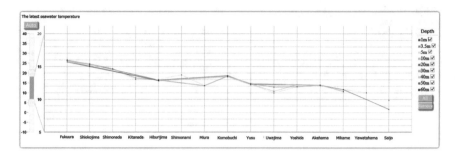

Fig. 3 Displaying the latest measurement data as a graph

displayed can be selected by unchecking the check box. The users can simultaneously grasp variation in seawater temperature between measurement points and variation in seawater temperature due to difference in water depth with this function.

It is possible to see the variation in seawater temperature considering spatial spread of the sea area by using these three functions in combination.

2.3.2 Visualization of Seawater Temperature Variation over Time

Function (d) is the function to display the variation of the past seawater temperature graphically for each measurement point, as shown in Fig. 4. In this function, the graph to be displayed is the line graph with seawater temperature on the vertical axis and time as the horizontal axis and its color of the line is divided like function (c) because it becomes possible to grasp the variation of seawater temperature due to the passage of time and the difference of water depth at the same time. In addition, like the function (c), in addition to making it possible for the user to arbitrarily select the depth of water to be displayed and to change the upper and lower limits of the sea water temperature to be displayed, the time of short-term or long-term seawater temperature. From the desires of fishery researchers who want to be able to grasp the change, the display period can be selected from 24 h, 48 h, 7 days, 30 days, 60 days, and the display period is selected as a slider type. So that it can be switched by a simple operation. It is possible to visualize the time change of the seawater temperature at each depth at each measurement point with this function.

2.3.3 Storing Past Information

As shown in Fig. 5, specifies the measurement place and the measurement period and presses the button written as download so that the measurement data within the measurement place and the measurement period is downloaded as a csv format file

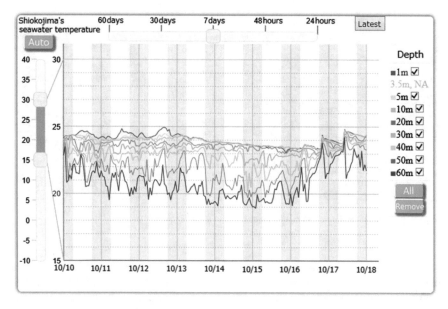

Fig. 4 Displaying seawater temperature variation over time

Fig. 5 Saving measured data as a file in csv format

in function (e). We set the location designation as the radio button method, designate the period as the text box, and display the calendar when entering the period so that the user can download by a simple operation. Past measurement data list can be stored arbitrarily by the user in the terminal with this function.

2.3.4 Saving Graph Display Settings

Function (f) is a function implemented to enhance the convenience of this system. When accessing this system, it is desirable that the graph can be displayed with the same display settings as at the previous access, so it was made possible to display at the previous display setting at the time of re-access. In addition, since it is desirable that the user can arbitrarily create several display setting patterns and it is desirable that the display setting pattern can be easily called, it is possible to register the display setting pattern and call by the button.

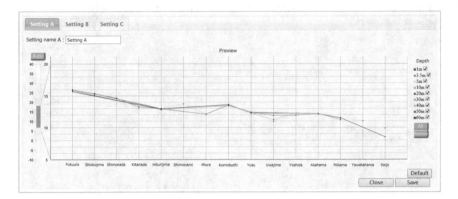

Fig. 6 Saving graph display setting pattern

The display setting pattern is registered using the UI as shown in Fig. 6. The display setting pattern can be separately registered in the graph of the latest measurement data and the graph of seawater temperature variation over time, and the user can register the setting pattern in the same sense as changing the display setting of the graph. It is possible to reflect the display setting pattern instantly registered on the current display setting by simply pressing the button after registering the display setting pattern.

3   Conclusion

We developed a system can visualize seawater temperature information for support prediction of red tide occurrence.

We introduced request from fishery researcher and display measured data as table and graph, making easy data analysis possible. In addition, in developing this system, we strive to be able to display graphs in a simple operation and to display in a short time and devised about the layout, example, buoys that are out of order in the display of the latest measurement data are known for the measurement data of the buoys, the background color in the "latest measurement date" column of the table is made red, the display can be switched between the current situation and the past situation by press the button, etc.

However, the additional functions such as display of three-dimensional seawater temperature distribution maps and improvement of existing functions are sought from fishery researchers, so it is necessary to respond to those requests in the future.

Acknowledgements I would like to thank Mr. Hidetaka Takeoka of South Ehime Fisheries Research Center and Mr. Akihiko Takechi of Ehime Research Institute of Agriculture, Forestry and Fisheries for their cooperation in this research.

This work was partially supported by Strategic Information and Communications R&D Promotion Programme (SCOPE) (152309003), the Ministry of Internal Affairs and Communications, Japan.

Reference

1. Fisheries Agency Seto Inland Sea Fisheries Adjustment Office: The red tide of the Seto Inland Sea in Heisei 27(Heisei 27 nendo Seto-naikai no akashio), pp. 1 (2015)

Part IV
Decision Support Systems

Support Product Development Framework by Means of Set of Experience Knowledge Structure (SOEKS) and Decisional DNA

Muhammad Bilal Ahmed, Cesar Sanin and Edward Szczerbicki

Abstract In this paper, we propose a framework to support product development activities by utilizing Set of Experience Knowledge Structure (SOEKS) and Decisional DNA (DDNA). This idea will provide a new direction to researchers working on product development, especially designers and manufacturers. They will be working on the same platform and this will be reducing their communication gap. Once the final idea is perceived about product development, it will be easy to design and manufacture it quickly and efficiently. Early consideration of manufacturing issues can shorten product development cycle time, minimize overall development cost, and ensure a smooth transition into production. In product development process knowledge of previous products and processes is very important, as product development requires both knowledge and experience. This framework will store knowledge in the form of experiences of past decisions, and the system will update itself after every decision is taken.

Keywords Product development · Product design and manufacturing
Set of experience · Decisional DNA · Virtual engineering objects (VEO)
Virtual engineering processes (VEP)

M. B. Ahmed (✉) · C. Sanin (✉)
The University of New Castle, Callaghan, NSW, Australia
e-mail: Muhammadbilal.ahmed@uon.edu.au

C. Sanin
e-mail: cesar.sanin@newcastle.edu.au

E. Szczerbicki (✉)
Gdansk University of Technology, Gdansk, Poland
e-mail: edward.szcerbicki@newcastle.edu.au

© Springer International Publishing AG, part of Springer Nature 2018
A. Sieminski et al. (eds.), *Modern Approaches for Intelligent Information and Database Systems*, Studies in Computational Intelligence 769,
https://doi.org/10.1007/978-3-319-76081-0_22

1 Introduction

In today's competitive global market, companies fight vigorously to provide good products having greater value and low cost, and therefore, employ best product development strategies. There is a huge rise in market growth and technological changes in the past years, and this tempestuous environment requires new methods and techniques to bring successful new products to the marketplace [1]. There is need to reduce all the non-useable repetitive processes and mistake proofing at early stages of product development, as world market requires shorter product development times [2]. It is very important for companies with short product life cycles, to quickly and safely develop new products and new product platforms that fulfill reasonable demands on quality, performance, and cost. Classical methods such as Stage-Gate model process [3], Integrated Product Development (IPD) [4], Concurrent Engineering (CE), and Simultaneous Engineering (SE) were developed for re-engineering of existing products and new products. Unfortunately, there are reports and project reviews as feedback principles, which for all types of development means fragmented information, delayed information, and reactive management [5]. Thus, there is a need for an effective decisional support system to process a customer order, enquiries as they should be equipped with satisfactory information and knowledge of manufacturing processes to enable quick and accurate response to the customer. The sales department working in alliance with the design and production department should also ensure that they can deliver what has been assured to the customer within the due date [6]. It has been proven that ICT improves the new product development process by limiting distances between designers and manufacturers working on different platforms, and saving on costs and time [7]. Similarly, with the beginning of a fourth Industrial revolution (Industry 4.0), which will be focusing on generating smart products, procedures and processes, it requires increasing efforts to understand a smart factory environment and also the product development has to perform a significant transformation [8].

Analysis of Cyber-Physical Systems, and Virtual Engineering Object, Virtual Engineering Process and Virtual Engineering Factory has revealed that there are fundamental similarities among these concepts, both at theoretical as well as practical levels [9]. There are different factors on which new product development process gets success, and knowledge and experience are some of them [10]. Due to the enormous increase in knowledge and experience base decisional activities, it is very difficult to store all the relevant information [11]. There are few companies which produce their product development manuals, but they are not enough to store decisional experiences of each and every step. This work proposes a concept of supporting product development activities by using a knowledge representation technique called Set of Experience Knowledge structure and Decisional DNA.

The structure of the paper includes the background in Sect. 2, which presents the concepts of product development, Set of experience knowledge structure and Decisional DNA, Virtual engineering objects, Virtual engineering processes and Virtual engineering factory. Product development process by using SOEKS and the

Decisional DNA is discussed in Sect. 3, along with the smart products in Industry 4.0 and proposal for framework and its working. Finally, the concluding remarks are presented in Sect. 4.

2 Background

2.1 Product Development

Generally, a product is defined as a good, service, place, organization or an idea. In this research, products are objects which are manufactured for the end users. Product development is a concatenation of different processes and sub-processes [12]. Kusar et al. [13] summarized different stages of new product development, in which at the earlier stages the aim is to make an initial market, business, and technical assessment, whereas at the later stages they recommend to actually design and develop the product. This research contributes towards later stages.

The main aim of product development is to provide the product at lower production costs, good quality and quick access to the market, so that it may contribute to customer satisfaction [14]. Most organizations are forced to move from traditional face-to-face teams to virtual teams or adopt a combination between the two types of teams [2]. In the past, various product development processes or techniques have been engaged by different companies, i.e. Stage-gate model process, Development funnel product model process, Integrated product development process, Toyota NPD, and Product development process by Ulrich and Eppinger [4]. Product development by using set of experience knowledge structure and Decisional DNA and virtual engineering objects will make this process systematic, smarter and faster.

2.2 Set of Experience Knowledge Structure and Decisional DNA

Set of experience knowledge structure (SOEKS) is a smart knowledge structure that collects and analyses formal decision events and uses them to represent experiential knowledge. A formal decision is defined as a choice (decision) made or a commitment to act that was the result (consequence) of a series of repeatable actions performed in a structured manner. A set of experience (SOE, a shortened form of SOEKS) has four components: Variables (V), functions (F), Constraints (C) and Rules (R). Each formal decision is represented and stored in a unique way based on these components. Variables are the basis of the other SOEKS components, whereas functions are based upon the relationships and links among the variables. The third SOEKS component is constraints, which, like functions, are connected to variables. They specify limits and boundaries and provide feasible solutions. Rules are the fourth components and are conditional relationships that operate on variables. Rules are relationships between a condition and a consequence connected by the statements 'if/then/else' [15]. The four components of a SOE and its structural body can be defined by comparing it with some important features of human DNA. First, just as the combination of its four nucleotides (Adenine, Thymine, Guanine, and Cytosine) makes DNA unique, the combination of its four components (Variables, Function, Constraints, and Rules) makes an SOE unique.

Each formal decision event is deposited in a structure that combines these four SOE components. Several interconnected elements are visible in the structure, resembling part of a long strand of DNA, or a gene. Thus, a SOE can be associated to a gene and, just as a gene produces a phenotype, a SOE creates a value for a decision in terms of its objective function. Hence, a group of SOEs in the same category form a kind of chromosome, as DNA does with genes. Decisional DNA contains experienced decisional knowledge and it can be categorized according to areas of decisions. Further, just as assembled genes create chromosomes and human DNA, groups of categorized SOEs create decisional chromosomes and DDNA. In short, a SOEKS represents explicit experiential knowledge which is gathered from the previous decisional events [16].

2.3 Virtual Engineering Objects, Virtual Engineering Processes and Virtual Engineering Factory

Virtual engineering object (VEO) is a novel knowledge representation technique of an engineering object associated with its experience and formal decision [17]. It gathers and reuses its own experience knowledge; this is attained by gathering information from six different aspects of an object, viz., (i) Characteristics, (ii) Functionality, (iii) Requirements, (iv) Connections, (v) Present State, and (vi) Experience. Virtual engineering process (VEP) is a knowledge representation of manufacturing process/process-planning of artifacts having all shop-floor-level information with regard to required operations, their sequence and resources needed to manufacture it. To encapsulate knowledge of the aforementioned areas, the VEP is designed having the following three main elements or modules (i) Operations, (ii) Resources, and (iii) Experience. Whereas, Virtual engineering factory (VEF) is a knowledge representation of a complete manufacturing process and it is an extension of the VEO-VEP concept to a factory level. The VEF elements include: (i) VEF-Loading/Unloading, (ii) VEF-Transportation, (iii) VEF-Storage, (iv) VEF-Quality Control, (v) VEF-Experience.

The relationship between physical and virtual processes in terms of VEO and VEP is shown in Fig. 1, where three levels of virtual manufacturing are depicted briefly. The bottom level is VEO. It is a representation of an artifact at machine level, such as machining parameters, tolerances, and surface conditions etc.

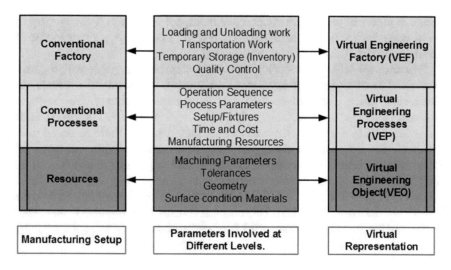

Fig. 1 Relationship between physical and virtual world

The middle-level VEP deals with the information at the process or shop floor level, such as operation sequences, process parameters, time and costs etc. Finally, the top level is VEF. It stores the experience or formal decisions related to the various different aspects involved at the system level, such as material handling, storage quality control, and transportation etc. The technological base for VEO-VEP-VEF is a powerful knowledge representation technique, which is already explained in the previous section i.e. set of experience knowledge structure (SOEKS) and Decisional DNA [9].

3 Product Development by Using SOEKS and Decisional DNA

3.1 Smart Products in Industry 4.0

Industry 4.0 highlights the idea of reliable digitization and linking of all productive units in a manufacturing set-up, and creates a real-world virtualization into a huge information system. It is an integration and assimilation of smaller concepts such as "Cyber-Physical Systems (CPS)", "Internet of things (IoT)", "Internet of services (IoS)", and "Smart Products", etc. [18]. It is not only limited to the direct manufacturing in the company, but also includes a complete value chain from providers to customers. Industry 4.0 is a specialization of IoT, applied to the manufacturing and complete industrial environment. It can be also perceived as a natural transformation of the industrial production systems triggered by the digitalization trend. This hypothesis has been effectively supported by comparison of 'conventional' topics in industrial production systems and Industry 4.0 topics. It is obvious that the main issues/topics did not really change, just the technology and approaches for tackling the connected issues are new [19].

Industry 4.0 will be producing new type of products, which will be smart products. These products are implanted with sensors, identifiable components, and processors which carry information and knowledge to convey the functional guidance for the customers. They will be transmitting the user's feedback to the manufacturing system. With these elements, many functions could be added to the products, for example, measuring the state of products or users, carrying this information, tracking the products, and analyzing the results depending on the information. The smart products are not only smart during the manufacturing process, but they continue to provide the data about their state, knowledge, and experience during their lifetime. This data can be used for preventive maintenance; it can provide the manufacturer useful information about the lifetime and reliability of their products [20].

3.2 Framework for Product Development Process

This framework follows all the basic rules of the traditional Stage-Gate model process used for product development. The Stage-Gate model process can be recognized as a conceptual and operational model employed to move the new product, from idea generation to product launch. Every stage is made as to reduce the uncertainties and risks. At the end of each stage, the Go/Kill gates are given to gauge and decide whether to move to the next stage or not [3]. Framework to support the product development process by using SOEKS and Decisional DNA is shown in Fig. 2.

It consists of user interface, diagnoser, prognoser, solution and decisional DNA. The working detail of each section is explained as follows:

User Interface/Acquisition Phase: In this phase, all relevant information, such as requirements and constraints are collected. This is a particularly difficult stage, as it requires manual inputs from users. It can be a simple query or an innovation, for example, a query for selection of suitable material for a relevant product development, or addition of new materials which can be used in the future.

Diagnoser: It receives data from the user interface and gathers information through various applications. It involves two main activities: problem breakdown and relationship characterization. A large and complex problem is normally decomposed into smaller tractable sub-problems, which can be solved separately and/or in parallel. In order to solve the problem accurately and efficiently, it is also very important to precisely characterize the relationships between input parameters and

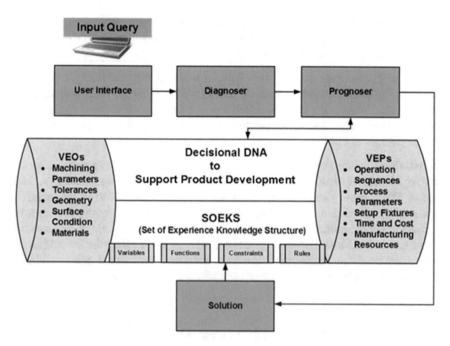

Fig. 2 Framework for product development by using set of experience knowledge structure and decisional DNA

objectives. Therefore, we can say that synthesis for decision-making is developed in this process. Each sub-problem may concern only certain aspects, i.e. material selection process will consider about material related properties, and machine selection process will concern about machining parameters. It is therefore necessary to integrate them to solve the problem as a whole, taking all the objectives and constraints into consideration. It creates a set of experience (SOE) from the query.

Prognoser: Firstly, it creates sub-solutions according to customer's requirements; for example, in material selection, it can require hardness of material, percentage of various metallic ingredients among others. In the next step, it transforms the information into a unified language or a measuring system. It gathers and organizes data to create various sets of experiences, which are described by a specific set of variables, functions, constraints, and rules. After this, it interacts with Decisional DNA to find the similar sets of experiences, and provides actions to be performed. Decisional DNA also contains various VEOs and VEPs of relevant products or products of the same family. Such VEOs and VEPs are some sorts of prepared recipes for incoming enquires; they already contain various successful sets of experiences in them. A new set of models is prepared by considering the measurement of uncertainty, incompleteness, and imprecision. Therefore, the final set of proposed solutions is created and sent to the solution layer of the system.

Solution: The solution layer allows the user to select the best solution out of the proposed solutions. Based on a set of priorities defined by the user; it chooses the best solution among the possible solutions provided. The decisional event is then sent to the Decisional DNA of product development, so that it can be stored for future decisional events.

Decisional DNA of Product Development: It is the brain of the system, where sets of experience are stored and managed. One set of experience represents a Decisional DNA gene. Genes of the same category are grouped together, which is collectively called as a decisional chromosome. A group of such chromosomes, for instance, product chromosome, process chromosome, and technology chromosome constitute a Decisional DNA of product development. It also interacts with other components of the framework during the solution process and presents similar experiences that help in finding a reliable solution in less time.

3.3 Process Flow Chart of Framework to Support Product Development by Using Decisional DNA

How the framework extracts the right knowledge from decisional DNA of product development system is presented in Fig. 3. Design and development of a threading tap (a tool to create screw threads, which is called threading) is chosen as a case study. It gives a clear idea about how knowledge of all the activities within the product development are collected and utilized to provide decisional support throughout the product life cycle. Working phenomena of few steps of the proposed framework are explained below:

Material Selection: In first step, user tells about its requirements regarding material selection, i.e. required hardness, tensile strength and various metallic ingredients etc.

   Proper material selection and its availability check are our first goals; therefore, the information goes into Decisional DNA of product development. It tries to find any appropriate VEO, within the family of the same product. Two events could happen:

(a) It finds the relevant VEO, which consists of all required SOEKS for current enquiry. These VEOs can be used for further action.
(b) It cannot find relevant VEO, so it will search for any appropriate SOEKS. If it finds SOEKS, they will be used to form a new VEO and can be used for current enquiry. It will also store this VEO for future correspondence. In case, it cannot find appropriate SOEKS, the relevant department is informed.

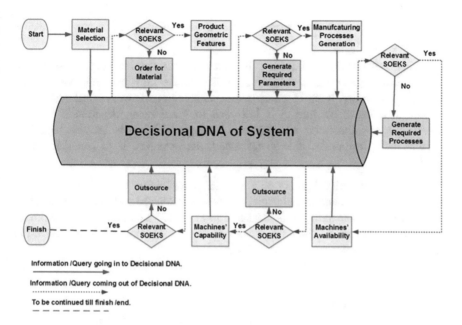

Fig. 3 A process flow chart of framework for explaining an example

It will also look for any available VEP, as all the manufacturing resources need to be known for material selection.

Product Geometric Features: When analyzing the component for design, it needs to be assessed at the system level. This makes it easy to understand its overall functional requirements, its method of operation and the components with which the component of interest interfaces. The functional requirements of the individual features and their physical attributes need to be identified, including the form, the feature relationships, and associations [21]. Therefore keeping this in mind, as the user provides inputs regarding minimum geometric features information, it goes into Decisional DNA of product development and it tries to find any further geometric features within the family of the same product. Two events could happen:

(a) It finds the relevant SOEKS in terms of design parameters. This information is generated on the bases of rules stored in Decisional DNA. These SOEKS can be used for further action.
(b) It cannot find any relevant SOEKS for design parameters, a totally new design is generated and saved for future reference.

Manufacturing Processes Generation: Once the geometric features are generated, the next step is to generate a complete list of manufacturing processes. In case of a screwing thread, we have to see whether it will require casting, forging, molding or machining (turning, milling, grinding, and heat treatment etc.). These parameters can be simply recalled from existing VEP, if no VEP is found, a new SOEKS is generated.

Machines' Selection/Availability: In this step, all the machines required for manufacturing processes which are identified in the previous step, are checked for their availability. If any machine is not available in terms of physical availability or due to a busy schedule, it is informed to the relevant department for outsourcing or any alternative arrangement. A simple VEP can be recalled regarding manufacturing resources, if it is not found, the next step is to generate a new SOEKS.

Machines' Capability: It is not only machine's availability which is enough to be known by designers and manufacturers, they should also be well aware of each and every machine's capability. It is measured in terms of its maximum and minimum limits. If it is not found, the new machine selection information is generated in the form of SOE and is structured as follows:

If selected machine is conventional Lathe Machine, THEN
Available machines' capacity is:

- Maximum Diameter = 250 mm and Maximum Length = 1200 mm.
- Minimum Diameter = 1.5 mm and Minimum Length = 25.4 mm.

Once it is confirmed that product can be easily manufactured in an existing facility, the next steps are to develop new drawings, order for tooling and fixture, selection of measuring equipment and prototype manufacturing respectively.

4   Conclusion and Future Work

In this research, we presented a framework for supporting product development by using SOEKS, Decisional DNA, Virtual Engineering Objects and Virtual Engineering Processes. The system is capable of storing knowledge as well as experiences of past decisional events. Future work will involve the following aspects:

- Refinement of design parameters which can be recalled in terms of variables, functions, constraints, and rules.
- Further development of an approach to show, how manufacturing collective intelligence will be achieved for Cyber-Physical Production Systems (CPSS), operating in Industry 4.0 environment.

References

1. González, F.J.M., Palacios, T.M.B.: The effect of new product development techniques on new product success in Spanish firms. Ind. Mark. Manage. **31**, 261–271 (2002)

2. Starbek, M., Grum, J.: Concurrent engineering in small companies. Int. J. Mach. Tools Manuf. **42**, 417–426 (2002)
3. Cooper, R.G.: Stage-gate systems: a new tool for managing new products. Bus. Horiz. **33**, 44–54 (1990)
4. Härkönen, J.: Improving Product Development Process Through Verification and Validation, vol. C, p. 327 (2009)
5. Ottosson, S.: Dynamic product development—DPD. Technovation **24**, 207–217 (2004)
6. Oduoza, C.F., Xiong, M.: A decision support system framework to process customer order enquiries in SMEs. Int. J. Adv. Manuf. Technol. **42**, 398–407 (2009)
7. Ozer, M.: Information technology and new product development: opportunities and pitfalls. Ind. Mark. Manage. **29**, 387–396 (2000)
8. Group, I.W.: Recommendations for implementing the strategic initiative INDUSTRIE 4.0. Final Report, Apr 2013
9. Shafiq, S.I., Sanin, C., Szczerbicki, E., Toro, C.: Virtual engineering factory: Creating experience base for industry 4.0. Cybern. Syst. **47**, 32–47 (2016)
10. Ernst, H.: Success factors of new product development: a review of the empirical literature. Int. J. Manage. Rev. **4**, 1–40 (2002)
11. Waris, M.M., Sanin, C., Szczerbicki, E.: Framework for product innovation using SOEKS and decisional DNA. In: Asian Conference on Intelligent Information and Database Systems, pp. 480–489. Springer (2016)
12. Bauch, C.: Lean Product Development: Making Waste Transparent (2004)
13. Kušar, J., Duhovnik, J.e., Grum, J., Starbek, M.: How to reduce new product development time. Robot. Comput.-Integr. Manuf. **20**, 1–15 (2004)
14. Cagan, J., Vogel, C.M.: Creating Breakthrough Products: Innovation from Product Planning to Program Approval. Ft Press (2002)
15. Sanin, C., Szczerbicki, E.: Set of experience: A knowledge structure for formal decision events. Found. Control Manage. Sci. 95–113 (2005)
16. Sanin, C., Szczerbicki, E.: Experience-based knowledge representation: SOEKS. Cybern. Syst. Int. J. **40**, 99–122 (2009)
17. Shafiq, S.I., Sanin, C., Szczerbicki, E., Toro, C.: Decisional DNA based framework for representing Virtual Engineering Objects. In: Asian Conference on Intelligent Information and Database Systems, pp. 422–431. Springer (2014)
18. Lee, E.A., Seshia, S.A.: Introduction to Embedded Systems. Cyber-Phys. Syst. Approach **2** (2015)
19. Rojko, A.: Industry 4.0 concept: background and overview. Int. J. Interact. Mobile Technol. (iJIM) **11**, 77–90 (2017)
20. Abramovici, M., Stark, R.: Smart Product Engineering: Proceedings of the 23rd CIRP Design Conference, Bochum, Germany, 11th–13th Mar 2013. Springer Science & Business Media (2013)
21. Urbanic, R.J., ElMaraghy, W.: A design recovery framework for mechanical components. J. Eng. Des. **20**, 195–215 (2009)

Actual Situation and Development in Online Shopping in the Czech Republic, Visegrad Group and EU-28

Libuše Svobodová and Martina Hedvičáková

Abstract The article is focused on the development of the number of people shopping on the Internet. Customer groups will be further analyzed in terms of age, gender and economic activity. The aim of the article is to confirm or refuse one hypothesis and answer three scientific questions. Hypothesis is that there is no gender difference in shopping on the Internet in the Czech Republic. Scientific questions are buying over the Internet grows in total over the past seven years. Young people (16–24 years old) buy over the Internet more than the Czech population older than 55 years. The most economically active group in online shopping is employed people. The hypothesis was confirmed in two analyzed years. One scientific question was positive, two were negative. Since 2014, there has been a decline in the number of students purchasing online. Gender shopping also changed trend in the last monitored period. Secondary data was used for the article.

Keywords Development · Gender · Internet · Online shopping
Statistics · Use

1 Introduction

Online shopping has an increasing tendency in the Czech Republic, but also in other EU countries. With the approaching Christmas time, even the motivation of individual businesses to increase their profits is growing. The Czech Republic has been inspired abroad, especially in the USA. To increase their sales in pre-Christmas time, Free Shipping Days will be available in the Czech Republic. These newly built on the traditional Day of free shipping, which organizes online

L. Svobodová (✉) · M. Hedvičáková (✉)
University of Hradec Králové, Rokitanského 62, 500 03 Hradec Králové, Czech Republic
e-mail: libuse.svobodova@uhk.cz

M. Hedvičáková
e-mail: martina.hedvicakova@uhk.cz

© Springer International Publishing AG, part of Springer Nature 2018   269
A. Sieminski et al. (eds.), *Modern Approaches for Intelligent Information and Database Systems*, Studies in Computational Intelligence 769,
https://doi.org/10.1007/978-3-319-76081-0_23

shopping mentor Heureka.cz. Their start is scheduled for November 13th to November 15th. On days free shipping then, after more than a week to build on-purchase "holidays" imported from the USA and Great Britain. Black Friday this year falls on November 24, Cyber Monday on November 27th.

The Black Friday brand is used in Czech e-commerce quite often, even outside the world's usual autumn term. In the US, this is an unofficial shopping feast, during which the first Christmas discount events start. The Terminus follows Thanksgiving, always the first Friday. Cyber Monday is the equivalent of "Black Friday", which is reserved for online stores abroad, and which falls on the following Monday. Increasingly, however, these two terms coincide in one. Also thanks to these events, the number of Internet purchases in the Czech Republic is growing. Annual revenue growth of e-commerce should be a 15% increase compared to last year [1].

2 Literature Review

The rapid expansion of the Internet has spawned a growing body of literature on the impact of online shopping (here called e-shopping) on physical shopping. Weltevreden's results indicate that in the short run, e-shopping is unlikely to have a significant effect on purchases at city center stores. In the long run, however, e-shopping may well substitute for going to actual stores [2].

In 2015, more than half of Poles took advantage of the electronic shopping form [3]. The reasons for the partial abandonment of purchasing products in a traditional way in favour of on-line shopping are convenience, savings and a richer range of products. The increase in the number of Internet shop users also contributes to a number of promotional activities organized by online shops [4].

Many authors also examine differences in the behavior of men and women when shopping. The topic of shopping orientation is widely analyzed in theoretical and empirical studies of consumer behavior. According to Gehrt et al. [5] shopping orientation describes general consumers' inclination to shopping. It is expressed through several variables such as information search, assessing alternatives, and choice of a product (Brown et al. [6]). Brown et al. [6] claim that shopping orientation also reflects a general shoppers' attitude to the process of shopping [7].

There is a lot of research on online shopping (Al-Maghrabi et al. [8], Limayem et al. [9], Jamil and Mat [10], Orapin [11], Tseng et al. [12], Xie et al. [13], Lim et al. [14]).

Another very important factor is shopping for individual age groups. For examples Hedvicakova and Svobodova [15] or Pascual-Miguel et al. [16].

Internet users who bought or ordered goods or services for private use over the Internet in the previous 12 months by age group is presented in Fig. 1. Figure 1 shows a growing trend for all age groups in EU-28. However, there is a slowdown in growth rates.

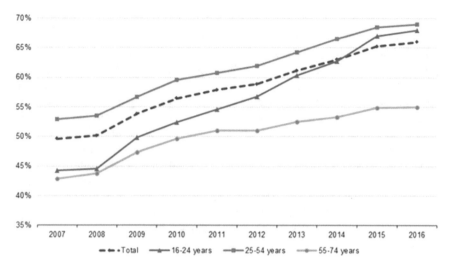

Fig. 1 Internet users who bought or ordered goods or services for private use over the Internet in the previous 12 months by age group, EU 28 [17]

3 Methodology and Goal

The focus of the article is to describe the development and current situation focused on making of the purchases on the Internet by individuals.

The structure of the article is follows. Firstly, it describes the situation throughout the world. The main part of the article is dedicated on selected activities carried out by individuals on the Internet and on individuals making purchases on the Internet. Most of data are taken from the Czech Republic and selected statistics from the Visegrad Group. The goal of the article is to analyze on-line shopping by selected groups of population in the Czech Republic and individuals who ordered goods or services over the Internet for private use in the countries of Visegrad group.

The following hypothesis has been established:

- There is no difference in gender shopping on the Internet in the Czech Republic in the last seven years.

The following scientific questions have been established:

- The volume of Internet purchases has increased in the last seven years in the Czech Republic. At the same time, the percentage of people who shop online is increasing.
- Young people (16–24 years old) purchase over the Internet more than the Czech population older than 55 years.
- The most economically active group in online shopping is employed people.

The article is based on secondary sources. The secondary sources provide information about online shopping and use of the Internet. Information and data were taken from professional literature, information collected from professional press, web sites, discussions and previous participation at professional seminars and conferences related to the chosen subject. The foundation of the information was gained from the Czech Statistical Office, Eurostat and APEK (Association of E-commerce in the Czech Republic). Then it was necessary to select, classify and update accessible relevant information from the numerous published materials that provide the basic knowledge about the selected topic. Data from the Czech Statistical Office was based on the sample survey on ICT usage in households and by individuals, which was obtained from a sample survey of the labour force since 2005. The survey is conducted through personal interviews using a personal computer (Computer Assisted Personal Interviewing—CAPI) from a sample of about 10 000 individuals aged 16 years and over. Conversion of the results to the entire population of the Czech Republic took place consistently with the methodologies of the two abovementioned investigations [18].

4 Results

The next chapter focuses on the two core areas from statistics data done by the Czech Statistical Office and Eurostat that will be analyzed. Firstly, we will focus on selected activities carried out by individuals on the Internet in the Czech Republic, secondly on individuals making purchases on the Internet in the Visegrad group.

4.1 Selected Activities Carried Out by Individuals on the Internet

Most people seek information about goods and services, 68.3% in total in the Czech Republic in 2016. 47.0% of the population are interested in travelling and accommodation, and 42.7% in health. In 2015 was monitored a decrease by 1.2% in finding information about goods and services compared to last year, and even by 4.5% in the search for health compared to 2014. From the perspective of Internet services, most people use the Internet for internet banking, 47.4% in total, and 43.6% of people use the Internet for shopping. The data in Table 1 shows that although over 68.3% of the individuals search information about goods and services, only 43.6% really buy via the Internet. In on-line shopping a slight progression in the monitored period is seen.

Communication is the most often done activity by Czech population on the Internet. 71.8% sending/receiving emails. 41.4% individuals marked in 2016 active participation in social networks. This value is still rising. Searching information on goods and services is the second often carry activity on the Internet by individuals. 68.3% of individuals are looking for those information. Leisure activities, especially

Table 1 Selected activities carried out by individuals on the internet (as a %) [18]

Indicator	2010	2011	2012	2013	2014	2015	2016
Communication							
Social networking	9.3	24.6	31.0	34.3	36.9	37.4	41.4
Searching information on							
Goods and services	49.8	51.6	58.2	60.0	64.4	63.2	68.3
Internet services							
Internet banking	21.0	27.4	32.3	39.3	42.6	44.9	47.4
Internet shopping	25.4	28.0	30.6	34.4	39.3	41.9	43.6
Selling of goods (e.g. in auctions)	7.9	10.1	11.3	15.9	14.3	12.5	11.1

reading online new, newspapers and magazines is in 2016 with the 62.2% the third activity done on the internet. Internet banking is presented by 47.4% and travel and accommodation by 47% followed by Internet shopping with 43.6%. Selling of goods (on auctions) is the least presented activity in the article. In a deeper analysis and use of statistical functions, it was subsequently found that social networking has an interesting trend in the monitored period. This specifically relates to a shift from rapid growth to stagnation. Demand saturation may be the reason of the situation. The logarithmic trend explains 97% of the variance. The logarithmic trend for Internet shopping explains also 97% of the variance. The logarithmic trend for searching information of goods and services explains 92% of the variance.

4.2 Individuals Making Purchases on the Internet

The popularity of on-line shopping continues to grow, which is evident from Fig. 1. In the next section, we will focus on different groups of people shopping via the Internet. In 2016, a total of 43.6% of the Czech population aged 16+ shopped on-line via the Internet. There was monitored the interesting shift in the gender shopping on the Internet in 2016. While in 2015 women bought less than men by 1.4% in 2016 was firstly monitored that woman purchase more on the Internet by 3.7% than in the previous year and by 2.6% more than man. People aged 25–34 years are the most active buying age group, where a total of 72.0% of this group used the Internet for shopping in 2016. The next group conclude by 59.4% people aged 35–44 years and 58.7% was monitored for age 16–24. The purchasing on the internet is falling by age from 25–34 years when it takes a peak. All groups rose in the all monitored years instead of the comparison between 2015 and 2016 for the youngest group aged 16–24 years (see Table 2).

On the basis of the recorded data it is possible to accept the first scientific question. Buying over the Internet grow in total over the past seven years.

Table 2 Individuals making purchases on the Internet (as a %), own creation based on data [18]

Indicator	2010	2011	2012	2013	2014	2015	2016
Total	*25.4*	*28.0*	*30.6*	*34.4*	*39.3*	*41.9*	*43.6*
Males	28.0	29.0	31.5	35.6	40.5	42.6	42.3
Females	23.0	26.9	29.8	33.4	38.1	41.2	44.9
Age group							
16–24 years	38.8	40.5	46.3	53.9	62.2	60.6	58.7
25–34 years	44.5	48.0	54.3	58.3	63.2	66.9	72.0
35–44 years	34.7	39.3	43.1	46.9	52.6	59.2	59.4
45–54 years	22.1	25.3	27.9	32.4	40.1	41.2	46.6
55–64 years	10.5	13.1	15.7	19.8	21.7	25.7	28.3
65+ years	3.0	3.7	3.9	4.5	7.6	8.0	9.7

Gender purchases via Internet

The results will be again analyzed by using statistical methods. There is a statistically significant difference between men and women. Groups show growth with an increase of 2.8% by males and 3.6% by females per annum. The linear trend explains over 95% of the variance, esp. 95% by males and 99% by females (see Fig. 2). The trend has changed in the last year in the comparative period. While men tended to shop online more than women in all monitored years, the trend changed in the last year 2016. Women started shopping more than men.

According to Fig. 2 and Table 3 in 2012 and 2015, the H0 hypothesis was confirmed. It means that men buy on the Internet just as women. In 2010, 2011, 2013, 2014 and 2016, the zero hypothesis was rejected and it was confirmed by H1 that men are buying significantly more than women on the Internet.

Although more men than women (by 5%) purchased in the first year, it was not the same in other years. The difference ranged around 2.4% in 2014 to the lowest value of 1.4% in 2015 and 1.7% in 2012. The change occurred in the last monitored year. Women buy by 2.6% more than man.

Fig. 2 Gender purchasing on the Internet, own creation based on data [18]

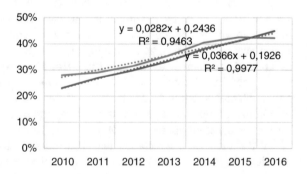

Table 3 Gender purchasing on the Internet and statistics, own creation based on data [18]

Year	2010	2011	2012	2013	2014	2015	2016
p-value	0.000	0.019	0.065	0.021	0.014	0.156	0.009
alfa	0.050	0.050	0.050	0.050	0.050	0.050	0.050

Age groups and purchases via Internet

Regarding age groups the trend is approximately the same. The differences are statistically significant (again due to the sample size). Consistently, the highest values gain the age group 25–34, followed by the 16–24 group and 35–44.

All indicators decrease with age in all analyzed years with one exception. Age group 16–24 is the only one exception that fell with purchasing on the Internet by almost 2% in the last monitored period. In the evaluation of the purchases focused on the individual age groups we can say that outside of one specific group of very young people 16–24 years, which is below the purchase of 25–34 years, online purchasing decrease with age. Thus, the older, the less individual groups of people are buying online. The situation is shown in Fig. 3.

It is possible to confirm the scientific question that young people (16–24 years) purchase via the Internet more than the Czech population older than 55 years.

Economic activity status and purchasing on the Internet

A visible trend is that the most active shoppers on the Internet were women on maternity or parental leave and students followed by employed individuals. Data in 2010 and 2011 for woman on maternity leave were not monitored. Another finding is that the employed bought more than the unemployed in 2015 by 19.5% and in 2016 rose the difference on 24.5% (see Fig. 4).

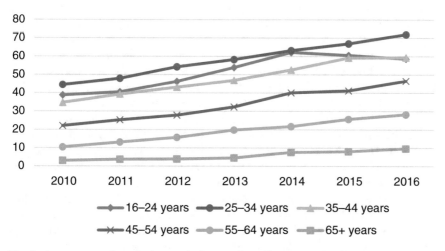

Fig. 3 Age groups and purchasing on the Internet (as a %), own creation based on data [18]

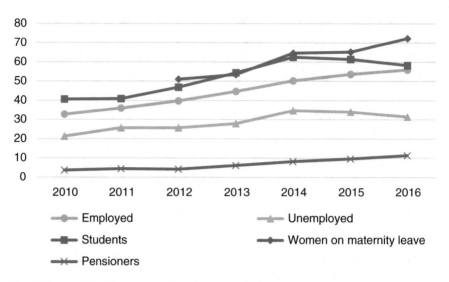

Fig. 4 Economic activity status and purchasing on the Internet (as a %), own creation based on data [18]

4.3 Individuals Who Ordered Goods or Services Over the Internet for Private Use in Visegrad Group and EU-28

Individuals who ordered goods or services over the Internet for private use in 2012–2016 are presented in Fig. 5. Data from EU-28 and countries concluded in the Visegrad Group were involved into the comparison. These results also display that on-line purchases in each monitored year gaining in popularity between individuals. The most active buyers via the Internet in the whole period are Slovaks, followed by Czechs, Poles and Hungarians. Values of ordered goods or services are in all countries below EU-28. The biggest shift was recorded with 12% increase in the Czech Republic in the whole period. The smallest change was recorded with 5% in Slovakia and with 7% in Poland. Despite was in Hungary monitored increase by 11% it is still with 36% in 2015 on the last position in the comparison.

The largest share of the online purchases in the last 12 months was recorded in Europe by United Kingdom. 83% of all individuals purchase online at least once in the year. United Kingdom was followed by Denmark with 82%. Luxembourg and Norway recorded 78%. In Sweden was the value a little bit smaller, 76%. Germany and Netherlands gained 74%. All other countries recorded less than 67% that monitored Finland. Conversely, the least on-line shopping was recorded in 2016 in Romania with 12%, Macedonia 15%, Bulgaria and Turkey with 17% and Cyprus and Italy monitored 29% followed with 31% in Portugal and Greece [19].

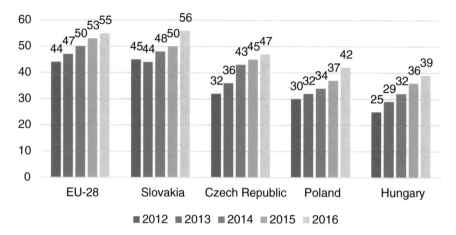

Fig. 5 Individuals who ordered goods or services over the Internet for private use in the 12 months prior to the survey from 2012 to 2016 (as a %), own creation based on data [19]

5 Discussion and Conclusion

A peep into the exponential growth of the main players in this industry indicates there is still a large reservoir of market potential for e-commerce. The conveniences of online shopping make it an emerging trend among consumers, especially Gen Y [21].

This trend has also been confirmed in the Czech Republic. The most active buyers via the Internet in the whole period in Visegrad Group are Slovaks, followed by Czechs, Poles and Hungarians. Values of ordered goods or services are in all countries below EU-28. Sales of e-shops in the Czech Republic over the 2011–2017 period have an increasing tendency. E-shop sales in 2017 will be record-breaking to 115 billion CZK [1]. The first scientific question was confirmed. The growth rate of % of the shoppers' population is slowing, but the overall volume of sales is growing faster. There was monitored the interesting shift in the gender shopping on the Internet in 2016. While in 2015 women bought less than men by 1.4% in 2016 was firstly monitored that woman purchase more on the Internet by 3.7% than in the previous year and by 2.6% more than man. Second scientific question, that young people (16–24 years old) purchase over the Internet more than the Czech population older than 55 years, was confirmed. But only young people (16–24 years old) decreased purchases on the Internet in 2016. The third scientific question that the most active group in online shopping is employed people was not confirmed. The most economically active group in online shopping are women on maternity leave and students. However, the number of student online shoppers is declining since 2014. The hypothesis, that there is no gender difference in the online shopping on the Internet in the Czech Republic, was accepted only in 2012 and 2015.

The question in the discussion is whether to continue the trend of a decreasing number of students' shoppers on the Internet. Due to aging of the population that use ICT in personal life or at work, the question is whether the number of people shopping on the Internet in older age groups will gradually increase in next years.

Acknowledgements The paper was written with the support of the specific project 2018 granted by the FIM UHK, Czech Republic and thanks to help students Veronika Domšová and Martin Král.

References

1. APEK (2017) https://www.apek.cz/. Accessed 21 Oct 2017
2. Weltevreden, J.W.J.: Substitution or complementarity? How the Internet changes city centre shopping. J. Retail. Consum. Serv. **14**(3), 192–207 (2007)
3. Gemius dla e-Commerce Poland: E-commerce in Poland (2015). https://www.gemius.pl/files/reports/E-commerce-w-Polsce-2015.pdf. Accessed 21 Oct 2017
4. Małecki, K., Wątróbski, J.: The classification of internet shop customers based on the cluster analysis and graph cellular automata. Proced. Comput. Sci. **112**, 2280–2289 (2017)
5. Gehrt, K.C., Guvenc, G.A., Lawson, D.: A factor-analytic examination of catalog shopping orientations in France. J. Euromark. **2**, 2, 49–69 (1992)
6. Brown, M., Pope, N., Voges, K.: Buying or browsing? An exploration of shopping orientations and online purchase intention. Eur. J. Mark. **11**(12), 1666–1684 (2003)
7. Banytė, J., Rūtelionė, A., Jarusevičiūtė, A.: Modelling of male shoppers behavior in shopping orientation context. Proced. Soc. Behav. Sci. **213**, 694–701 (2015)
8. Al-Maghrabi, C., Dennis, S.V,. Halliday.: Antecedents of Continuance Intentions towards e-shopping: the case of Saudi Arabia. J. Enterp. Inf. Manage. **24**(1), 85–111 (2011)
9. Limayem, M., Khalifa, M., Frini, A.: What makes Consumers Buy from Internet? A Longitudinal Study of Online Shopping, pp. 421–432 (2000)
10. Jamil, N.A., Mat, N.K.: To Investigate the drivers of online purchasing behavioral in malaysia based on the theory of planned behavior (TPB): a structural equation modeling (SEM) approach. In: International Coference On Management, pp. 453–460 (2011)
11. Orapin, L.: Factors influencing internet shopping behavior: a survey of consumers in Thailand. J. Fash. Mark. Manag. **13**(4), 501–513 (2009)
12. Tseng, Y.F., Lee, T.-Z., Kao, S.-C., Wu, C.: An extension of Trust and Privacy in the Initial Adoption of Online Shopping: An Empirical Study, pp. 159–164 (2011)
13. Xie, G., Zhu, J., Lu, Q., Xu, S.: Influencing Factors of Consumer Intention towards Web Group Buying, pp. 1397–1401 (2011)
14. Lim, Y.J., Osman, A., S.N., Romle, A.R., Abdullah, S.: Factors influencing online shopping behavior: the mediating role of purchase intention. Proce. Econ. Financ. **35**, 401–410 (2016)
15. Hedvicakova, M., Svobodova, L.: internet use by elderly people in the Czech Republic. In: Book: Digital Nations—Smart Cities, Innovation, and Sustainability, pp. 514–524 (2017)
16. Pascual-Miguel, F.J., Agudo-Peregrina, A.F., Chaparro-Peláez, J.: Influences of gender and product type on online purchasing. J. Bus. Res. **68**(7), 1550–1556 (2015)
17. Eurostat: E-commerce statistics for individuals (2017). http://ec.europa.eu/eurostat/statistics-explained/index.php/E-commerce_statistics_for_individuals. Accessed 21 Oct 2017
18. Czech statistical office: Statistical Yearbook of the Czech Republic—2016, 21. Information society. https://www.czso.cz/csu/czso/21-information-society. Accessed 15 Oct 2017

19. Eurostat: Information society—households and individuals (2017). http://ec.europa.eu/eurostat/web/digital-economy-and-society/data/database. Accessed 15 Oct 2017
20. Eurostat: E-commerce statistics for individuals (2017). http://ec.europa.eu/eurostat/statistics-explained/index.php/E-commerce_statistics_for_individuals. Accessed 15 Oct 2017
21. Goyal, A.: A study of psychological perspective of customers w.r.t. rising digital retailing. Int. J. Recent Sci. Res. **8**(6), 17708–17718 (2017)

How Product Brand Effects Consumer Decision

Vaclav Zubr, **Hana Mohelska** and **Marcela Sokolova**

Abstract The aim of the paper is to evaluate the effect of product brand on consumer decision regarding purchase and in connection with this, to answer the specified research questions using statistical methods. The partial goal is to find out what influence a close person has on consumer decision-making. The theoretical part of the study defines the basic concepts of the given issue. The application part of the paper is first devoted to the characteristics of the analysed markets. The main part of the paper is our own research focused on the brands of laptops, which was conducted through online polling in the Czech Republic and Taiwan. The data obtained from respondents in the 19–26 age group is then analysed and compared. The data showed that there is a relationship between the brand that the consumer owns and whether it is the first brand that comes to his mind. There are some other interesting results such as that the "high cost = high quality" opinion is not related to the results of the survey.

Keywords Taiwan · Czech republic · Laptop · Product brand
Consumers

1 Introduction

The Internet has been a part of the life of almost everyone throughout the world in the last few years. The available statistics from 2017 show that approximately 50% of Internet users come from Asia, 17% from Europe [1]. In 2016, 71% of the population of the European Union connected the Internet every day, 59% of the same population used mobile devices to connect to the Internet. The most popular mobile devices were mobile phones, laptops and tablets [2].

V. Zubr (✉) · H. Mohelska · M. Sokolova
Faculty of Informatics and Management, University of Hradec Kralove,
Rokitanskeho 64, 500 03 Hradec Kralove, Czech Republic
e-mail: vaclav.zubr@uhk.cz

© Springer International Publishing AG, part of Springer Nature 2018
A. Sieminski et al. (eds.), *Modern Approaches for Intelligent Information
and Database Systems*, Studies in Computational Intelligence 769,
https://doi.org/10.1007/978-3-319-76081-0_24

Each individual's behaviour is influenced by their upbringing, their surroundings and their own opinions. Very important is also the environment in which the individual lives as well as the family and friend relationships and social and economic events taking place in their surroundings (factors from cultural, social, personal and psychological categories [3–5].

Recently, the use of laptops and mobile phones mostly ensure to young individuals the connection with friends and workplace, laptops are also needed when use smart connected devices [6–8]. According to the Czech Statistical Office sources, 76.5% of Czechs aged over 16 use the Internet and two thirds connect via a laptop [9]. According to Consumer Barometer survey for 2016, the most numerous group that uses these devices are people under the age of 25. When purchasing a laptop, Czechs mostly give consideration to two brands (31%) [10].

In Taiwan, 83.8% of the population use the internet [11]. Specifically, for example students can tablet devices and laptop devices for effective mobile learning [12]. Laptops are also still one of the most productive devices for work out of the office [13]. According to the Consumer Barometer survey results in Taiwan, in 2016, 69% of consumers used a computer and most users are also under the age of 25. When purchasing a laptop, Taiwanese mostly give consideration to two brands (32%) or three brands (22%) [10].

In terms of laptops, some of their manufacturers (1st Apple, 4th Microsoft, 7th Samsung, 48th HP, 58th Sony, and 99th Lenovo) have also been ranked among the world's most valuable brands for 2016, brands constructed by Interbrand [14]. This data can be compared with IDC's 2015 results, and according to those the Lenovo PC brand is the world-wide brand, another best-selling brand is HP, Dell is the third best-selling brand. Apple is directly after and the fifth best-selling brand is Acer [15].

For the purposes of our study, the Czech Republic was elected as the representative of European Internet users, Taiwan being elected as the representative of Asia. These two countries were elected due to the long-standing cooperation between the University of Hradec Kralove and three Taiwanese universities: Chang Jung Christian University, National Taiwan University of Science and Technology, National Chin-Yi University of Technology.

1.1 Czech Republic

The population of the Czech Republic in 2017 is 10,555,815. The Czech Republic's currency is the Czech crown. The official language is Czech. In 2017, the number in the economically active population is 76.1% and the unemployment rate in April 2017 is 3.3%. The average wage for the first quarter of 2017 is 27,889 CZK [9]. The Czech Republic is characterised by the automotive industry (Škoda Auto, TPCA).

1.2 Taiwan

The population of Taiwan in 2017 is 23,405,309. The state currency is the Taiwanese dollar—the average rate is: 1 TWD = 0.8179 CZK. The official language is Mandarin Chinese. In 2015, the number in the economically active population is 11,200,000 and the unemployment rate is 3.78%, which has had decreasing trend in the last five years. The average wage for 2015 is 48,490 TWD (= 40,523 CZK). Taiwan "is one of the world's largest suppliers of computer chips, LCD and OLED panels, DRAM computer memory, devices for electronic networks and consumer electronics." [16] Laptop brands such as Acer, Asus, GIGABYTE and UMAX also come from Taiwan.

2 Methodology

The study is aimed at respondents in the 19–26 age group, where two countries (Czech Republic and Taiwan) are first analysed individually and then the comparison of these two countries is carried out.

Based on the literature research [4, 6–8, 12, 13] the questionnaire was compiled. The Czech questionnaire was distributed in the Czech language. The Taiwanese questionnaire was distributed in English. To ensure the validity of the questionnaire, translation to the English and back-translation to Czech language was performed by two independent translators. The distribution of the questionnaire in Chinese (official language in Taiwan) was not possible due to the lack of translators in Chinese.

The aim of this paper is to assess the impact of the product brand and close person on consumer purchasing decisions. The impact of close person can be including answers to the following research questions:

1. Is there a dependency between gender and influence by a negative attitude from a close person in connection with buying a laptop?
2. Is there a dependency between earnings and "high cost = high quality" opinion?
3. Is there a relationship between the brand that the consumer owns and whether it's the first brand that comes into their mind?
4. Do the Czechs and Taiwanese differ in the importance of branding when choosing a laptop?
5. Is there a dependency between nationality and the influence of a negative attitude from a close person in connection with buying a laptop?

In the application part the market research was carried out, namely online polling. The questionnaire was created using Google application in both Czech and English and was distributed in the Czech Republic and Taiwan. First, the pilot study was carried out among 10 Czech and 10 Taiwanese students. After pilot study was the Czech questionnaire disseminated through the Facebook social network on

University of Hradec Kralove groups, other universities and companies [17]. It was filled out by a total of 586 respondents. The Taiwanese questionnaire was also distributed via the Facebook social network through four Czech colleagues currently studying at Taiwan University or participating in a Taiwanese exchange stay. The questionnaire was further disseminated through the Interpals.net site. In total, we were able to get answers from 163 respondents. For the evaluation of five research questions, statistical tests were used—the $\chi2$ Independence Test and the Mann-Whitney Test. The significance level was set to $\alpha = 0.05$ for all tests.

3 Results

A total of 447 respondents from the Czech Republic (256 women, 191 men) and 121 respondents from Taiwan (91 women, 30 men) aged 19–26 participated in the survey. Most respondents from Czech Republic were those with a secondary school graduation with a leaving examination (57%) and with a university education (40%). Most respondents from Taiwan were with a university education (75%), university education with a higher qualification, respondents with higher professional education or secondary education with leaving examination. In contrast to the Czech Republic, there were no respondents with basic education or an apprenticeship certificate. In the Czech Republic, the distribution of respondents according to the income group is also responsible for the large number of students, with the largest number of respondents with a net income of 5,001–10,000 CZK per month (27.5%). Then the respondents with the income of up to 5,000 CZK a month (25.7%) and respondents with income between 10,001–20,000 CZK a month (21.5%) are the most represented. 2% of respondents declined to answer this question.

The Taiwanese respondents in many cases have no income (38.8%) or have income (equivalent to the Czech crown) of 16,717–25,074 CZK per month (14.9%) or 16,716 CZK per month (33%). 6.6% of respondents declined to answer this question.

A laptop is owned by 96% of Czech respondents and 90% of Taiwanese respondents.

The most common brand among Czechs is Lenovo, followed by Acer and HP. Taiwanese most often own brands such as Asus, Acer and Apple. Only 3% of the respondents from both countries aren't happy with their laptop. Most commonly, they are owners of Acer, Asus and MSI brands. Half of the Czechs and 74% of Taiwanese also buy or usually buy other products of the same brand as their laptop. Czechs and Taiwanese most often mark the Apple brand as a symbol of quality.

Czechs ranked the product quality (97%), followed by positive reviews and customer support among the three most important aspects that build the good brand name of a laptop. Taiwanese also chose product quality (94%), customer support and quality added value services.

The most important aspects influencing Czech respondents in selecting a laptop include the quality of the laptop, their own experience and reviews. For Taiwanese respondents, it's also the quality of the laptop, their own experience and after-sales service. The brand affects or usually affects 47% of Czechs and 53% of Taiwanese.

3.1 Evaluation of the Research Questions

Selected tests in the SPSS statistical program were used to evaluate the research questions. For each research question, first there is a classical pivot table, which is compiled on the basis of input data and supplemented by sums, followed by a table with the research questionnaire solution with the results of the respective test.

Is there a dependency between gender and influence by a negative attitude from a close person in connection with buying a laptop?

It's possible that women are more likely to be influenced by a negative attitude from a close person than men. To evaluate this question, the χ^2 independence test developed in SPSS statistical program and data from Czech respondents were used.

The p-value 0.004 at the significance level of 0.05 confirms the relationship between gender and a negative attitude influence from a close person when selecting a laptop brand (Tables 1 and 2).

Is there a dependency between earnings and "high cost = high quality" opinion?

The respondents, who had no income and who didn't want to report their income, were removed. The other respondents were divided into two income groups: up to 10,000 CZK per month and above 10,001 CZK per month. The χ^2 independence test developed in SPSS statistical program and data from Czech respondents were used (Tables 3 and 4).

Table 1 Solving research question No. 1 (1)

		Yes	Preferably yes	Preferably no	No	Total
Gender	Men	13	66	84	28	191
	Women	23	127	83	23	256
Total		36	193	167	51	447

Table 2 Solving research question No. 1 (2)

	Value	df	Asymptotic significance (2-sided)
Pearson chi-square	13,385[a]	3	0.004
Likelihood ratio	13,440	3	0.004
N of valid cases	447		

[a]0 cells (0.0%) have an expected count less than 5. The minimum expected count is 15.38

Table 3 Solving research question No. 2 (1)

		Yes	Preferably yes	Preferably no	No	Total
Income group	Up to 10 000 CZK	16	106	95	21	238
	Over 10 001 CZK	9	65	55	15	144
Total		25	171	150	36	382

Table 4 Solving research question No. 2 (2)

	Value	df	Asymptotic significance (2-sided)
Pearson chi-square	0.347[a]	3	0.951
Likelihood ratio	0.345	3	0.951
N of valid cases	0.382		

[a]0 cells (0.0%) have an expected count less than 5. The minimum expected count is 9.42

The p-value 0.951 at the significance level of 0.05 does not confirm that the "high price = high quality" opinion depends on the amount of income.

Is there a relationship between the brand that the consumer owns and whether it is the first brand that comes into their mind?

It can be assumed that consumers will first think of the laptop brand which they currently own. For the evaluation, the respondents who didn't own a laptop, were removed, and only six of the most common brands owned by Czech respondents: Lenovo, Acer, HP, Asus, Apple, and Dell were used. To evaluate this question, the χ^2 independence test developed in SPSS statistical program and data from Czech respondents were used (Tables 5 and 6).

The p-value 0.001 at a significance level of 0.05 confirms the strong dependence between the brand that the consumer owns and whether it's the first brand that comes into their mind.

Table 5 Solving research question No. 3 (1)

		Yes	No	Total
Owner of a laptop brand of:	Lenovo	89	27	116
	Acer	40	38	78
	HP	60	16	76
	Asus	51	20	71
	Apple	28	14	42
	Dell	18	13	31
Total		286	128	414

Table 6 Solving research question No. 3 (2)

	Value	df	Asymptotic significance (2-sided)
Pearson chi-square	20,333[a]	5	0.001
Likelihood ratio	19,825	5	0.001
Linear-by-linear association	0.892	1	0.345
N of valid cases	414		

[a]0 cells (0.0%) have an expected count less than 5. The minimum expected count is 9.58

Table 7 Solving research question No. 4 (1)

	Country	N	Mean rank	Sum of ranks
How does the brand affect you when choosing a laptop?	Czech Republic	447	283.11	126,550.00
	Taiwan	121	289.64	35,046.00
	Total	568		

Do the Czechs and Taiwanese differ in the importance of branding when choosing a laptop?

To evaluate this question, the Mann-Whitney test developed in the SPSS statistical program was used (Tables 7 and 8).

The p-value 0.683 at a significance level of 0.05 does not confirm a statistically significant difference between Czechs and Taiwanese regarding the importance of brand when selecting a laptop.

Is there a dependency between nationality and the influence of a negative attitude from a close person in connection with buying a laptop?

The $\chi 2$ independence test developed in the SPSS statistical program and data of Czech and Taiwanese respondents were used (Tables 9 and 10).

The p-value 0.000 at a significance level of 0.05 confirms the strong dependence between nationality and a negative attitude from a close person. The results suggest that the Taiwanese are influenced more than the Czechs.

Table 8 Solving research question No. 4 (2)

	How does the brand affect you when choosing a laptop?
Mann-Whitney U	26,422.00
Wilcoxon W	126,550.000
Z	0.408
Asymp. Sig. (2-tailed)	0.683

Table 9 Solving research question no. 5 (1)

		Yes	Preferably yes	Preferably no	No	Total
Country	Czech Republic	36	193	167	51	447
	Taiwan	31	50	19	21	121
Total		67	243	186	72	568

Table 10 Solving research question No. 5 (2)

	Value	df	Asymptotic significance (2-sided)
Pearson chi-square	41,282[a]	3	0.000
Likelihood ratio	39,243	3	0.000
N of valid cases	568		

[a]0 cells (0.0%) have an expected count less than 5. The minimum expected count is 14.27

4 Discussion

Nowadays, consumers are overwhelmed by the large number of brands and products on the market. They are primarily influenced by their individual attitudes, accepted values and the environment in which they grow up and live. However, the perceived value of the trademark also plays an important role, which helps makers to influence consumers' perceptions and attitudes towards the brand, it's necessary to understand the consumer's purchasing behaviour and buyer's decision-making process. However, not only the brand, but also the price, friends, past experiences and the warranty as well as laptop parts can contribute to the perception of the quality of the laptop [18].

The aim of this work was to assess the impact of a product brand and close person on consumer purchasing decisions, including answers to the research questions. Since there was overwhelming representation of students in both groups of respondents, it was possible to compare these groups. The results show that 96% of Czech respondents and 90% of Taiwanese own a laptop, which positively influences the survey's worthiness. A number of Taiwanese laptop users may reflect the production (and therefore better availability) of some laptop brands in Taiwan. The most common brand that Czechs own is Lenovo, Acer and HP. Taiwanese most often own brands such as Asus, Acer and Apple. As can be seen, the Taiwanese support local production. These results are in line with the 2014 study, according to which most consumers preferred buying laptops from local retailers [19]. In another studies focused on brand awareness and purchase intention have been shown that the high awareness of brand name lead to increase of brand loyalty and customers will buy the well-known brand rather than the unknown one [20, 21]. More than half of Czechs are loyal to their laptop brand. The respondents would probably not have bought Acer again, which may be related to possible negative experiences with this brand and with customers' expectations that exceed product performance [22]. As shown by the labelled best laptop brands for 2017, Acer products often gain a mid-range rating [23]. More than half of Taiwanese are also loyal to their laptop brand, the most loyal are owners of the Apple brand. Toshiba owners wouldn't buy this brand again. When compared to another studies dealing with the factors affecting the choice of the car or cellular phone, the results are similar [20, 24].

It has been confirmed that the "high price = high quality" view doesn't depend on the level of income. This result may also be related to the results of other studies which have shown that consumers tend to use price as a quality indicator, especially if they're not sure about the product's purchase or its quality. It should be noted that price influence as a quality indicator is different, depending on the presence of other factors [25].

When selecting a laptop, the Taiwanese can be more influenced by a close person when selecting a laptop [26].

We can consider the distribution of the questionnaires between the Taiwanese in English language and not in their native language as limitation of this study.

5 Conclusion

Consumers currently have the choice of many laptop brands that are rated at different levels. Producers try to advertise their products as the best; they can also use a well-known person to promote their product. The question remains, how strong product brand influence on consumer decision-making. The research suggests that other attributes of a laptop are far more important for respondents than the brand. Despite the fact that this influence is very individual, the brand does have certain influence.

Interbrand annually announces the 100 most valuable brands in the world. For the year 2016, the Apple brand, which is valued at 178.1 billion US dollars, was ranked in first position [14]. Taiwanese respondents most often recalled this brand first, and the respondents of both countries most often identified it as a quality symbol, and they also chose this brand as the brand with the most compelling logo. The research [27] conducted by Tomáš Kovářík in connection with the bachelor's thesis shows that Apple's consumers refer to the processing of their products as high quality.

The best-selling brand for 2015, according to IDC, was the Lenovo brand whose total market share is 20.7%. The Czech respondents most often recalled this brand first and also this is the most common brand of laptop owned by Czechs. Other best-selling brands include: HP, Dell, Apple and Acer.

At the same time, there is a relationship between the brand that the consumer owns and whether it is the first brand that comes to mind. Also, no difference was found between the Czech and Taiwanese respondents in terms of brand importance when selecting a laptop. The amount of income and the "high cost = high quality" opinion are not related according to the results of the survey.

Consumer decision-making could make be easier with use of decision support systems to help the consumer evaluate individual selection criteria and help him to choose a laptop brand. Application of these methods could be a subject of future research.

In the future, it would certainly be appropriate to carry out similar quantitative research on a representative sample of the entire population.

Acknowledgements The paper was written with the support of the Specific project 2018 granted by the University of Hradec Králové, Czech Republic and thanks to help of student Iveta Sixtová.

References

1. Miniwats Marketing Group: Taiwan (2017). http://www.internetworldstats.com/asia/tw.htm. Accessed 15 July 2017
2. Eurostat: Digital Economy and Society Statistics—Households and Individuals (2017). http://ec.europa.eu/eurostat/statistics-explained/index.php/Digital_economy_and_society_statistics_-_households_and_individuals. Accessed 2 Aug 2017
3. Kotler, P., et al.: Moderní Marketing. Grada Publishing, Praha (2007)
4. Dědková, J., Honzáková, I.: The customer behavior analysis in the Czech-German part of the Euroregion Neisse-Nisa-Nysa. Sci. J. Econ. Manage. Trade České Budějovice **13**(4), 13–20 (2010)
5. Pescher, Ch., Reichhart, P., Spann, M.: Consumer decision-making processes in mobile viral marketing campaigns. J. Interact. Mark. **28**(1), 43–54 (2014)
6. Belgiawan, P.F., Schmöcker, J.D., Fujii, S.: Understanding car ownership motivations among Indonesian students. Int. J. Sustain. Trans. **10**, 295–307 (2014)
7. Mohelska, H., Sokolova, M.: Smart, connected products change a company's business strategy orientation. Appl. Econ. **48**(47), 4502–4509 (2016)
8. Van, H.T., Fujii, S.: A cross Asian country analysis in attitudes toward car and public transport. J. East. Asia Soc. Trans. Stud. **9**, 411–421 (2011)
9. Czech Statistical Office (2017). https://www.czso.cz. Accessed 20 July 2017
10. Google: Consumer Barometer with Google (2017). https://www.consumerbarometer.com/en/. Accessed 30 July 2017
11. Miniwats Marketing Group: Internet Usage and Population Statistics (2017). http://www.internetworldstats.com/stats.htm. Accessed 9 Aug 2017
12. Cheung, S.K.S.: A Study on the use of mobile devices for distance learning. In: International Conference on Hybrid Learning, Guangzhou, China 13–15 Aug 2012
13. MakeUseOf: 4 Reasons Why You Don't Need a Laptop Anymore (2017). http://www.makeuseof.com/tag/4-reasons-dont-need-laptop-anymore/. Accessed 15 July 2017
14. Interbrand: Best Global Brands 2016 Rankings. (2017). http://interbrand.com/best-brands/best-global-brands/2016/ranking/. Accessed 12 July 2017
15. IDC: PC Market Finishes 2015 as Expected, Hopefully Setting the Stage for a More Stable Future, According to IDC (2016). https://www.idc.com/getdoc.jsp?containerId=prUS40909316. Accessed 12 July 2017
16. Czech Economic and Cultural Office Taipei: Taiwan (2017). http://www.mzv.cz/taipei/cz/second_article$1325.html?action=setMonth&year=2017&month=2&day=1. Accessed 15 July 2017
17. Tahal, R., Formánek, T., Mohelska, H.: Loyalty Programs and Personal Data Sharing Preferences in the Czech Republic. E + M Ekonomie a Management **20**(1), 187–199 (2017)

18. Nazari, M., Arab, R.R.: The influence of external signals on perceived quality and purchase intention of high involvement products (case study: laptop). New Mark. Res. J. **4**(2), 223–241 (2014)
19. Arora, R., Chawla, A.: Mapping of consumer perceptions for laptops: a case study. Int. J. Adv. Res. Manage. Soc. Sci. **3**(7), 357–372 (2014)
20. Chi, H.K., Yeh, H.R., Yang, Y.T.: The impact of brand awareness on consumer purchase intention: the mediating effect of perceived quality and brand loyalty. J. Int. Manage. Stud. **4** (1), 135–144 (2009)
21. Shahid, Z., Hussain, T., Zafar, F.: The impact of brand awareness on the consumers' purchase intention. J. Acc. Mark. **6**(1), 223 (2017)
22. Zhang, Y.: The impact of brand image on consumer behavior: a literature review. Open J. Bus. Manage. **3**(1), 58–62 (2015)
23. Laptop: Best & Worst Laptop Brands 2017 (2017). https://www.laptopmag.com/articles/laptop-brand-ratings. Accessed 9 Aug 2017
24. Šefara, D., Franěk, M., Zubr, V.: Socio-psychological factors that influence car preference in undergraduate students: the case of the Czech Republic. Technol. Econ. Dev. Econ. **21**(4), 669–685 (2015)
25. Shirai, M.: Impact of "High Quality, Low Price" appeal on consumer evaluations. J. Promot. Manage. **21**(6), 776–797 (2015)
26. Gajjar, N.B.: Factors affecting consumer behavior. Int. J. Res. Humanit. Soc. Sci. **1**(2), 10–15 (2013)
27. Kovářík, T.: Vnímání značky Apple. Bachelor´s thesis, Tomas Bata University in Zlín (2015)

Investments Decision Making on the Basis of System Dynamics

Galymkaiyr Mutanov, Marek Milosz, Zhanna Saxenbayeva and Aida Kozhanova

Abstract The rapid increase in the volume of incoming and processed information in oil companies has led to a change not only in the automation of the process of data processing and research, but also in the intellectualization of informational and organizational processes, building and implementing effective methods and intellectual supporting technologies of decision-making. Oil companies have always paid great attention to making the scientific and reasonable decisions about the investment scale and structure in the extraction sector to enable them to minimize business risks and make high profit. According to the theories and methods of system dynamics, a dynamic model for analyzing forecasting the scale and structure of investments for the oil industry has been built and presented in this article. As well as the problem of data extraction in intellectual information systems is described. The formulated model can be applied to analyze and predict the structure and size of the investment process as a new method and provide a basis for decision-making.

Keywords System dynamics · Oil company · Investment · Causal loop diagrams

G. Mutanov · Z. Saxenbayeva · A. Kozhanova (✉)
Al-Farabi Kazakh National University, Almaty, Kazakhstan
e-mail: aida_8304@bk.ru

G. Mutanov
e-mail: rector@kaznu.kz

Z. Saxenbayeva
e-mail: zhanna.saksenbaeva@kaznu.kz

M. Milosz
Lublin University of Technology, Lublin, Poland
e-mail: m.milosz@pollub.pl

© Springer International Publishing AG, part of Springer Nature 2018   293
A. Sieminski et al. (eds.), *Modern Approaches for Intelligent Information and Database Systems*, Studies in Computational Intelligence 769,
https://doi.org/10.1007/978-3-319-76081-0_25

1 Introduction

The main purpose of information systems is the timely presentation of the necessary information to decision-makers for making effective decisions when managing with investment capital. However, during the process of the development of information technologies, intellectual informational systems have taken on a significant part of routine operations, as well as the functions of preliminary analysis and assessments. A decision-making based on the analysis of data, their behavior over time only improves and increases the effect, especially in investing.

The basis of the information system is consisted of blocks "database—DB", "rule base", "machine of logical inference—MLI". The database stores the original data. The knowledge and experience of the expert are fixed in the rule base. MLI outputs the result, interacting with the database and rule base. All three blocks should be described mathematically [1]. Therefore, the methodology of system dynamics, allowing to model complex systems at a high level of abstraction, without taking into account small details: the individual properties of individual products, events or people is the basis of an intellectual information system.

System dynamics (SD) was first developed in the late 1950s at the Massachusetts Institute of Technology under the leadership of Jay Wright Forrester in 1958 [2]. In his work, the author computatively analyzed the supply chain of the whole system. This system consists of three inventories (factory, distributors and retailers), as well as several ordering and delivery processes [3]. The system dynamics method is often chosen for complex systems, since this method is specially developed for modeling and studying complex systems with multi-parametric, nonlinear and dynamic characteristics [4]. This refers to the dynamic problems arising in complex social, managerial, economic or ecological systems, literally any dynamic systems characterized by interdependence, mutual interaction, information feedback and circular causality. The work of Jay Wright Forrester "Industrial Dynamics" [5] is still a significant exposition of philosophy and methodology in this field. At the present time, the system dynamics is applied in the economy, public policy, environmental research, protection, the construction of theory in the social sciences and other areas of management.

In more recent modeling environments, more complex integration schemes are available (although the equation written by the user may look like a simple Euler integration scheme), and temporary scenarios may not be in evidence. Important modeling environments include Vensim, Stella, iThink, PowerSim and AnyLogic.

Diagrams of feedback loops with information and circular causality are tools for conceptualization of a complex system structure and transmission of data based on the model. In the methodology of the dynamic system, a problem or a system (e.g., ecosystem, political system or mechanical system) is first represented as a causal loop diagram [6]. The causal loop diagram is a simple map of the system with all its constituent components and their interactions. Capturing interactions and consequently the feedback loops, causal loop diagram reveals the structure of the system.

Understanding the structure of the system, it becomes possible to determine the behavior of the system over a certain period of time [7].

2   Prehistory and Related Work

2.1   Modeling in the Oil Industry

In this part, we will give an overview of a wide range of different models based on fossil resources based on system dynamics. Ford [8] best demonstrates the importance of models of fossil energy resources for the United States. He uses system dynamics to model and influence the strategic fuel reserve in California. As part of this study, he describes the simulation analysis that he developed for the California Energy Commission. The analysis of the simulation estimates the impact of the strategic fuel reserve (SFR), designed to limit the growth of gasoline prices during the days following the disruption of oil refining. In addition, the modeling method is characterized by a clear display of the dynamics of prices and storage, its representation of long delays that limit responsiveness to both demand and requirement, and the inclusion of unintended and presumed impacts within the same model.

In a certain analogy, Fan et al. [9] developed a model of system dynamics to capture the dynamics of investment in coal in China. The model of system dynamics simulates the behavior of the entire system caused by investments in mines, and the influence of investments in fields on the coal system is studied.

Moreover, the potential for coal production is projected in various scenarios in 2020, on the basis of which policy recommendations are proposed. Summing up, the model examines the impact of investments in state fields and geological exploration, and suggests an optimal investment size.

Tang et al. [10] predicts the oil reserves and oil production at the Daqing Oil Field in China until 2060. In particular, this document examines the state of oil fields in Daqing, the largest field in China, and forecasts its ultimate recoverable reserves by use of the SD model. Chinese politicians should pay attention to whether oil production in new oil fields effectively compensate for the reduction in the production of large brownfields. Close speaking, Li et al. [11] predict a rise in natural gas consumption in China until 2030. They estimate gas consumption in China by the method of sectoral division, which shows a different trend of growth in gas consumption in various industries. From a general point, a forecast results give some reasonable preconditions for the development of the gas industry in China.

Hosseini et al. [12] to use the SD methodology for modeling of a peak of oil production in Iran and the assessment of impacts. In this study they consider the main factors influencing the peak of oil production in Iran, using the SD approach. The developed model can help practitioners, especially politicians, in the oil sector to obtain a systematic and comprehensive understanding of the influencing factors and relationships that led to the peak of oil in Iran.

2.2 Tools for Modelling and Simulation

For the analysis of economic processes, simulation models developments are created and are being created currently on the basis of special environments. At this point in time, the most common environment of simulation models developments such as Stella (Ithink), Anylogic Vensim, Powersim are well-known. They allow not only quickly create simulation models using simple visual tools, but also to analyze the work of created models and to use these models to assess the impact of management decisions on the course of economic processes in modeled systems.

The Stella approach to systems modeling has some common features with its predecessor, the simulation language Dynamo. Dynamo clearly defined "stocks" (tanks), and flows (inputs and outputs) as key variables in the system vocabulary, which owns Stella [13]. Stella users are provided a graphical user interface where they can create graphical models of the system using four basic principles: stocks, flows, converters and connectors [14]. Communication between the transmitters (which transmit the transforming variables) and other elements can be done using converters. Users can enter values for stocks, flows and converters (including using the many built-in functions) [15]. Stella does not distinguish between the external and intermediate variables within the system; All of them are converters [16].

AnyLogic is a multi-model simulation modeling tool developed by the Any-Logic company. It supports agent-based modelling methodology, discrete events and system dynamics.

Simulation modelling Language Powersim can be used for modelling both simple and complex systems. It is known that complex systems are characterized by a multiplicity of descriptions. So for them it is impossible to build the only true model, but you can only describe their behavior with the help of those or other models, reflecting the characteristic behavior of the simulated systems in specific situations.

3 System Dynamics Model of Oil Company

3.1 Model Development Methodology

Let's consider a design pattern of a system dynamic model, which will have a form as shown in Fig. 1. The model elements and the relationships between them define the structure of the model. System dynamics is based on the theory of feedback and

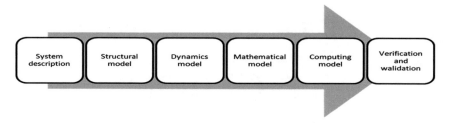

Fig. 1 Methodology of dynamics model development

integrates such subjects as information theory, management science and decision theory [17]. With principles defining the system function of the system, the whole system can be a model as a graph structure and functional relations, it then constructs a feedback loop with the theory of feedback control, finally, establishes the dynamic model of the system and models it using a computer [18]. Oil company is engaged in exploration, development and production of oil and gas. As a result, investments in exploration and production are to be divided into four areas, including investment in exploration, investment in development and investment in operational services. Investments in the exploration operations consist of investment in geophysical and geochemical exploration, investment in exploration wells and other relative investments. Development investments consist of investments in wells, investment in the construction of onshore facilities and other related investments. Investment in operational services encompass investments in power generation and communication, as well as investments in security and protection of the environment [19].

3.2 Description of Oil Company and Its Structural Model

In the proposed model (Fig. 2), the nature of the main variables is expressed by the following:

- expenses for the provided resources, characterizing the subsystem of oil prospecting and exploration;
- transport logistics, reflecting the process of oil transportation;
- operation and modernization of oil field equipment as characteristics of current and major repairs;
- oil exploration, as the basis for profit of the oil company.

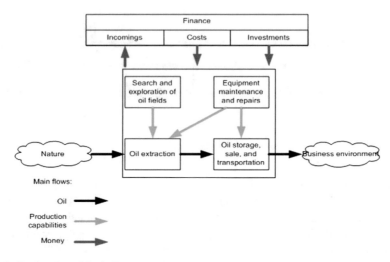

Fig. 2 Structural model of oil company

3.3 Development of Causal Loop Sub-models

Sub-system: **Oil extraction, search and exploration of oil fields**

The model of the subsystem of a field prospecting and exploration is based on an equation that reflects the costs of explored reserves of resources. In this case, we will assume that the assessment of the field is characterized by a relative capital investment.

A relative capital investment for each year of field development is the ratio of accumulated investments to the annual oil production. According to Zheltov [19] it is "the expenditure of labor and material resources in monetary terms for the creation of fixed assets of the enterprise, i.e. expenses for drilling of the wells, construction of commercial oil transport facilities, separation of hydrocarbons, demineralization and demulsions of extracted products, treatment of process water and its utilization, etc.".

In mathematical expression, the equation will have the following form:

$$\frac{dR}{dt} = \frac{K_v}{V} - S_m \tag{1}$$

where R—the costs of proven reserves of resources, K_v—capital investment (thousand tenge), V—volume of exploration, (thousand barrels per day), S_m—costs of the conservation of marginal fields (thousand tenge per thousand bar).

Imagine K_v as:

$$K_v = \sum_{i=1}^{n} a_i S_i n_i \tag{2}$$

where a_i—coefficient of a proportional value of fixed assets and cost of wells of the i-th field $(i = \overline{1, n})$, S_i—cost of one well of the i-th field (thousand tenge), n—number of wells of the i-th field.

At the same time, production volumes depend on the type of production:

$$V = K_r \sum_{i=1}^{n} V_i \tag{3}$$

where K_r (from 0 to 1)—coefficient, characterizing the complexity of hydrocarbon production (type of production, the closer to 1 the more complex), V_i—volume of production of the i-th field (thousand barrels per day).

Thus, the general view is as follows:

$$\frac{dR}{dt} = \frac{\sum_{i=1}^{n} a_i S_i n_i}{K_r \sum_{i=1}^{n} V_i} - S_m \tag{4}$$

Sub-system: Oil storage, sale, and transportation

Dynamics of expenses for oil transportation will depend on the volume of consumption of customers, such as refineries. In the model we assume that the transportation costs from the field will be reflected in the cost of storage and it requires more detailed consideration. Of course, accounting and capacity of the transport system is necessary, which also requires a more careful analysis.

The general view of a transport logistics model as follows:

$$\frac{dT}{dt} = C_0 \frac{D}{q} + C_1 \left(1 - \frac{D}{q}\right) \tag{5}$$

where C_0, cost of orders (thousand tenge), C_{1j}—cost for storage in a warehouse (thousand tenge), D—consumption of orders (thousand barrels per day), q—volumes of orders, (thousand barrels per day), T—transportation cost.

Given that there can be several orders we, for k—number of stocks, get (Fig. 3):

$$\frac{dT}{dt} = \sum_{j=1}^{k} \left(C_{0j} \frac{D_j}{q_j} + C_{1j}\left(1 - \frac{D_j}{q_j}\right)\right). \tag{6}$$

Sub-system: Equipment maintenance and repairs

During oil and gas extraction there are technological and organizational features that affect the formation of production costs, the organization of management accounting and the level of the cost of production.

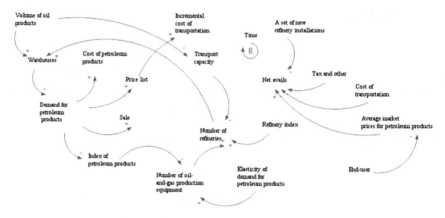

Fig. 3 Model oil storage, sale, and transportation

In the process of oil production, such works as installation and dismantling of mechanical and power equipment, underground and aboveground well repair, maintenance of reservoir pressure, collection and transportation of oil and gas, etc., are being done. The costs associated with the operation of downhole motors are included in the cost of a day (hour) of operation of the rig as time-dependent. But, only the depreciation of the downhole motor kit depends on all these costs from the time of drilling.

The operation of the equipment is accompanied by continuous and irreversible changes in the parts and connections caused by wear, deformation, corrosion and other factors, the accumulation and overlap of which lead to a decrease in working characteristics and failure. Works on maintenance and repair of equipment allow you to reduce the possibility of malfunctions and maintain the performance of products at the proper level.

The size of the overhaul includes all work types related to maintenance and routine maintenance; replacement and restoration of all worn out parts and assemblies, including basic ones; the determination of the state of the foundation, the magnitude and nature of its draught. With all this a complete disassembly of the product, washing, defectoscopy and substitution of units, parts, with the following assembly, adjustment, testing of repaired equipment, painting and marking is being done.

The maintenance and improvement of oilfield equipment involves variable and fixed costs. Variables (current) reflect the urgent repairs of equipment, and permanent (capital) is the cost of constant updating of technical equipment of the oil company.

A mathematical equation reflecting the monetary cost of repairs is:

$$\frac{dP}{dt} = P_r + P_k + P_0 \tag{7}$$

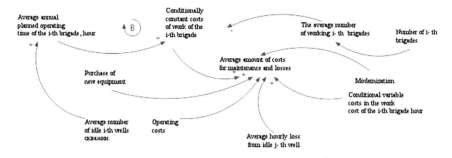

Fig. 4 Model of implementations of maintenance and repairs

where: $P_r = g_{0i}(t_{qi}V_{mi} + V_{zi})$—current repair of wells and oilfield equipment; g_{0i}—coefficient, reflecting the category of repair complexity on the i-th field, t_{qi}—time of repair work on the i-th field (час.), V_{mi}—cost of repair works on i-th deposits per unit of time (thousand tenge per hour), V_{zi}—t e cost of spare parts (thousand tenge); $P_k = g_{1i}V_d$—costs on overhaul of oilfield equipment; g_{1i}—share of the total profit for the overhaul of equipment; $P_o = g_{2i}V_d$—cost of maintenance; g_{2i}—share of total profit for technical support of equipment, V_d—amount of profit per year (thousand tenge) (Fig. 4).

Sub-system: **Finance**

Oil companies pay great attention to the analysis and forecast of investment in exploration and production, in terms of scale and structure, and can then take appropriate action to adjust according to changing situations. Many methods are available as methods of analysis and forecasting such as time series analysis, regression analysis and econometric methods [12] (Fig. 5).

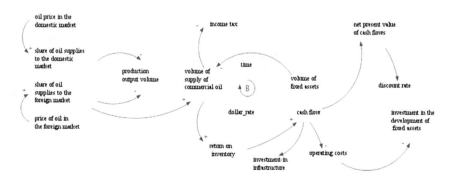

Fig. 5 Model of implementations of cash flows

4   The Problem of Data Extraction in Intellectual Information Systems

Intellectual information system is based on the concept of using the knowledge base to generate algorithms for solving investment problems in an oil company will allow: to diagnose the company's condition; help in crisis management; choose the optimal solutions according to the company's development strategy and its investment activities; economic analysis of the company's activity; strategically plan; to analyze the investment, evaluate the risk; to form a portfolio of securities, etc.

Every intelligent information system, taking into account adaptability to the changes in the subject area and information needs of users, performs the following functions: perceives user-entered information requests and necessary initial data, processes entered and stored data in the system in accordance with the known algorithm, and generates the required output information.

When creating systems of managing with knowledge base, already known and specially developed data models are used. For example, in a number of works, knowledge bases, realized by means of relational DBMS are described. Therefore, today the development of systems requires the realization of easier and more convenient access to databases. It is important to make access and manipulation easier in complex databases.

And it means that intelligent databases differ from conventional databases by the possibility of retrieving the required information upon request, which can not be exactly stored, but can be extracted from the database.

The technology of extracting the right data from large databases is the main problem. Optimization is important for users, since with this ability they need to know only a few rules and commands for using the database.

It is required to perform a search on the condition, which must be further defined in the course of solving the problem. Intellectual system without the help of the user on the structure of the database in itself builds the path to access the data files. The query is formulated in a dialogue with the user, the sequence of steps of which should be performed in the most convenient form for the user. This task is set in designing an information system of supporting decision-making of investment in an oil company.

5   Conclusions and Future Work

Management of investment activity of oil companies in modern conditions is connected with the adoption of complex and expensive management conditions. Oil development is the mono-productive production process and is highly specialized; on the other hand, it had to concentrate a wide range of multidisciplinary scientific capabilities.

The main advantage of the methodological approach based on modeling of model building is that the model is a complex component of the building, where the functions of actual and expert information on accounting, analysis, planning and management represent the union, indivisible, interdependent process. Another advantage of the system is that it operates with resources of any type and destination, automatically simulating the dynamics of their transformation in accordance with the input information. A dynamic model in which different individual characteristics should be in a particular hierarchy properly describes the development of the company. The article presents the general issues of interaction of the structural model subsystems of the oil company. The next step will be to develop criteria for each subsystem and identify components of the investment portfolio for the oil company.

References

1. Migas, S.S.: Intelligent information systems, abstract of lecture. St. Petersburg (2009)
2. Forrester, J.W.: Industrial dynamics: a major breakthrough for decision makers. Harv. Bus. Rev. 36(4), 37–66 (1958)
3. Leopold, A.: Energy related system dynamic models: a literature review. CEJOR 24(1), 231–261 (2016)
4. Zhang, B., Wang, Q.: Analysis and forecasts of investment scale and structure in upstream sector for oil companies based on system dynamics. Pet. Sci. 8(1), 120 (2011)
5. Forrester, J.W.: Industrial Dynamics. MIT press, Cambridge (1961)
6. Sterman, J.D.: Business Dynamics: Systems Thinking and Modeling for a Complex World. McGraw, New York (2000)
7. Meadows, D.: Thinking in Systems: A Primer. Earthscan (2000)
8. Ford, A., Vogstad, K., Flynn, H.: Simulating price patterns for tradable green certificates to promote electricity generation from wind. Energy Policy 35(1), 91–111 (2007)
9. Jeong, S.-J., Kim, K.-S., Park, J.-W., Lim, D.-S., Lee, S.-M.: Economic comparison between coal-fired and liquefied natural gas combined cycle power plants considering carbon tax: Korean case. Energy 33(8), 1320–1330 (2008)
10. Tang, X., Zhang, B., Höök, M., Feng, L.: Forecast of oil reserves and production in Daqing oilfield of China. Energy 35(7), 3097–3102 (2010)
11. Li, J., Dong, X., Shangguan, J., Hook, M.: Forecasting the growth of China's natural gas consumption. Energy 36(3), 1380–1385 (2011)
12. Hosseini, S.H., Kiani, B., Mohammadi Pour, M., Ghanbari, M.: Examination of Iran's crude oil production peak and evaluating the consequences: a system dynamics approach. Energy Explor Exploit 32(4), 673–690 (2014)
13. Bossel, H.: Modeling and Simulation. A K Peters, Ltd., Wellesley, MA (1994)
14. de Souza, R., Huynh, R., Chandrashekar, M., Thevenard, D.: A comparison of modelling paradigms for manufacturing line. In: IEEE International Conference on Systems, Man, and Cybernetics, Beijing, pp. 1253–1258 (1996)
15. Hannon, B., Ruth, M.: Modeling Dynamic Systems. Springer, New York City (1997)
16. Yan, G.: System Dynamics, pp. 4–30. Intellectual Press, Shanghai (1991)
17. Kathryn, M., Bartol, D., Matin C.: Management. The McGraw-Hill Companies (1998)
18. Chen, X., Yu, B., Chang, G., Gong, Y.: Research on investment size of onshore oilfield development. Oper. Res. Manag. Sci. 11(2), 122–123 (2011)
19. Zheltov, U.P.: Development of oil field. Textbook, 2nd edn. Nedra, Moscow (1998)

Dynamic Configuration of Same-Day Delivery in E-commerce

Arkadiusz Kawa, Bartlomiej Pieranski and Wojciech Zdrenka

Abstract The main disadvantage of e-commerce when compared to brick-and-mortar stores is time that is needed to deliver ordered products to clients. That is why a speed of delivery has a great potential to create a value for e-shoppers. Based on this assumption the concept of same-day delivery was developed and introduced into business activity. At the same time the scientific investigation of this concept is on its infancy. For that reason, this paper aims to propose a conceptual model of same-day delivery in which a crucial part plays a delivery platform based on the hub and spokes (H&S) model and the co-opetition concept.

Keywords E-commerce · Same-day delivery · Value creation

1 Introduction

E-commerce has been growing rapidly over the last years. On a global scale, in 2016 retail sales over the Internet reached the value of 2.7 billion dollars, representing an increase of 124% compared to 2012. Despite this impressive growth, on-line sales account for only 7% of total sales [1]. This means that there is ample space for further dynamic developments in this type of retailing [2]. However, the main barrier to the development of e-commerce includes the terms of delivery which involve the necessity to incur additional costs and a time lag between making a purchase and the delivery of the goods [3]. That is why a very important part in

A. Kawa · B. Pieranski (✉) · W. Zdrenka
Poznan University of Economics and Business, al. Niepodleglosci 10,
61-875 Poznań, Poland
e-mail: bartlomiej.pieranski@ue.poznan.pl

A. Kawa
e-mail: arkadiusz.kawa@ue.poznan.pl

W. Zdrenka
e-mail: wojciech.zdrenka@ue.poznan.pl

© Springer International Publishing AG, part of Springer Nature 2018 305
A. Sieminski et al. (eds.), *Modern Approaches for Intelligent Information and Database Systems*, Studies in Computational Intelligence 769,
https://doi.org/10.1007/978-3-319-76081-0_26

e-commerce plays a logistics activity. It has a huge potential to create a great value for end-customer by delivering an ordered product to their homes or offices. However, on the other hand, logistics can also destroy a value for customer by its inflexibility (for instance in terms of time and place of delivery). What seems to be the most disadvantageous for customers is a time lag between time of ordering product and time of delivering it. This point of view is highly supported by recent research [4]. According to it the most important delivery features for end-customer are both (low) price of delivery and speed of delivery. This is the reason why both e-shops and logistics service providers have been struggling to find a way to minimize the time of shipment at zero extra costs. One of the most promising solution of this problem is so called same-day delivery. This approach already has and will have in the future a great ability to create a value both e-shoppers and e-retailers. Consumers will contribute from convenience when buying online and multiple delivery options to choose from. This is the reason why same-day delivery is an increasingly popular delivery form of products ordered online. On the other hand, for online retailers reducing delivery time increase immediate product access, which significantly improves their competitive advantage over brick-and-mortar retailers. With same-day delivery, e-retailers will be able to increase its share of total retail and foster the sale of product categories typically are not sold online (i.e. DIY products such as tools, for example, that are usually purchased for immediate use). However one has to keep in mind that same-day delivery being a big opportunity for e- retailers at the same time requires for instance very short fulfilment lead-times and flexible last-mile delivery [4].

The same-day delivery approach has drawn an attention of academic researchers as well. This article aims to contribute in developing a theoretical background for this type of delivery option. To reach this end the authors propose their own conceptual model of the-same day delivery. What marks it out from similar models existing in literature is a unique combination of two concepts, namely: hub-and-spokes and co-opetition. This is assumed that these two concepts by complementing each other can lead to increase the efficiency of proposed model. The structure of the article is organized as follow: in the first section the possibilities of creating value in e-commerce by logistics activity are investigated. Then— existing in literature—models of same-day delivery are presented. These two sections are followed by development of same-day delivery model based on delivery platform (Sect. 3) and exemplification of it (Sect. 4). The paper is summarized by conclusion (Sect. 5).

2 Logistic Source of Value in E-commerce

The main role of logistics is to deliver right product, in the right quantity and the right condition, at the right place, at the right time, for a right customer, at the right costs. These are so called "Seven R's of Logistics" [5]. The importance of logistics isn't questionable but for many years logistics was considered only as a cost generation center with no capacity of differentiation [6]. Now no one have doubts that logistics is one of key elements creating value for customers [7]. In e-commerce logistics is especially important due to the fact that end-customers have the direct contact with logistics services. Logistics providers' performance have the direct impact on the customers' satisfaction [8]—ergo e-retailers' success.

In e-commerce logistics activity can create value for customers in five areas: availability, delivery time, place of delivery (or reception), packaging and communication of information. Delivery method is one of the key areas that determines the level of value delivered to the customers. Among many different delivery methods in e-commerce, the most common are: courier, post, pick-up drop-off (PUDO) and parcel lockers. Each of these methods can create different level of value for customers. Couriers are the most popular and the most convenient delivery method due to the fact, that parcels are delivered at the door of the customer's home or work, but on the other hand they are one of the most expensive delivery methods. Post is just as convenient as couriers and cheaper but the delivery time is in most cases much longer. Both of these methods require from the customers to be present at home or work. This requirement can be a disadvantage which may force some of the customers to seek for alternatives delivery methods. One of such a method is PUDO. It allows the customers to send or collect the parcels in dedicated places such as shopping malls, petrol stations or shops. For customers which are often out of home and have no possibility to receive parcels at work, PUDO may create higher value than couriers or post. The value can be also increased by the fact that in most cases PUDO is a cheaper solution than couriers and post. In this method however parcels collection is limited by the opening hours of the dedicated points. This issue is solved by the last of the above-mentioned

Tab.1 Value creation in e-commerce by different delivery methods—costs and benefits comparison

Delivery method	Benefits	Costs
Courier	• Door-to-door delivery	• High price • Customer presence required to collect the parcel • Customer cannot choose the time when the parcel will be delivered
Post	• Door-to-door delivery • Low price	• Customer presence required to collect the parcel • Customer cannot choose the time when the parcel will be delivered
Pick-up drop-off (PUDO)	• Low price • Customer may choose when to collect the parcel	• Limited by the opening hours of the pick-up points • Customer has to reach the pick-up point
Parcel lockers	• Low price • Customer may choose when to collect the parcel • Accessibility 24/7	• Customer has to reach the pick-up point

methods—parcel lockers. Parcel lockers are booths designed for self-service collection and dispatch of parcels which are localized in easy accessible 24/7 places. This delivery method is convenient for people who seek for value in being independent from the couriers, postman or the opening hours of reception points. Price and time of delivery is comparable with PUDO [9] (see Table 1).

3 Models of Same-Day Delivery

The main disadvantage of methods mentioned in the previous section: are time of delivery and the static delivery parameters (date and time of delivery and delivery localization). These methods were designed to deliver products in couple of days to the predefined localization. In many cases this leads to "not-at-home syndrome" and the "ping-pong" effect (when agreed-upon delivery times are not met by customers), what generates extra costs (economic and environmental ones) incurred due to the extra miles driven—especially in areas of low consumer density [10].

These days more and more customers expect to have ordered online products delivered even at the same day when the order was placed and have a possibility to dynamically change the delivery conditions (e.g. time or localization). Traditional models either can't fulfill these expectations or if they do, the cost of such as services is so high that the customers aren't interested in it. There are companies which offer deliveries at the same day. However, currently this is possible only by the global companies and available only in limited number of cities.

The most basic model of same-day delivery is called point-to-point. It is characterized by transporting products from one place to another without loading and unloading them anywhere on the way. However, such delivery service is quite expensive. Their costs are comparable to or even higher than those of domestic delivery service. It is hard to find savings in such a delivery form, because most of the costs are consumed by the working time of the couriers.

Figure 1 illustrates the processes of ordering products online and then delivering them. Customer (C1) and customer (C2) buy products from the same e-tailer (E). Both of them are located in different parts of the same city. In the case of (C1), the delivery service is provided by courier (S1). Initially, (S1) goes to the warehouse (E), where the ordered products are picked up. In the next step, courier delivers them directly to (C1). In general, the order goes along the path from points (S1) to (C1) through (E). The delivery route of the ordered products in the case of (C2) is similar. This shipment is moved from point (S2) to (C2) through (E). Such direct (point-to-point) delivery service is very quick but not cost-efficient. Couriers must travel long distances, very often with only one product. Moreover, the path between the couriers (C1, C2) and e-tailer (E) is "empty" (without shipments). Relatively high freight rates of local transport may deduce a value for end-customers and as a result discourage them from choosing the same-day delivery service when buying products online. In addition, this ineffective transport (in the case of vehicles) is very harmful for the environment. Fortunately, many city deliveries are carried out by bicycles.

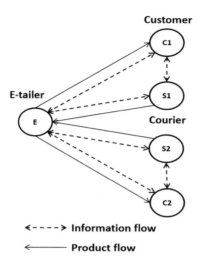

Fig. 1 Point-to-point model in same-day delivery services

The point-to-point model is generally used when number of requirements are met. Namely: there are single large consignments of goods, a high level of the use of the means of transport is ensured, the transport distances are short and there are few possibilities of obtaining additional loads on the transport route [11]. As mentioned above, the direct flow of shipments between e-tailers and customers (see Fig. 1) is ineffective. The relatively high costs of product delivery to customers appear due to the underutilization of the cargo space. Therefore, a solution is needed to overcome this problem, which will enable logistics enterprises to cooperate, especially to gain access to data about logistics services and supply capacities, and as a result, it will consolidate shipments from different e-tailers and reduce the costs of same-day deliveries.

To overcome the problem of high costs of same-day delivery approach, researchers all-over the world have been proposing different models over the past years. These researches focus on three main alternative methods of delivering goods within one day: pickup points and click and collect, crowdsourcing and dynamic routing systems. In case of pickup points and click and collect methods, these models focus either on the offering the parcel lockers localized in large retailers [12] or pickup point networks in urban and suburban areas [13]. Some authors, like T. Cherrett et al., focuses on combination of the pickup point methods with transport consolidation. Charret solution for example offers same-day delivery for students which live in campuses [14].

Other type of models are based on the crowdsourcing. In literature, there are two main models which enables same-day delivery based on the crowdsourcing. One of them is a delivery by exploiting the social networks of retail store customers. This model assumes a usage of Social Transportation to transport the parcels and a usage of IT system to gather data from real-time sensors and GPS devices to compute optimal networks to connect individuals for optimized transportation applications [15]. Other crowdsource-based same-day delivery model considers the combination of truck operator and crowdsources as last-mile deliverers. This model was designed to operate in urban areas where cyclists and pedestrians collect parcels from the trucks parked in the predefined localizations to deliver the parcels to the customers. The truck operator focuses on the truck planning and the crowdsources place the bids to execute the delivery to the end-customer [16].

The third group of models are models based on dynamic routing systems. These models are based on the advanced IT systems which combines anticipatory algorithms and GPS devices. An example of such as model is a model proposed by G. Ghiani et al. which assumes to use anticipatory algorithms for the dynamic vehicle dispatching problem. This model presents a solution that anticipate future demands through a Monte Carlo sampling procedure, to manage in an unified way several kinds of decisions, including vehicle dispatching, route scheduling and vehicle relocation, by simulating near-future demand [17]. Another model which is based on the dynamic routing approach is a model presented by the D. Reyes, M. Savelsbergh and A. Toriello. This model basis on the approach where the parcels are delivered to the trunk of customers' cars. The model proposed by the authors tries to resolve a new type of vehicle routing and scheduling problem—roaming delivery locations. In this model, the authors present algorithms which anticipates routes based on the number roaming delivery locations visited by the customer, maximum distance of roaming delivery location from the home location [18].

4   Assumptions of the Same-Day Delivery Platform

To find a remedy to the presented drawbacks of presented same-day delivery approaches a new model is proposed. A crucial part of it is a delivery platform which is based on the hub and spokes (H&S) model and the co-opetition concept. In contrast for instance to the point-to-point deliveries, hubs that connect the individual places by line-hauls are used (see Fig. 2). The H&S model minimizes storage costs and reduces the individual costs of transportation. Although a single

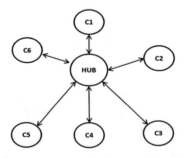

Fig. 2 Hub and spokes model

consignment is transported over a long distance, the total distance for all shipments counted separately is shorter than in the case of direct deliveries. This solution works very well for a large number of items that are posted and received at multiple locations [19]. The second, vital component of the proposed platform is the co-opetition concept (a neologism from cooperative competition). Briefly speaking, it relies on cooperation between different entities which are competitors [20].

The proposed solution assumes establishing cooperation between CEP (courier, express, and postal) operators, small transport companies and local carriers. They are a part of the platform, which is dedicated to dynamic configuration of logistics chains for the needs of same-day delivery processes. Such a platform brings together freight forwarders for transport (e-tailers) and the abovementioned logistics service providers. The platform is designed to group and systematize information on services and prices and to assist in the decision-making process concerning carrier selection. Based on predefined criteria, such as the place of shipment and delivery, shipment size and weight, delivery time and cost, logistics minima, supported directions and supply (load capacity) offered by all members of the enterprise network, the system adapts the loads to specific transport companies. For this purpose, it uses the resources of other organizations.

In the case of last mile deliveries (from hub to customers) not only couriers, but also private drivers, who have their own means of transport, can be used to convey the shipments. This is done according to the earlier indicated crowdsourcing delivery model (described in Sect. 3) in which the society and its resources are involved to provide logistic services. The driver delivering shipments in this model receives a special smartphone application, which is used to accept orders, fulfill, monitor and settle them.

The same-day delivery platform is connected to e-shops, auction platforms and price search engines. After the customer decides to purchase a given product, s/he receives an offer that meets the required criteria. On this basis, s/he can choose the best offer and place the order for delivery of the goods. For the customer, this means saving money and having a greater impact on the shipment, i.e. a higher perceived value.

The idea behind this platform is very similar to flight search engines (such as Skysanner, Google Flights), but the former includes additional features. The system automatically recommends the shipping options that are adjusted to the ordered products to the customer of the online store. For example, for a larger package courier or mail services are suggested rather than delivery to a parcel locker. Depending on the planned date of delivery, the system may offer different prices. Express deliveries will be more expensive than economical transport [9].

An important feature of the logistics platform is flexibility of the pricing policy. In the case of traditional courier companies, the cost of shipping is fixed at a given time and it is independent of the current market demand. On the logistics platform, this cost is dynamically determined and depends on the number of packages sent by/to the different companies. For example, if there is a high demand for a particular transport connection, the rate may be lower [9].

The basis of the same-day delivery platform is mainly a well-developed operating system which consists of working people and infrastructure. The latter includes hubs, local branches, means of transport and information about them. To ensure a fast and correct flow of information between individual entities of the operating system, a logistics firm has to use appropriate information technologies. Information technologies (IT) are now so closely connected with the operating system that one cannot exist without the other [21]. For this reason, the construction of the same-day delivery platform has to be based on an advanced IT tool which enables to browse, analyze and choose the best of all the available offers meeting the predefined requirements in a short time. Such IT should ensure automatic communication among particular entities and efficient and quick capture of data from the market.

An important feature of the proposed platform is interoperability. It is understood as an ability of different information systems to cooperate, safely exchange data with a predefined structure, as well as to use this data mutually in order to create information. Communication between IT systems of different logistics enterprises is possible independently of the programming languages and operation systems that they use or the applied information exchange standard [22]. What is important is that the access to the proposed system does not require application of any specialized IT systems [23].

The main component of the proposed solution is the Track and Trace (T&T) module which enables to monitor the vehicles in real time. Thanks to that, the system can freely and rationally coordinate the operation of the business. It observes the path of the consignment for all the time and on that basis, selects the optimal route; in case of a loss of the shipment appropriate action can be taken. T&T can also get complete statistical information about the quality of the services provided and, if necessary, the company may seek to raise their standards. With T&T, logistics firms have a clearly defined system of accountability for the delivery and know where the consignment is at every moment [24]. Thanks to T&T, the customer has a possibility to dynamically change the time and place of delivery.

In addition, standardization of the processes and the used infrastructure is needed. For example, shipments are transported in certain loading units, and the barcode labels describing the shipment (details of the sender and recipient, terms of delivery, etc.) must be processed by the various entities dealing with the shipments [9].

5 Exemplification of the Model

In Fig. 2 the local delivery distribution system within a city using the H&S system is illustrated based on the example from Sect. 4. The processes of online product orders are the same as in Fig. 1. Customers (C1) and (C2) buy products at e-tailer (E). In this case, city delivery processes are modified by an adaptation of the H&S system. The local transport companies have access to the hub which is managed by

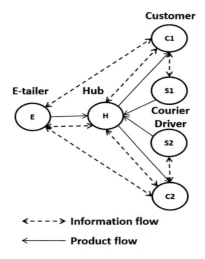

Fig. 3 Hub and spoke model in same-day delivery services

separate logistics firms. This hub is connected to the local warehouses of e-tailers by local line-hauls. These line-hauls use light commercial vehicles which pick up the products from e-tailers from the same area of the city. Within the city there may be more than one hub. It depends on the demand for the same-day delivery services. In comparison with Fig. 1, the processes of deliveries presented in Fig. 3 are a little bit different. In this case, the orders placed by customers (C1) and (C2) are operated jointly by a local transport company which picks up and delivers goods from e-tailer E to hub H, where the products are unloaded, sorted and loaded again. From this hub the shipments are delivered by couriers (from CEP companies) or by drivers (from local transport companies or private drivers) to (C1) and (C2). The choice of an appropriate provider depends on the prices and availability at a given time.

Thanks to using the proposed platform, direct connections from e-tailers to customers with only one order are eliminated. As a result of that, it is possible to achieve the benefits in the form of shorter transportation distances. Moreover, the selection of offers competitive to those of the couriers or private drivers by the system leads to the reduction of the costs of transportation between the hub and the customers. In effect, the e-tailers and their customers can get a lower price for a same-day shipment. The drawbacks of this solution, on the other hand, are more handling and sorting operations and line-hauls [25]. They create the need for connections and sometimes extend the delivery time. Thanks to shipment consolidation, there is a greater risk of package losses or damages. However, with a large scale of the operations, the H&S system in the same-day delivery process is more profitable.

It is worth noting that in some cases it may also use point-to-point delivery—when the time and cost of delivery is less important for the customer or when there is a correspondingly large volume of shipments on the same direct connection.

6   Conclusions

In the paper same-day delivery platform which is based on the hub and spokes (H&S) model and the co-opetition concept was proposed. As mentioned earlier the H&S model minimizes storage costs and reduces the individual costs of transportation. The other component of the platform namely co-opetition relies on cooperation between competitors. The platform itself—being connected to e-shops, auction platforms and price search engines—brings together e-tailers and the logistics service providers. Thanks to utilization of appropriate information technologies the unique features of a platform in question can be distinguished: interoperability, flexibility of pricing policy and (thanks to use T&T module) possibility to track a vehicle in real time.

Beyond any doubts this conceptual model is at its infancy. The authors are fully aware that it requires a further development. As a next step the simulation will be provided to verify the model and compare it with already existing delivery methods (see Table 1). During the simulation process one will investigate how the IT technologies that are employed in the model contribute to minimizing not only cost but also the time of same-day delivery.

Acknowledgements This paper has been written with financial support of the National Center of Science [Narodowe Centrum Nauki]—grant number DEC-2015/19/B/HS4/02287.

References

1. Global B2C e-commerce global report 2016. www.ecommercewiki.org/wikis/, www.ecommercewiki.org/images/5/56/Global_B2C_Ecommerce_Report_2016.pdf
2. Pieranski, B., Szymkowiak, A.: Does the device matter? Differences between the behaviour of e-shoppers using smartphones, tablets and desktops, paper in press (2017)
3. Borusiak, B., Pierański, B.: Forms of food distribution. Food retailing. In: Klaus G. Grunert (Eds.) Consumer Trends and New Product Opportunities in the Food Sector, pp. 151–171. Wageningen Academic Publishers, Wageningen (2017)
4. Hausmann, L., Hermann, N., Krause, J., Netzer, T.: Same day delivery: The next evolutionary step in parcel logistics (2014). www.mckinsey.com/industries/travel-transport-and-logistics/our-insights/same-day-delivery-the-next-evolutionary-step-in-parcel-logistics
5. Mangan, J., Lalwani, C., Butcher, T.: Global Logistics and Supply Chain Management. Wiley (2016)
6. Gil-Saura, I., Servera-Francés, D., Fuentes-Blasco, M.: Antecedents and consequences of logistics value: and empirical investigation in the Spanish market. Ind. Mark. Manage. **39**, 493–506 (2010)
7. Kawa, A.: Orientacja sieciowa przedsiębiorstw branży usług logistycznych (Network Orientation of Logistics Services Branch). Wydawnictwo Uniwersytetu Ekonomicznego w Poznaniu (2017)
8. Liu, X., He, M., Gao, F., Xie, P.: An empirical study of online shopping customer satisfaction in China: a holistic perspective. Int. J. Retail Distrib. Manage. **36**(11), 919–940 (2008)

9. Kawa, A.: Logistyka jako instrument kreowania wartości dla klienta w handlu elektronicznym (Logistics as an instrument of customer value creation in e-commerce). Przedsiębiorczość i Zarządzanie. **18**(4), 357–372 (2017)
10. Slabinac, M.: Innovative solutions for a "Last-Mile" delivery—a European experience. Bus. Logist. Modern Management, 111–130 (2016)
11. Woxenius, J.: Alternative transport network design and their implications for intermodal transport technologies. Eur. Transp. **35**, 27–45 (2007)
12. Vissera, J., Nemotob, T., Brownec, M.: Home delivery and the impacts on urban freight transport: a review. Proc.-Soc. Behav. Sci. **125**, 15–27 (2014)
13. Morganti, E., Dablanc, L., Fortin, F.: Final deliveries for online shopping: The deployment of pickup point networks in urban and suburban areas. Res. Transp. Bus. Manage. **11**, 23–31 (2014)
14. Cherrett, T., Dickinson, J., McLeod, F., Sit, J., Bailey, G., Whittle, G.: Logistics impacts of student online shopping—evaluating delivery consolidation to halls of residence. Transp. Res. Part C **78**, 111–128 (2017)
15. Devari, A., Nikolaeva, A.G., He, Q.: Crowdsourcing the last mile delivery of online orders by exploiting the social networks of retail store customers. Transp. Res. Part E **105**, 105–122 (2017)
16. Kafle, N., Zou, B., Lin, J.: Design and modeling of a crowdsource-enabled system for urban parcel relay and delivery. Transp. Res. Part B **99**, 62–82 (2017)
17. Ghiani, G., Manni, E., Quaranta, A., Triki, C.: Anticipatory algorithms for same-day courier dispatching. Transp. Res. Part E **45**, 96–106 (2009)
18. Reyes, D., Savelsbergh, M., Toriello, A.: Vehicle routing with roaming delivery locations. Transp. Res. Part C **80**, 71–91 (2017)
19. Kawa, A.: Supply chains of cross-border e-commerce. In: Advanced Topics in Intelligent Information and Database Systems. Volume 710 of the series Studies in Computational Intelligence, pp. 173–183. Springer International Publishing (2017)
20. Brandenburger, A.M., Nalebuff, B.J.: Co-opetition. Crown Business (2011)
21. Kawa, A.: Application of cloud computing in logistics services. In: Grzybowska K. (Eds.) Logistics—Selected Concepts and Best Practices, pp. 9–22. Publishing House of Poznan University of Technology, Poznań (2012)
22. Kawa, A., Golinska, P., Pawlewski, P., Hajdul, M.: Cooperative purchasing of logistics services among manufacturing companies based on semantic web and multi-agent system. In: Demazeau, Y. et al. (Eds.) 8th International Conference on Practical Applications of Agents and Multiagent Systems, Advances in Intelligent and Soft Computing, vol. 71. Springer, Berlin, Heidelberg (2010)
23. Kawa, A.: SMART logistics chain. In: Pan, J.-S., Chen, S.-M., Nguyen, N.T. (Eds.) ACIIDS 2012, Part I. Lecture Notes in Artificial Intelligence LNAI 7196. Springer, Berlin, Heidelberg (2012)
24. Kawa, A., Ratajczak-Mrozek, M.: Cloud community in logistics e-cluster. In: Nguyen, N.T., Attachoo, B., Trawiński, B., Somboonviwat, K. (Eds.) 6th Asian Conference on Intelligent Information and Database Systems, ACIIDS 2014, Bangkok, Thailand, April 7–9, 2014, Proceedings, Part II, pp. 495–503. Springer, Berlin (2014)
25. Cook, G.N., Goodwin, J.: Airline networks: a comparison of hub-and-spoke and point-to-point systemsairline networks: a comparison of hub-and-spoke and point-to-point systems. J. Aviation/Aerosp. Educ. Res. **17**(2) (2008)

Lean and Agile Supply Chains of E-commerce in Terms of Customer Value Creation

Arkadiusz Kawa and Anna Maryniak

Abstract Among the reasons for shopping online are the following: availability of products 24 h a day, more attractive prices than in traditional shops, ease of finding rare or specialized products, a wide range of ways of having the purchases delivered or of receiving them, speed of order processing. It is not a simple task to reconcile all of these requirements, particularly because many different actors are involved in delivering value to the final customer. It also requires balancing between greater freedom, speed and low costs. The aim of this paper is to identify the nature of supply chains in terms of their lean and agile approach in the context of the type of product being moved (which can be sold via e-commerce), logistical solutions and supply chain management and configuration. The basic source of empirical materials was the author's survey. The studies were conducted via a direct visit of the interviewer in the enterprises.

Keywords E-commerce · Supply chain · Lean · Agile · Customer value

1 Introduction

Undoubtedly, e-commerce is now one of the most important trends in economy. According to eMarketer [1], e-commerce sales in the world in 2017 will grow by approx. 23.2% compared with 2016 and will amount to approx. $2.3 trillion (10% of total retail sales). By 2020, the global sale via the Internet is even expected to reach $3.9 trillion.

The dynamic development of e-commerce is driven by increased household access to the Internet, but also by growing mobility and popularity of portable

A. Kawa (✉) · A. Maryniak
Poznan University of Economics and Business, al. Niepodległości 10,
61-875 Poznań, Poland
e-mail: arkadiusz.kawa@ue.poznan.pl

A. Maryniak
e-mail: anna.maryniak@ue.poznan.pl

© Springer International Publishing AG, part of Springer Nature 2018
A. Sieminski et al. (eds.), *Modern Approaches for Intelligent Information and Database Systems*, Studies in Computational Intelligence 769,
https://doi.org/10.1007/978-3-319-76081-0_27

devices (e.g. smartphones, tablets), via which customers order goods and services at a convenient time and place more and more frequently. They do not only order things of greater value, but, more and more often, everyday products to which they want to have very fast access. This requires management tools suitably adapted to e-commerce. Internet retail sales is different from traditional channel sales in that a certain kind of promise to fulfill the order in the right place, time and cost is sold apart from the products themselves. One of the key tools of online trade, then, is logistics. It allows not only to attract new customers (by availability of goods, different forms of delivery and a low shipping cost), but also to retain those who have already placed an order (by timeliness, compliance of the goods with the order, no damages).

E-commerce has shifted the center point of the logistics system from retailer to consumer, a new set of expectations emerges. The consumers are seeking ways to maximize convenience, choice, and price—establishing a completely different shopping experience [2]. As more shoppers buy online, the demand for a seamless shopping experience lands on retailers. They have to look for new possibilities to meet the customer expectations.

Ensuring the expected level of customer service involves searching a balance (so-called trade-off) between the benefit anticipated by the customer and the level of the costs necessary to provide it. Creating an agile supply chain enables more freedom and flexibility to tailor the product to the current needs. Creating a lean supply chain makes it possible to offer goods at competitive prices [3]. Awareness of the extent to which the supply chain and the product being moved within it is agile and lean is likely to facilitate decision-making concerning the development of e-commerce and the choice of the tools to support its development.

Among the scientific papers combining the topics of e-commerce and supply chains, there are mainly studies that raise the subject of logistics customer service [4], challenges of the logistics industry related to the dynamic growth of online sales [5] and studies that deal with the behavioral and marketing issues connected to the creation of virtual sales channels (including cultural, economic and political conditions) [6]. However, there are no articles in which the impact of supply chain types on the ability to create virtual sales channels with the support of smart solutions is considered.

The aim of the study is therefore to diagnose the nature of the supply chains of manufacturing enterprises that are or may be involved in the development of e-commerce. The second objective is to identify the relationship between the type of supply chains (analyzed in the context of the type of the product being moved, logistics solutions and chain management and configuration) and the development of e-commerce and to indicate what kind of smart solutions can be used in this area.

The structure of the paper is as follows. Section 2 describes the lean and agile supply chain. Section 3 presents the research methodology. Section 4 shows the research results. In Sect. 5 smart e-commerce solutions in lean and agile supply chains are given. Section 6 summarizes the article and points to future directions of the research.

2 Lean and Supply Chain

This paper concentrates on two seemingly very different types of chain—the first one is based on creating a cost advantage (lean supply chain) and the second one is based on the individualization of the product and logistics offer (agile supply chain). The decisions regarding how much focus there is on one or the other solution seem to have the greatest impact on the way the supply chain is managed, the form of the flowing products and the configuration of the distribution channels.

The leanness and agility of supply chains depend on their characteristics. Harrison and van Hoek [7], for example, define the nature of chains (and their respective strategies) using the criteria of market presence and those of competing in it. Naylor et al. [8] assign an agile or lean character to chains depending on their resistance, ability to quickly reconfigure and the scheduling level. Christopher and Towill [9] classify chains according to the nature of the product (its type, life cycle, diversity), the purchasing policy, the level of sales forecasting or the profit margin. Goldsby et al. [10] divide chains on the basis of the criterion of costs (of warehousing, stockholding, inbound and outbound transport, production and sourcing of raw materials). However, the most popular distinction between agile chains and lean chains is based on the criterion of locating the decoupling point of goods [11]. It is the storage place for the major stocks in the supply chain. This point separates the part in which all activities are carried out according to the customer's order from the part determined by the demand forecast. In the research, a mathematical approach is used to adopt legality in a supply chain, as well as case studies, bibliometric tests and other research methods. These is research, which focuses on the relations in supply chains as well in the products moving along them.

3 Research Methodology

On the basis of the mentioned suggestions and the review of literature, activities characteristic of lean and flexible supply chains as well as their intensity in the Likert scale (from 1 to 5, where 1 meant "definitely yes" and 5—"definitely no") were identified.

The subject of the research were big and medium-sized enterprises, classified, according to the data of the Polish Central Statistical Office, in section C (manufacturing enterprises), excluding the enterprises which are not predestined for e-commerce. In Section C, 1087 medium and large entities were registered. At the first stage of the research, according to statistical procedures, 280 records were drawn. At the second stage, a non-probability sampling took place. The choice criterion was a declaration that the chains, in which the companies were functioning, to any extent were agile and lean. The respondents were decision makers dealing with supply chain management. The research took place in 2017.

The basic source of empirical materials was the author's survey. The studies were conducted via a direct visit of the interviewer in the enterprises. The used method involved structured interviews. Although the overall number of studied enterprises amounted to 115, materials from 71 enterprises were qualified for final analysis. The surveys which were only partially completed or in which there were inconsistencies with regard to the checklist questions were excluded from the studies.

The quality of the results were verified using validity and reliability measures (all convergent factor loadings and Cronbach's alpha coefficients of constructs were higher than 0.60).

4 Nature of Supply Chains—Research Results

In order to identify the nature of supply chains in which the studied enterprise participate, average values of thirty test items concerning the aspects connected with product, logistics and supply chain management were calculated. Fifteen questions concerned lean supply chains, and the other fifteen questions—agile supply chains. The graphs present the average values for all of the studied supply chains. Therefore, on the basis of the gathered material, one can expand the studies in the future and characterize particular supply chains with the use of the case study method.

Within the area of "product", the representatives of the studied enterprises indicated that the most important thing for them is designing products taking into consideration the following issues: the increase in production efficiency, the decrease of products' defectiveness and the creation of wide range of products in order to better match individual needs of the recipients (Figs. 1 and 2).

The respondents concluded that, within logistics activities, they, above all, limit unnecessary activities during the movement of products in supply chains, optimize transportation routes (mainly from the point of view of the costs), smoothly adjust

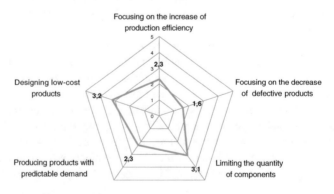

Fig. 1 Lean supply chain—nature of the product

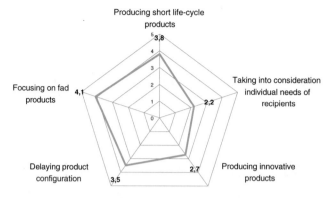

Fig. 2 Agile supply chain—nature of the product

to the current transportation needs and reduce the costs of warehousing infrastructure service (Figs. 3 and 4).

Within the process of supply chain management, the enterprises strive to eradicate problems "at root". At the same time, they try to reconcile two seemingly contrary strategies. On the one hand, they organize the flows of goods in supply chains according to a previously specified schedule, and on the other hand, they try to quickly respond to current needs and ensure a high level of products' availability. In addition, the studied entities participate in relatively static supply chains since they conduct only small changes during the reconfiguration (Figs. 5 and 6).

Taking into consideration average results for the studied aspects (Table 1): "products", "logistics" and "supply chain management", one can conclude that within each of these aspects the enterprises in the first place adopt a policy of lean supply chains creation, and further on—agile supply chains. Lean activities in supply chains are rather important for the enterprises, while the agile ones are of moderate importance.

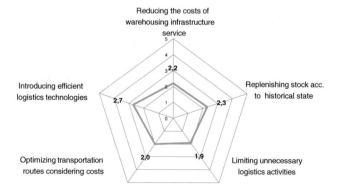

Fig. 3 Lean supply chain—logistics activities

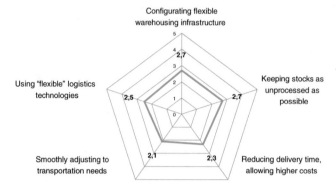

Fig. 4 Agile supply chain—logistics activities

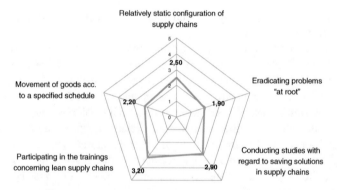

Fig. 5 Lean supply chain—supply chain management and its configuration

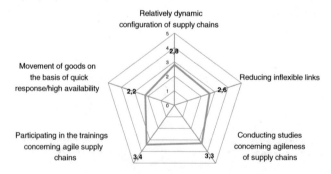

Fig. 6 Agile supply chain—supply chain management and its configuration

Table 1 Agile and lean activities

Dimension	Average values for lean activities	Average values for agile activities
Product	2.5	3.3
Logistics	2.2	2.5
Supply chain management	2.5	2.9

5 Smart E-commerce Solutions in Lean and Agile Supply Chains

On the basis of the research carried out, it was found that companies do not generally have chains unambiguously profiled as lean or agile. The type of the product being moved, the way in which the supply chain is managed, as well as stock, transport and storage management policies are not subject to a single, clear strategy. It is also difficult to say that it is a mixed strategy. It rather seems that, with a few exceptions, managers do not appear to have a clear vision of the maximum allowable costs of agility.

In such conditions, the creation of virtual sales channels entails a high risk. Introducing them on an ad hoc basis, without embedding them in the company's strategy and without determining their importance for the value for the customers may cause more profits to be lost.

The following are examples of smart solutions that are used in the context of the product type, logistics solutions and e-commerce chain management and configuration.

5.1 Product in E-commerce

Authors suggest that the introduction of new products into the supply chain is ever more frequent [12, 13], and the products themselves are less and less standard. As a result, it is becoming increasingly difficult to pursue a strategy of lean product economy and economies of scale. The development and introduction of online sales partly removes these barriers. Mass customization enables to generate product characteristics freely chosen by individual purchasers on a huge scale. The customer not only decides on the shape of this product, but also on the delivery method, type and frequency of the information received. The market of prosumers, who do not only engage in potential modifications of the final appearance of the product, but also design it almost from scratch, is also growing. The closer the decoupling point is to the lower part of the supply chain, the easier it is to adapt to individual customer requirements. It can therefore be assumed that the development of Internet distribution has an impact on the possibilities to create agile chains and at the same time facilitates the implementation of a strategy of saving at every stage of the product design. In the latter case, it is very important to pay attention to the quality of products, because in e-commerce customers return a large part of products (up to 70%), which generates additional transport costs and other operating costs.

E-commerce is also connected with product digitization or addition of digital information to products. From the perspective of logistics, in which the physical flow of things must be present, complete digitization, where the product is transformed into an intangible form, e.g. music, film, book, is not of interest. E-tailers are interested in the addition of digital information to the product, which provides the customer with an added value, e.g. the ability to compare products in an application available on the vendor's online platform, access to ratings and reviews of other users.

5.2 Logistics in E-commerce

Logistics in e-commerce is the main value creator for the customer. Without logistics, in particular without the delivery of goods to the customer, the online sales process would be very limited. It is also a source of costs that can reduce the value for the customer. For these reasons, within the framework of the logistic activity, the most organizational and technological solutions are introduced which improve the operational functioning of companies as well as their customer service.

In traditional trade, the retailer sells a product that the customer sees on the shelf at a certain moment, while in e-commerce the seller offers a kind of promise to fulfill the order. E-customers are not only interested in the product itself, but also in receiving up-to-date information about the shipment, in flexible and fast delivery and in simple and free returns of goods. If the product is not delivered to the customer in due time, is damaged or the driver's service is not satisfying, then the customer may not re-purchase from the given e-tailer.

Besides door-to-door delivery, the customer can pick up the shipment at a PUDO (Pick Up, Drop Off) point, from a self-service terminal (e.g. parcel lockers) or at a bricks and mortar store. Returns of the purchased goods may be carried out in a similar way. In addition, payment for the purchases can be made during the fulfillment of the order, but also upon collection of the shipment from the courier (cash on delivery) or the self-service terminal [14].

The important solution in e-commerce logistics affecting the agile supply chain is the Track and Trace (T&T) system which enables to monitor the vehicles in real time. It can also get complete statistical information about the quality of the services provided. Thanks to this system, the customer has a possibility to dynamically change the time and place of delivery.

Automation is another area that companies are developing in order to reduce costs and speed up processes. In e-commerce, sorter and conveyor systems, mezzanines and other similar solutions are implemented to save storage space. More and more companies are investing in carry pick solutions, where the rack is transported by a robot to the picker, and pouch sorter solutions, in which orders are sorted at a high speed using special pouches, as well as picking robots, whose performance has not been fully satisfactory yet, but their use during peak times, holidays and at night gives companies a great competitive advantage. These solutions affect positively the lean supply chain.

5.3 Supply Chain Management in E-commerce

In most online shops, the buyer has a free choice in the methods of purchase, testing, reception and payment, thus (s)he decides to create the value of his or her product. This has a major impact on the supply chains that are being set up, which are more and more often configured for individual transactions. If this is combined with the fast-growing cross-border trade, where consumers around the world buy billions of products from different countries every day, a complex network of links is created. Each of customer can thus be the creator of logistic processes.

The importance of end-customers in e-commerce supply chain is growing. This can be seen from the trends in the logistics services industry. More and more solutions are emerging that allow to configure the supply chain. Apart from the place where they can pick their order up, the customers can also choose a convenient time to do so. In addition, it is possible to change deliveries dynamically using a smartphone.

At the same time, the concept of sharing economy is developing in supply chain which assumes the use of resources from outside the logistics services industry with the participation of modern technologies. For example, cars belonging to other companies or private persons are used to transport consignments. Similarly, free storage space is made available. This concept is based on the assumption that "access is better than ownership". People possessing free resources, shoppers and online shops benefit from this. Customers can, then, simultaneously use and offer services to other market players. For the time being, these services are being developed mainly in larger cities where direct delivery is carried out without loading bays. Over time, more advanced solutions may emerge, resulting in even greater involvement of communities in logistics processes.

6 Conclusion

The approach to logistics has considerably changed as a result of the emergence of Internet technologies in business. This is not only about the digitalization of some products (e.g. music, films, books), where logistics is not needed, but also about the very dynamic development of e-commerce of other goods. On the one hand, the Internet has eliminated intermediary links in the supply chain and, on the other hand, new sales and distribution channels have been created. The central focus of interest has been moved to the final customer placing the order at any location and time. The route to the store has been replaced with home delivery. After online sales appeared, the customer has become an integral part of the logistics process and, often for the first time, has dealt with logistics services.

Thanks to our research we can conclude that despite these trends of creating agile chains in the market, traditionally structured chains, in which the main determinants of strategy are production costs and productivity, function to a large extent, and therefore e-commerce solutions should first of all take into account the cost-efficiency needs. However, it should be added that at the level of individual companies, it can be seen that the chains in which they operate are a kind of hybrids and rarely aspire to be only lean or agile.

As noted in this paper, in e-commerce the customer influences the supply chain configuration (using a lean and agile approach at the same time), in particular in the field of the delivery of the purchased products. The smart solutions that are used in the context of the product type, logistics solutions and e-commerce chain management and configuration are very important.

The research just like other research is limited. The companies came from a well-developed economic region. The results may have been different, if it took place in poorer regions. What is more, it is worth expanding the future research to include also the moderators, such as company size or production profile.

A potential aim of further research may be to present the perception of this impact from the perspective of the customers themselves. As a result of these studies, it will be possible to develop a model for configuring the e-commerce supply chain, which is lean or agile depending on the market conditions, customers preferences and smart solutions. The results may be an inspiration to design modern technologies, which will improve managing of hybrid chains.

Acknowledgements This paper has been written with financial support of the National Center of Science [Narodowe Centrum Nauki]—grant number DEC-2015/19/B/HS4/02287.

References

1. eMarketer, Worldwide Retail and Ecommerce Sales: eMarketer's Estimates for 2016–2021 (2017). https://www.emarketer.com/Report/Worldwide-Retail-Ecommerce-Sales-eMarketers-Estimates-20162021/2002090
2. Rigby, D.: The Future of Shopping. Harvard Bus. Rev. **89**(12), 65–76 (2011)
3. Reis, L., Varela, M.L.R., Machado, J., Trojanowska, J.: Application of lean approaches and techniques in an automotive company. Rom. Rev. Precis. Mech. Opt. Mechatron. **50**, 112–118 (2016)
4. Lin, Y., Luo, J., Cai, S., Ma, S., Rong, K.: Exploring the service quality in the e-commerce context: a triadic view. Ind. Manage. Data Syst. **116**(3), 388–415 (2016)
5. Joong-Kun Cho, J., Ozment, J., Sink, H.: Logistics capability logistics outsourcing and firm performance in an e-commerce market. Int. J. Phys. Distrib. Logist. Manage. **38**(5), 336–359 (2008)

6. Lawrence, J.E., Tar, U.A.: Barriers to e-commerce in developing countries. Inf. Soc. Justice J. **3**(1), 23–35 (2010)
7. Harrison, A., van Hoek, R.: Logistics Management and Strategy. Pearson Education, Harlow (2002)
8. Naylor, J.B., Naim, M.M., Berry, D.: Leagility: integrating the lean and agile manufacturing paradigms in the total supply chain. Int. J. Prod. Econ. **62**, 107–118 (1999)
9. Christopher, M., Towill, D.R.: Supply chain migration from lean and functional to agile and customized supply chain management. Int. J. **5**(4), 206–213 (2000)
10. Goldsby, J.T., Friffis, S.E., Roath, A.S.: Modeling Lean Agile, and Leagile Supply Chain Strategies. J. Bus. Logist. **27**(1), 57–80 (2006)
11. Chan, F.T.S., Kumar, V.: Performance optimization of a leagility inspired supply chain model: a CFGTSA algorithm based approach. Int. J. Prod. Res. **47**(3), 777–799 (2009)
12. Maryniak, A.: Zarządzanie zielonym łańcuchem dostaw (Green Supply Chain Management). Wydawnictwo Uniwersytetu Ekonomicznego w Poznaniu, Poznań (2017)
13. Simchi-Levi, D., Kyratzoglou, I., Vassiliadi, C.: MIT Forum for supply chain innovation. Supply Chain and Risk Management, Massachusetts Institute of Technology, Massachusettss (2013)
14. Kawa, A.: Supply chains of cross-border e-commerce. In: ACIIDS 2017: Advanced Topics in Intelligent Information and Database Systems, pp. 173–183. Springer International Publishing (2017)

Improvement of Community Bus Operation Management System

Kento Ando, Yu Fujihara, Takuya Fujihashi, Keiichi Endo,
Hisayasu Kuroda and Shinya Kobayashi

Abstract We renovated the community bus operation management system which is operated in Tsushima area of Ehime Prefecture Uwajima City. The bus operation management system was developed in 2013. It is possible to grasp the bus location and the number of passengers by using the application installed on the tablet and the web application. However, it did not satisfy the necessary functions for the system. In this paper, we interviewed the problems of the current situation from customers and organized the problems of customers. We reflected systematic problems in the design of the system and developed a more satisfying system for customers.

Keywords Community bus operation management system · Renovation · Bus location · B-Map application · B-Map administrator system · B-Map web service

1 Introduction

Uwajima City in Ehime Prefecture has a central urban area with the sea on the west side and the other consists of the mountains. Since the majority of the population in the mountain area of Uwajima City is elder people, occupied by the elderly and they are difficult to drive their car to the urban area. To transport elder people in the mountain areas to the urban area, Uwajima City provides a public bus service, namely, community bus service. On the other hand, since the community bus service is run by limited personnel and budgets of Uwajima City, the number of buses in a day is few, e.g., one bus in a day in a certain bus line. Due to a few number of buses, users of the community bus service often request the current location of the bus. However, due to the limitation of budgets for the community bus service, each bus does not send its own location, and thus it is difficult to clearly response the users' request. In addition to the above issue, there is another issue in terms of usage

K. Ando · Y. Fujihara (✉) · T. Fujihashi · K. Endo · H. Kuroda · S. Kobayashi
Graduate School of Science and Engineering, Ehime University, 3 Bunkyo-cho,
Matsuyama, Ehime 790-8577, Japan
e-mail: fujihara@ict.ehime-u.ac.jp

© Springer International Publishing AG, part of Springer Nature 2018
A. Sieminski et al. (eds.), *Modern Approaches for Intelligent Information
and Database Systems*, Studies in Computational Intelligence 769,
https://doi.org/10.1007/978-3-319-76081-0_28

management: more specifically, handwritten papers are used to record the number of users in the community bus service. It causes a high management cost and low preservability compared to the digital data.

People in Advanced Course for Information and Communication Technology Specialist, Graduate School of Science and Engineering, Ehime University (ICT course) solved the above-mentioned issues as the project-based learning (PBL) problem given by Uwajima City. Specifically, we developed a Tsushima area's community bus management system, namely, B-Map, in 2013. This community bus runs coastal and mountain areas in south part of Uwajima City. After the development in 2013, we improve B-Map system every year as a PBL problem [1].

In this paper, we describe a project of B-Map system improvement conducted by ICT course in 2016.

1.1 Community Bus

Private bus services avoid buses in depopulated areas where revenue cannot be expected. However, in the depopulated areas where the aging is remarkable, the existence of public transport is a very important means as a leg of the residents. Therefore it can not easily be abolished. For this reason, Public institutions such as city halls operate buses or consign them to bus operators. Such a bus is called a community bus, and a small busy 10–20 microbuses are used, such as being able to enter a narrow path mainly.

1.2 About PBL

While utilizing the knowledge and skills possessed by learners, they grasp the knowledge and skills necessary for practicing in society, pseudo tasks imitating them, devise and solve a problem-solving method.

2 B-Map System

Uwajima City Hall is servicing community buses for areas where it is difficult to maintain a line for private bus companies for the purpose of enrichment of public transportation in Uwajima City.

They had two problems. The first Bus users want to check the location of the bus.but, since the means of contact with the bus driver is only a telephone during bus operation, it is not desirable to contact for safe operation. Therefore, it is difficult for Uwajima City to grasp the location of the bus, and it is impossible to confirm whether the bus is operated normally. The second they record the number of passenger in a digitized state, but the bus driver uses record paper. For this reason, the recording paper submitted by the bus driver needs to be digitized manually, it takes time.

In the B-Map system, by introducing a tablet terminal equipped with a dedicated application on the bus, the bus user can know the current position of the bus in real time based on the GPS data transmitted from the tablet terminal. Furthermore, can digitally record the number of passengers. With this system, it is possible to work corresponding to the work of Uwajima City staff who digitizes paper media records or reduce the work of telephone users by telephone.

2.1 Stakeholders

The B-Map system is a system developed for the following people. The first, it is "administrator" which means city hall to browse bus driving information of Uwajima City Hall. Administrators spend considerable time digitizing paper from paper media when browsing and aggregating bus driving information on current paper media. The second, it is "bus driver" which means the staff of Uwajima City which operates Tsushima district community bus. Since bus drivers have to record the number of the passenger during bus operation, time-consuming recording means have a very bad influence on bus operation. The third, it is "bus user" to use Tsushima district community bus. Most bus users are elderly people who live in depopulated areas. Also, Community bus has few runs. Therefore, They inquiring the Uwajima City Hall about the Community bus operation situation.

2.2 Overview of B-Map System

As shown in Fig. 1, the B-Map system consists of a dedicated application installed on a tablet terminal used by a bus driver (B-Map application), a server that stores data such as a bus operation log and operation records, the Web page of the city staffs browsing information (B-Map management), the web page of bus users browsing information (B-Map web service). The B-Map application is an application installed on the tablet terminal of Uwajima City Hall, and the bus driver works. On the application, in addition to receiving information on bus operations such as bus names and bus names displayed on the screen by communicating with the server, you can do server bus location and number of users. In addition, when an abnormality that may interfere with the operation of the bus occurs, it is possible to declare an abnormal state to the server. The B-Map management system can output information used for document creation, such as the administrator totals the number of community bus users by using a Web browser. Also, when an abnormal situation declaration is made from the bus, it is also used for checking and responding based on the bus location. The B-Map web service is used by bus users to check the bus location using a web browser.

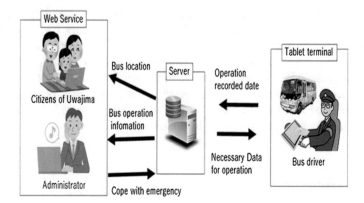

Fig. 1 Outline of B-Map system

3   Outline of Renovation

In 2016 B-Map system renovation project, nine development staff members took a project for two months. Based on the interview conducted by staff of Uwajima City Hall, we made the following renovation. This section explains the restoration contents from the three viewpoints of the B-Map application, the B-Map management system, and the B-Map Web service used in the B-Map system.

3.1   Renovation Concerning B-Map Application

The B-Map application has two requirements. The first is that the bus driver can collect the number of passengers and line information without operating the tablet as much as possible. The second is making it possible for applications to normally operate and collect data even in mountainous areas where radio does not reach.

1. The B-Map application receives information such as bus routes and bus stops at application startup and deletes it at the time of application termination. Therefore, it is necessary to start up in the communicable area. If you start up in an area where communication is impossible, application operation becomes impossible.
2. The B-Map application will run on the tablet, but since the characters are small and the buttons are concentrated, there is a possibility that the bus driver malfunctions.
3. In the B-Map application, each time you arrive at the bus stop, you need to select the bus stop you want to select the application and enter the number of passengers. However, this function may cause an input error or forget to change the bus stop.

4. It is difficult to understand the contents displayed on the screen and to check the number of the passenger and the total number of cargoes, as there is a problem with the layout configuration and character size of the B-Map application.

5. Because we have not dealt with the area where communication is impossible, the bus location information of the bus within the area where communication is impossible is not transmitted, and no data is left on the server.

6. The record of the number of passengers on the bus was sent to the server after bus operation. However, if you send in an area where communication is impossible, the record of the number of passengers will be sent to the server and disappear. The administrator needs to interview the bus driver when checking the fare and the number of the passenger.

Therefore, in this project, appropriate measures are taken and solved based on the reasons for problems 1–6.

1. In the B-Map application, line information such as the name of the route, service number and bus stop name used for bus operation is acquired at the time of application start, and it is discarded at the time of the application termination. Therefore, when launching an application in an area where communication is impossible, there is no such information at all and the application cannot be operated. However, information such as bus lines and bus stops are data that will not be changed suddenly. Therefore, this problem was solved by storing the line information in the area possessed by the B-Map application inside the tablet.

2. In the layout configuration of the B-Map application, as shown in Figs. 2 and 4, there are many margins and layout elements are concentratedly arranged, which causes a malfunction due to a push error. Therefore, as shown in Figs. 3 and 5, elements such as size and font buttons were solved by expanding them to the full-screen size of the tablet.

3. All the operations in the B-Map application are performed manually, and because of the layout configuration (2) problem, it has hindered the operation of the bus driver. Therefore, as shown in Fig. 5, based on the bus location acquired by the tablet terminal, the application automatically selects and displays the next bus stop and solved it.

4. Usability was not considered in the layout of the B-Map application. Records of the number of passengers taken and the bus fare etc. are displayed in a format as shown in Fig. 6. As a result, it is difficult to check the information displayed in lowercase letters. For this reason, as shown in Fig. 7, it was solved by increasing the character size and changing the display format.

5. In the B-Map application, current bus location acquired by the tablet every 3 s was sent to the server as it was and deleted from the tablet. Therefore, the bus location in the area incapable of communication is not stored but discarded, and the bus location as shown in Fig. 8 is generated. Considering that the capacity of the data based on the bus location is not large enough to affect the tablet terminal, all the bus location from the start to the end of the operation are saved like a file in the terminal. Furthermore, in order to save the complete bus operation log on the server side, after finishing the bus operation, we decided to send this file to

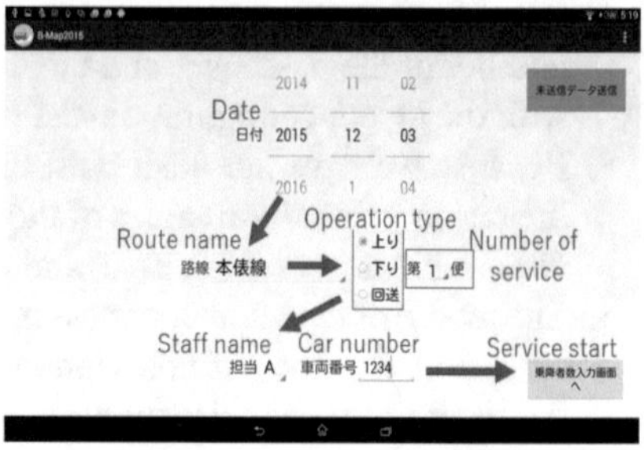

Fig. 2 Before renovation
line information page

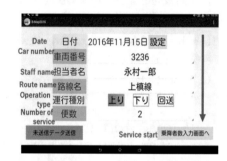

Fig. 3 After renovation line
information page

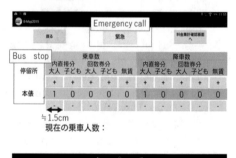

Fig. 4 Before renovation
count passenger page

the server in the communicable area. As a result of these two changes, as shown in Fig. 9, the bus location is also recorded in the area where communication is impossible.

6. The record of the number of passengers on the bus has been sent to the server after bus operation. However, there is no function to judge the availability of communication, and reception by the server is not confirmed. Therefore, when data transmission was performed in an area where communication is impossible, the transmission was performed even though transmission could not be performed. Therefore, we decided to return the reception completion from the server side to the application side, and we successfully recorded it by judging whether or not the transmission was completed normally on the application side.

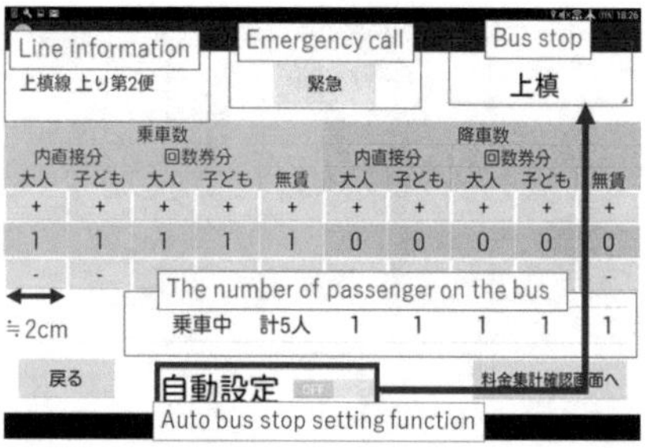

Fig. 5 After renovation
count passenger page

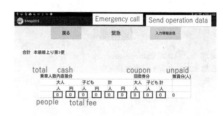

Fig. 6 Before renovation
confirm page

Fig. 7 After renovation
confirm page

3.2 Renovation Concerning B-Map Administrator System

The B-Map administrator system is a system that Uwajima officials who are administrators use for inquiries from bus users and grasp of bus users. Therefore, Uwajima officials need to display the bus location sent from the B-Map application.

However, as shown in Fig. 8, the movement speed of the bus calculated based on the time information associated with the position information of the bus stored as the operation log is a value which is inherently impossible. Also, it often does not match the tachometer value set up at Uwajima City Hall. This is a problem that occurs because the time information not related to the position information is stored by storing the position information transmitted from the B-Map application and the time information received by the server in association with each other.

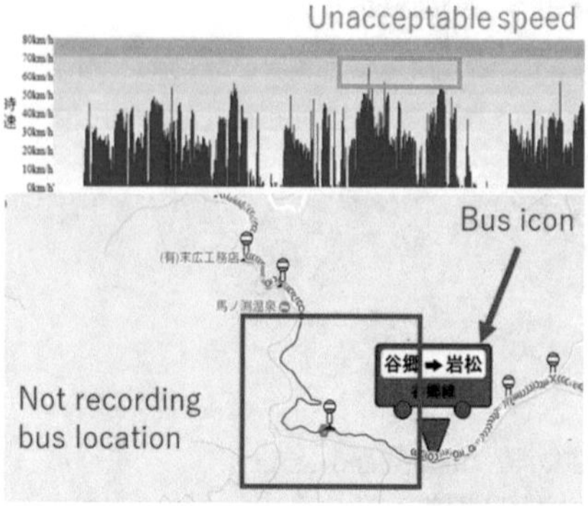

Fig. 8 Before renovation
administrator system page

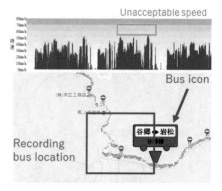

Fig. 9 After renovation
administrator system page

    In order to solve this problem, when acquiring the position information of the bus
at the tablet terminal of the bus driver, time information is acquired at the same time,
the time information is linked and transmitted. As a result, as shown in Fig. 9, the
speed error could be reduced.

3.3 Renovation Concerning to B-Map Web Service

The B-Map web service is a service that a bus user displays the current bus location
on Google map as an icon in order to know the current bus location. Therefore, like
the B-Map administrator system, it is necessary to be able to display the real-time
data sent from the B-Map application.

    However, in the current situation, when a bus enters an area where communication
is impossible, the bus icon does not change from the last transmission the bus location
of the communicable area. This is a problem that is occurring because there is no
bus operation processing that does not move for a certain period of time.

    Therefore, if the bus operation log has not been transmitted for more than one
minute, the bus icon becomes transparent when the next service on the same route is

Fig. 10   Clear the bus icon

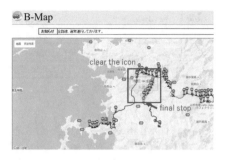

Fig. 11   Semi clear the bus icon

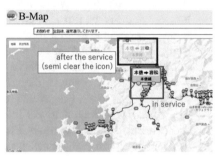

running or the bus is near the last stop. If the bus operation log is not sent for more than one minute, the next service on the same route will not be executed and the bus will not be near the last stop. The bus icon is translucent. This makes it possible to distinguish the displayed bus icons.

4   Evaluation

In this chapter, based on the results of the tests in the B-Map project and the results of the tentative introduction, we were able to take measures against the problems mentioned in Chap. 3 Sects. 1–3. we will state each thing.

4.1   Evaluation for B-Map Application

1. When launching the B-Map application, the tablet has changed to a mechanism to acquire and save information such as bus routes and bus stops. Table 1 shows whether or not the B-Map application can be operated according to the communication situation. From this result, we solved the problem that the B-Map application cannot be operated in areas where communication is impossible which is the problem shown in Chap. 3 Sect. 3 (1).

Table 1 Verification result of available operation on B-Map application

	Communication possible	Communication impossible
Saved data	Operation possible	Operation possible
No saved data	Operation possible	Operation impossible

2. As shown in Figs. 2, 3, 4, and 5, the layout of the B-Map application has been arranged in an easy to understand manner, eliminating useless margins as a whole. As a result, we were able to use buttons about 5 mm in size per button, and at the same time, we could use one size font. In addition, as shown in Fig. 3, it is possible to improve operability by adopting a layout that is operable from the top to the bottom.

3. In order to confirm that the display of the bus stop of the B-Map application was automatically changed, the verification was carried out separately in two times. The first of all, I registered the bus route as a test route, actually ran and confirmed. Next, we tentatively introduced to Uwajima City's community bus and confirmed by actual route. From this result, it was possible to eliminate the inability to keep the consistency of the bus operation information due to forgetting to change to the bus stop shown in Sects. 3 and 1 (3).

4. In the same way as the layout change of the B-Map application in (2), we eliminated useless margins, on the whole, changed the layout to be easy to understand, and changed buttons and character fonts to large ones.

5. In Figs. 8 and 9, the bus operation record collected by the B-Map application is usually sent to the server and displayed by the B-Map management system. As a result, it became possible to store the bus operation record in the area where communication which was completely lost until now cannot be done.

6. The bus operation record will not be lost due to the actual communication status between the temporary introduction of the B-Map system and the transition to the project report meeting. As shown in Table 2, even when the operation is completed in the area where communication is impossible, the bus operation record can be transmitted to the server by communicating again in the communicable area. As a result, we were able to prevent the loss of bus operation records. As a result, the administrator can save time and effort to contact the driver to confirm the information.

Table 2 Verification result of available bus operation record transmission on B-Map application

	Communication possible	Communication impossible
Before renovation	Send completely	Send error
After renovation	Send completely	Send error

4.2 Evaluation on Repair of B-Map Administrator System

In Figs. 8 and 9, the speed information derived from the bus operation log collected by the B-Map application is within the commonsense range. This allows administrators to browse nearly accurate bus routes and speed information. Also, bus drivers can save time and effort to change the bus stop name to make the bus easier.

4.3 Evaluation for B-Map Web Service

In Figs. 10 and 11, in the B-Map Web system, the icons of the buses in areas where communication cannot be made are made transparent or translucent, and differentiation from the bus icons in the communicable area can be aimed. This made it possible for Uwajima citizens to distinguish bus location when using B-Map web service.

5 Discussion

In 2016 B-Map the system remodeling project, we decided the function to solve the problem from the request of Uwajima City Hall and implement it. As a result, it was a project that got a very high evaluation. Especially the reputation of the two functions was high. The first one is the automatic change function of the bus stop. This function reduces input errors and change mistakes, and furthermore, reduces the driver's work. The second is the change of the layout. This change improves the convenience of the driver and further makes it easier to teach the operation of the tablet to the newly appointed driver.

6 Conclusion

In this paper, we implemented a project to renovation "B-Map" operation management system of Uwajima City Tsushima district community bus which is ICT course.

In the project, three subjects affected by systematization were defined as "managers" "bus drivers" "bus users" in order to solve the two necessary problems. After that, we met city officials based on the system renovation plan based on the request of Uwajima city, formulated the project plan, and renovated the system. we improved the correspondence and layout of areas that can not communicate mainly in coastal areas and mountainous areas, implemented new functions in the B-Map application proposed from the user's point of view, and received the high evaluation.

Reference

1. Hirokawa, T., Hayashi, M., Miyase, Y., Watanabe, K., Endo, K., Kuroda, H., Kobayashi, S.: Development of management system for community bus operation. In: Proceedings of International Conference on Business and Industrial Research (ICBIR 2016) (2016)

Part V
Computer Vision Techniques and Applications

Novel Human Action Recognition in RGB-D Videos Based on Powerful View Invariant Features Technique

Sebastien Mambou, Ondrej Krejcar⬤, Kamil Kuca⬤
and Ali Selamat⬤

Abstract Human action recognition is one of the important topic in nowadays research. It is obstructed by several factors, among them we can enumerate: the variation of shapes and postures of a human been, the time and memory space need to capture, store, label and process those images. In addition, recognize a human action from different view point is challenging due to the big amount of variation in each view, one possible solution of mentioned problem is to study different preferential View-invariant features sturdy enough to view variation. Our focus on this paper will be to solve mentioned problem by learning view shared and view specific features applying innovative deep models known as a novel sample-affinity matrix (SAM), able to give a good measurement of the similarities among video samples in different camera views. This will also lead to precisely adjust transmission between views and study more informative shared features involve in cross-view actions classification. In addition, we are proposing in this paper a novel view invariant features algorithm, which will give us a better understanding of the internal processing of our project. We have demonstrated through a series of experiment apply on NUMA and IXMAS (multiple camera view video dataset) that our method out performs state-of-the-art-methods.

S. Mambou · O. Krejcar (✉) · K. Kuca · A. Selamat
Faculty of Informatics and Management, Center for Basic and Applied Research,
University of Hradec Kralove, Rokitanskeho 62, 500 03 Hradec Kralove, Czech Republic
e-mail: ondrej.krejcar@uhk.cz

S. Mambou
e-mail: sebastien.mambou@uhk.cz

K. Kuca
e-mail: kamil.kuca@uhk.cz

A. Selamat
e-mail: aselamat@utm.my

A. Selamat
Faculty of Computing, Universiti Teknologi Malaysia, 81310 Johor Baharu, Johor, Malaysia

© Springer International Publishing AG, part of Springer Nature 2018   343
A. Sieminski et al. (eds.), *Modern Approaches for Intelligent Information
and Database Systems*, Studies in Computational Intelligence 769,
https://doi.org/10.1007/978-3-319-76081-0_29

Keywords Action recognition · View point · Sample-affinity matrix
Cross-view actions · NUMA · IXMAS

1 Introduction

Data taken from human body are pervasive and represent an important interest to
computer vision communities [1, 2] and machine learning [3, 4]. Sometimes, these
data are used for action detection from multiple Views, e.g. a set of dynamic actions
capture by several cameras views, as on (Fig. 1). In Cross-View scenario a classification on such set of action data is difficult as every original data sources are
recorded by different cameras at various locations, with a possibility to appear
totally different. Figure 1a, is a good example where an action recorded from side
view looks different compared to the one recorded from a top view. Consequently,
features extracted from one view cannot be used as sufficient discriminative factors
in different angle. Several papers have focused their attention on how to develop
view invariant images for action recognition [5, 6]. Images are normally taken from
video based on frame rate per second. Also, there is a Self-Similarity Matrix
(SSM) descriptor in several approaches [5, 7] to detect and restore user actions in
different views which can illustrate quality of their solution. Every detected action is
shared among different angles and also available in each other views [7, 8]. They
suggest to have same quality detected actions from shared features as from original
one taken from different views. Unfortunately this prediction cannot be valid as the
discriminative parameters in one view, may be incredibly far away from those in
other views (an example can be the top view in Fig. 1a). Thus, indubitably confusion of classifiers might be the cause, as they did not restrict information sharing
among action categories, which could lead to the consequence of producing same
detected actions for different movies taken from same point of view.

Within this paper, we are proposing developed deep networks which can learn
data for detection of human actions in different view angles. We are declaring that

Fig. 1 Illustration of a
multi-view scenarios where
a illustrates how human
activities are record from
multiple viewpoints,
b however represents how
multiple sensors are attached
on the human body with the
purpose of collecting
sufficient action data

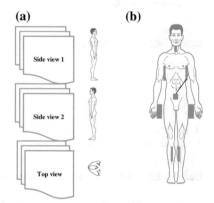

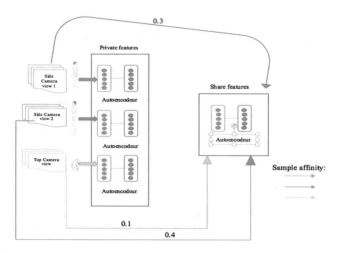

Fig. 2 An overview of our feature's extractor

we are able to use our computed human action data as a substitution of original data in different views which are not originally available. For this goal we are developing a new Sample Affinity Matrix (SAM) which is able to provide a good measurement of the similarities among available data from different camera views [9]. We will be able to accurately adjust transfer for different views.

Performance of our approach can be further improved by setting of layers labels and information of features. Authors develop this feature learning problem in a diminished autoencoder framework (Fig. 2) [10], which we are trying to overcome with our approaches. SAM is gracefully included in the learning of shared features, with a goal to evaluate several different solutions to detect a human action recognition from video samples [10]. In contrast of SSM [5], SAM focuses its attention on similarity of samples. SSM is based on the similarity of different video frames. We tried our approach on three freely available multi-view datasets, while we come to know that our approach significantly outperforms state-of-the- art approaches. The remaining part of this paper is structured as follows. We will present related work on our topic, follow by a sharp study of views invariant features, immediately after will be presented our deep architecture where our approach is deeply explained and our algorithm is shown, and never the less, we will prove our method to be better than state-of-the-art methods by the help of a series of experiments.

2 Problem Definition of Our Research

Many View invariant techniques struggle on how to extract the similarity among several frames recorded at the same time and from different angles of a RGB Camera. This is not the case of the RGB-D camera which offers the depth property, powerful enough to allow us to extract the target area from the plan. However another Struggle is noticed on RGB-D, when comes the time to create an artificial Views which content all the features relevant to understand the action. We will introduce in this paper a novel algorithm base on Simple Affine Matrix (SAM) and a several Autoencoders (Fig. 2) powerful enough, to help us to extract share and private features relevant to the identification of the action being perform by the subject.

3 Related Work

As referred in [4], Multi-view study which focus on finding similarities between data of two individual views. It is not the only method, several have been published with a particularity of focusing, their attention on discriminative and expressive features from low level observation [11–17]. Co-training approach is one of those methods, it trains several learning algorithms for each view, and looks for the specific connections between two different data sets and different views. Authors in [18, 19] came respectively with Canonical correlation analysis (CCA), which was used to keep track of the common space among multiple views, and secondly the two projection matrices applied onto a common feature space so that they map multi-modal data. Also in [20], a study of an incomplete view problem came out with an assumption which is "from shared subspace, different views are generated". Mian [21] presented another method known as generalized multi view analysis (GMA) which is an extension of CCA. As Liu et al. [13] which presented a matrix factorization apply in multi-view clustering, [12] came with a collective matrix factorization (CMF) approach which learns correlations among relational feature matrices. In the same way, Ding et al. [22] solved the multi-view learning scenario problem, by introducing low rank constrained matrix factorization model. Many studies attempt to solve the issues, related to view-invariant action recognition so that, given multiple view samples, they generate activities labels. Authors in [23] presented a method which achieves a hierarchical classification of the 3D Histogram of Oriented Gradients (HOG), as well as local partitioning with the goal of representing consecutive images. Shan [24] and Camps [25] show SSM-based method which retrieves view-invariant descriptors within a log-polar block on the matrix by computing frame-wise similarity matrix in a given video. However, the representation power of SSM can be improved by using a multitask learning approach [7], which consist of sharing discriminative SSM features among views as investigating in [6, 8, 26, 27] where more specific methods can be found. Authors in [6] presented MRM-Lasso approach which consists on keep latent corrections through different views. It is done by studying a low-rank matrix which consists of pattern-specific weights. Jiang [8] and Zhang [28] however, came with transferable dictionary pairs which favor the common feature space to be sparse. As a comparison with another approach: the indifference of common multi-view learning methods [17, 19, 20, 29], our approach gives us the ability to store several layers of learners, to study in a better manner view-invariant features. Furthermore, to capture complex motion information uniquely found in specific views, we use private features in our methods; also we favor the incoherence between shared features and private features. If we make a comparison between our approach and the knowledge sharing approaches [6, 8, 26, 30, 31], our method equipoises information sharing among views, based on sample similarities. By considering the possibility that a data samples can appear similar in some views, our approach gives us the possibility of better differentiate various categories. As [29] compute within-class and between-class Laplacian matrices, SAM Z method directly takes between class and with-in class between-view information [9]. In addition, the distance between two views of the same sample is measured in our work with the help of SAM Z, in the difference of [29] which doesn't encode such distance Table 1.

Table 1 Pro and cons of some methods used so far

Propose solutions	Pros	Cons
Multi-view learning methods [11–16]	Attention on discriminative and expressive features	Low interest on private feature of the views
Cooperative approach [17]	Goal is to try several learning algorithms at each selected view. Consequently find some connection between two sets of different views data	Can deal with more than 2 views at the same time
[18, 19]	Keep track of the common space among multiple views, and use the two projection matrices to apply onto a common feature space so that they map multi-modal data	Don't pay sufficient attention on private features
[23] method	It achieves a hierarchical classification of the 3D Histogram of Oriented Gradients (HOG)	Doesn't stored sufficient layers of learners
[29] method	Computes within-class and between-class Laplacian matrices	The distance between two views of the same sample is not measured
Our approach	–It has the ability to store several layers of learners, to study in a better manner view-invariant features –our method equipoises data shared between views –the distance between two views of the same sample is measured with the help of SAM Z	–As a limited number of view to process at the same time, due to the high amount of computation required –required a lot of computer resources

4 Design of Proposed Solution—Deep Architecture

By going through the papers [10] and [32], we came out with a deep model made by stacking several layers of features learners discussed before in Single Layer Feature Learning. Layer by Layer, we applied a nonlinear feature mapping (defines here as a nonlinear squashing) function $\theta(.)$ on an output of each layer, so that: $C_g^v = \theta(X^v G^v)$ and $C_w = \theta(XW)$, we come out with a sequence of hidden features matrices. To train our networks $\{G_k^v\}_{v=1,k=1}^{V,K}$ and $\{W_k\}_{k=1}^{K}$ containing K layers, here we use a scheme "layer wise" training. It is good to mention that the input of the $(k+1)^{th}$ layer is the outputs of the nth layer C_{kg}^v and C_{kw}, which is the training input of our Matrix $\{G_{k+1}^v\}_{v=1}^{V}$ and W_{k+1}. Also by considering $k = 0$ (meaning the first layer as we have K layer), X and $X^v X^v$ have for raw features C_{0g}^v and C_{0w} respectively.

4.1 Flow Chart of Our Project

Our project can be resume in few simple steps explicated as follow:

- **At the same time, take 3 Videos from different View point**: Here, we attempt to have a clear picture of the subject from different angle around him. These images will be submit to the next component of the project as input.
- **let's extract important features from those images**: As those images are taken at the same time for the same subject, they have in common some features known as share feature and each image has its one feature called private feature. These features will be submit to the next component of the project as input.
- **Using a Novel Invariant feature Algorithm**: at this point, we are in the learning phase, relevant for the process.
- **Generate the target Views**: illustrate by (Fig. 3), we will compute the simple affinity matrix Z of each arrow, also the W mapping matrix and we will generate the target view containing each relevant feature useful to understand the action.
- **Assign a label and a description of the action being perform.**

4.2 Novel View Invariant Features Algorithm

We compute with our **supervising** approach, the similarity between source views and our target view (Fig. 3). This can be achieved by the help of our view shared mapping matrix W, which help us to learn share feature so that we compute weight value (z). The arrow with a biggest weight will the one between the more similar source view and our target view.

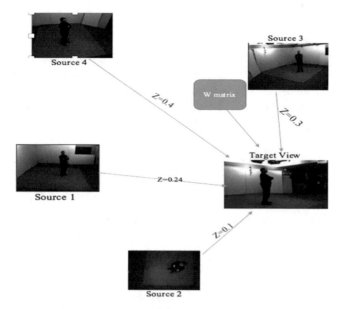

Fig. 3 Extraction of features from sources images and generation of the target view. Image provided from the public Dataset IXMAS

5 Experiment

Using three multiple view datasets as the Daily and Sports Activities (DSA) dataset [3], multi-view IXMAS dataset [33] and Northwestern- UCLA Multiview Action 3D (NUMA) dataset [9], we assess our approach. It is good to remain that all the datasets use here, are used in several papers like [3, 6, 8, 26, 27]. One-to-One and Many-to-One are considerate here as two cross view classification scenarios. The first one is trained on One view and assesses on another view while, the second is trained on V-1 views and assesses on the remaining view. We also adopted an intersection kernel support vector machine (IKSVM) as classifier with parameters $C = 1$, $\gamma = 1$, $\beta = 1$, $\alpha = 1$, $p = 0$ and $K = 1$ are default parameters for our approach with 1 as default number of layers. We consider for our experiment **NUMA** and **IXMAS**, which are multiple camera view video dataset. The first one obtained from 3 Kinect sensors in 5 environments, consists of 10 human actions and the second one obtained from 1 top view camera and 4 side view camera. As in [33] and [34–36] respectively, a bag of words is used here and, a set of global frame-based and local spatiotemporal trajectory based descriptors described an Action video. Furthermore, to build a video words and quantize those descriptors, we used k-means clustering method which resulted to the possibility of represent a given video with a histogram of the feature vector. Also it is good to know that a V feature vectors is constituted a representation of an action recorded by V camera views points.

5.1 Use of IXMAS Dataset

As on [10], a histogram of oriented optical flow and dense trajectory are obtained from the videos and by applying K-means, a dictionary (size 2000) is produced for each type of features. Also, to convert each video into a feature vector and encode those feature, we used the bag of words model. Similar to [8, 27], we use "leave one

Table 2 One-to-one cross-view recognition results of various supervised approaches on ixmas dataset. Numbers represent results of a particular methods [8, 28] and our supervised approach

	Test case 0	Test case 1	Test case 2	Test case 3	Test case 4
Training case 0	(71, 99.4, 99.1, **100**)	(82, 96.4, 99.3, **100**)	NA	(76, 97.3, **100**, 100)	(72, 90.0, 96.4, **100**)
Training case 1	(80, 85.8, 99.7, **100**)	(77, 81.5, 98.3, **100**)	(73, 93.3, 97.0, **100**)	(72, 83.9, 98.9, **100**)	NA
Training case 2	(75, 98.2, 90.0, **100**)	(75, 97.6, 99.7, **100**)	(73, **99.7**, 98.2, 99.4)	NA	(76, 90.0, 96.4, **100**)
Training case 3	(72, 98.8, **100**, 100)	NA	(74, **99.7**, 97.0, **99.7**)	(70, 92.7, 89.7, **100**)	(66, 90.6, **100**, 99.7)
Training case 4	NA	(79, 98.8, 98.5, **100**)	(79, 99.1, 99.7, **99.7**)	(68, 99.4, 99.7, **100**)	(76, 92.7, 99.7, **100**)
Average	(74, 95.5, 97.2, **100**)	(77, 93.6, 98.3, **100**)	(76, 98.0, 98.7, **99.7**)	(73, 93.3, 97.0, **100**)	(72, 92.4, 98.9, **99.9**)

Table 3 One-to-one cross-view recognition results of various unsupervised approaches on ixmas dataset. Numbers represent results of a particular methods [8, 26–28], and our unsupervised approach, respectively

	Test case 0	Test case 1	Test case 2	Test case 3	Test case 4
Training case 0	(79.6, 92.1, 99.4, 82.4, 72.1, **100**)	(76.6, 89.7, 97.6, 79.4, 86.1, **99.7**)	NA	(79.8, 94.9, 91.2, 85.8, 77.3, **100**)	(72.8, 89.1, **100**, 71.5, 62.7, 99.7)
Training case 1	(82.0, 83.0, 87.3, 57.1, 48.8, **99.7**)	(68.3, 70.6, 87.8, 48.5, 40.9, **100**)	(74.0, 89.7, 92.1, 78.8, 70.3, **100**)	(71.1, 83.7, 90.0, 51.2, 49.4, **100**)	NA
Training case 2	(73.0, 97.0, 87.6, 82.4, 82.4, **100**)	(74.1, 94.2, 98.2, 80.9, 79.7, **100**)	(74.0, 96.7, 99.4, 82.7, 70.9, **100**)	NA	(66.9, 83.9, 95.4, 44.2, 37.9, **100**)
Training case 3	(81.2, 97.3, 97.8, 95.5, 90.6, **100**)	NA	(75.8, 96.4, 91.2, 77.6, 79.7, **99.7**)	(78.0, 89.7, 78.4, 86.1, 79.1, **99.4**)	(70.4, 81.2, 88.4, 40.9, 30.6, **99.7**)
Training case 4	NA	(79.9, 96.7, 99.1, 92.7, 94.8, **99.7**)	(76.8, 97.9, 90.9, 84.2, 69.1, **99.7**)	(76.8, 97.6, 88.7, 83.9, **98.9**)	(74.8, 84.9, 95.5, 44.2, 39.1, **99.4**)
Average	(79.0, 94.4, 93.0, 79.4, 74.5, 99.9)	(74.7, 87.8, 95.6, 75.4, 75.4, **99.9**)	(75.2, 95.1, 93.4, 80.8, 72.5, **99.9**)	(76.4, 91.2, 87.1, 76.8, 72.4, **99.9**)	(71.2, 84.8, 95.1, 50.2, 42.6, **99.7**)

action class out" training scheme for equitable juxtaposition. We use one action class each time we want to test. In addition, all the videos in this action are removed from the feature learning steps including k-means, in order to assess the information transfer's potency of our approach. Also let keep in mind that we can see those videos in training the action classifier. Let's evaluate our approach Table 2.

In our experiment only one view is utilized as test view and the other views are utilized as training views. This experiment helps us to evaluate our approach which use learned private and shared features.

Our supervised approach is compared with well-known approach [23, 24, 27, 28]; with the incoherence in equation [8], and with the importance of SAM Z. We have observed an average of 99.9% achieved by our supervised approach as shown in Table 3.

6 Conclusions

Our Goal in this paper was to propose a method which is able to recognize human action from a cross-view. For this purpose, we have focused our attention on two novel approaches which have been utilized efficiently on shared and private features to accurately characterize human activities. These characteristics also cover appearance variations and wide range of angles in different views. We have also introduced in this paper, the sample affinity matrix to allow a computation of equivalence for different views using sample data. As for the need of precise computation of the quality of every sample to the shared features, our matrix is gracefully included in the process of keeping track of shared features, and to weight data transfer. We have also performed several experiments on the IXMAS, and NUMA datasets where results from experiment confirm that in case of cross-view action classification, our solution outperform state-of-the-art approaches. We have seen through this paper the potential of our approach, we are planning for the next step to instead of taking image extract during a time t, define manually in the experiment, we will deal with the space time and extract flow of action of a subject so that we can deal with activities. We will modify our algorithm and our model so that we can detect with accuracy a big number of activities.

Acknowledgements This work and the contribution were supported by The Faculty of Informatics and Management, University of Hradec Kralove, Czech Republic.

References

1. Fu, Y., Kong, Y.: Bilinear heterogeneous information machine for RGB-D action recognition. In: Proceedings of the IEEE Conference on Computer Vision and Pattern Recognition (CVPR), pp. 1054–1062 (2015)
2. Fu, Y., Kong, Y.: Max-margin action prediction machine. IEEE Trans. Pattern Anal. Mach. Intell. **38**, 1844–1858 (2016)

3. Barshan, B., Tunçel, O., Altun, K.: Comparative study on classifying human activities with miniature inertial and magnetic sensors. Pattern Recognit. **43**, 3605–3620 (2010)
4. Nanopoulos, A., Schmidt-Thieme, L., Grabocka, J.: Classification of sparse time series via supervised matrix factorization. In: Proceeding of the AAAI, pp. 928–934 (2012)
5. Dexter, E., Laptev, I., Pérez, P., Junejo, I. N.: Cross-view action recognition from temporal self-similarities. In: Proceeding of the ECCV, pp. 293–306 (2008)
6. Gao, Y., Shi, Y., Cao, L., Yang, W.: MRM-lasso: a sparse multiview feature selection method via low-rank analysis. IEEE Trans. Neural Netw. **26**, 2801–2815 (2015)
7. Ricci, E., Subramanian, S., Liu, G., Sebe, N., Yan, Y.: Multitask linear discriminant analysis for view invariant action recognition. **23**, 5599–5611 (2014)
8. Jiang, Z., Zheng, J., Phillips, J., Chellappa, R.: Cross-view action recognition via a transferable dictionary pair. In: Proceeding of the British Machine Vision Conference, pp. 125.1–125.11 (2012)
9. Kong, Y., Ding, Z., Li, J., Fu, Y.: Deeply learned view-invariant features for cross-view action recognition. IEEE Trans. Image Process. (2017)
10. Xu, Z., Weinberger, K., Sha, F., Chen, M.: Marginalized denoising autoencoders for domain adaptation. In: Proceeding of the ICML, pp. 1627–1634 (2012)
11. Guo, Y., Zhou, J., Ding, G.: Collective matrix factorization hashing for multimodal data. In: Proceeding of the CVPR, pp. 2075–2082 (2014)
12. Daumé, H., Kumar, A.: A co-training approach for multi-view spectral clustering. In: Proceeding of the ICML, pp. 393–400 (2011)
13. Zhang, K., Gu, P., Xue, X., Zhang, W.: Multi-view embedding learning for incompletely labeled data. In: Proceeding of the IJCAI, pp. 1910–1916 (2013)
14. He, R., Wang, W., Wang, L., Tan, T., Wang, K.: Learning coupled feature spaces for cross-modal matching. In Proceeding of the ICCV, pp. 2088–2095 (2013)
15. Tao, D., Xu, C., Xu, C.: Multi-view learning with incomplete views. IEEE Trans. Image Process. **24**, 5812–5825 (2015)
16. Kumar, A., Daume, H., Jacobs, D.W., Sharma, A.: Generalized multiview analysis: a discriminative latent space. In: Proceeding of the CVPR, pp. 2160–2167 (2012)
17. Fu, Y., Ding, Z.: Low-rank common subspace for multi-view learning. In Proceeding of the IEEE International Conference on Data Mining (ICDM), pp. 110–119 (2014)
18. Özuysal, M., Fua, P., Weinland, D.: Making action recognition robust to occlusions and viewpoint changes. In: Proceeding of the ECCV, pp. 635–648 (2010)
19. Dexter, E., Laptev, I., Pérez, P., Junejo, I.N.: Cross-view action recognition from temporal self-similarities. In: Proceeding of the ECCV, pp. 293–306 (2008)
20. Dexter, E., Laptev, I., Perez, P., Junejo, I.N.: View-independent action recognition from temporal self-similarities. IEEE Trans. Pattern Anal. Mach. Intell. **33**, 172–185 (2011)
21. Mian, A., Rahmani, H.: Learning a non-linear knowledge transfer model for cross-view action recognition. In: Proceeding of the CVPR, pp. 2458–2466 (2015)
22. Shah, M., Kuipers, B., Savarese, S., Liu, J.: Cross-view action recognition via view knowledge transfer. In: Proceeding of the CVPR, pp. 3209–3216, June 2011
23. Jiang, Z., Zheng, J.: Learning view-invariant sparse representations for cross-view action recognition. In: Proceeding of the ICCV, pp. 3176–3183 (2013)
24. Shan, S., Zhang, H., Lao, S., Chen, X., Kan, M.: Multi-view discriminant analysis. IEEE Trans. Pattern Anal. Mach. Intell. **38**, 188–194 (2016)
25. Camps, O.I., Sznaier, M., Li, B.: Cross-view activity recognition using hankelets. In: Proceeding of the CVPR, pp. 1362–1369 (2012)
26. Wang, C., Xiao, B., Zhou, W., Liu, S., Shi, C., Zhang, Z.: Cross- view action recognition via a continuous virtual path. In: Proceeding of the CVPR, pp. 2690–2697 (2013)
27. Salakhutdinov, R.R., Hinton, G.E.: Reducing the dimensionality of data with neural networks. Science **313**, 504–507 (2006)
28. Zhang, T., Luo, W., Yang, J., Yuan, X., Zhang, J., Li, J.: Sparseness analysis in the pretraining of deep neural networks. IEEE Trans. Neural Netw. Learn. Syst. (to be published). https://doi.org/10.1109/tnnls.2016.2541681

29. Weinberger, K., Sha, F., Bengio, Y., Chen, M.: Marginalized denoising auto-encoders for nonlinear representations. In: Proceeding of the ICML, pp. 1476–1484 (2014)
30. Larochelle, H., Lajoie, I., Bengio, Y., Manzagol, P.A., Vincent, P.: Stacked denoising autoencoders: learning useful representations in a deep network with a local denoising criterion. J. Mach. Learn. Res. **11**, 3371–3408 (2010)
31. Ronfard, R., Boyer, E., Weinland, D.: Free viewpoint action recognition using motion history volumes. Comput. Vis. Image Understand. **104**, 249–257 (2006)
32. Tabrizi, M.K., Endres, I., Forsyth, D.A., Farhadi, A.: A latent model of discriminative aspect. In: Proceeding of the ICCV, pp. 948–955 (2009)
33. Martinez, J., Little, J.J., Woodham, R.J., Gupta, A.: 3D pose from motion for cross-view action recognition via non-linear circulant temporal encoding. In: Proceeding of the CVPR, pp. 2601–2608 (2014)
34. Nie, X., Xia, Y., Wu, Y., Zhu, S.C., Wang, J.: Cross-view action modeling, learning and recognition. In: Proceeding of the CVPR, pp. 2649–2656 (2014)
35. Rabaud, V., Cottrell, G., Belongie, S., Dollar, P.: Behavior recognition via sparse spatio-temporal features. In: Proceeding of the VS-PETS, pp. 65–72, Oct 2005
36. Kläser, A., Schmid, C., Liu, C.L., Wang, H.: Dense trajectories and motion boundary descriptors for action recognition. Int. J. Comput. Vis. **103**, 60–79 (2013)

Study of CNN Based Classification for Small Specific Datasets

Huu Ton Le, Thierry Urruty, Marie Beurton-Aimar, Thi Phuong Nghiem, Hoang Tung Tran, Romain Verset, Marie Ballere, Hien Phuong Lai and Muriel Visani

Abstract Recently, deep learning and particularly, Convolutional Neural Network (CNN), has become predominant in many application fields, including visual image classification. In an applicative context of detecting areas with hazard of dengue fever, we propose a classification framework using deep neural networks on a limited dataset of images showing urban sites. For this purpose, we have to face multiple research issues: (i) small number of training data; (ii) images belonging to multiple classes; (iii) non-mutually exclusive classes. Our framework overcomes those issues by combining different techniques including data augmentation and multiscale/region-based classification, in order to extract the most discriminative information from the data. Experiment results present our framework performance using several CNN architectures with different parameter sets, without and with transfer learning. Then, we analyze the effect of data augmentation and multiscale region based classification. Finally, we show that adding a classification weighting scheme allows the global framework to obtain more than 90% average precision for our classification task.

Keywords CNN · Image classification · Small and dedicated dataset

H. T. Le (✉) · T. P. Nghiem · H. T. Tran · H. P. Lai
ICT Lab, University of Science and Technology of Hanoi (USTH),
VAST, 18 Hoang Quoc Viet, Cau Giay, Hanoi, Vietnam
e-mail: le-huu.ton@usth.edu.in

T. Urruty
XLIM, UMR CNRS 7252, University of Poitiers, Poitiers, France

M. Beurton-Aimar
LaBRI, University of Bordeaux, Bordeaux, France

R. Verset · M. Ballere
ENSEEIHT, Toulouse, France

H. P. Lai
M&S lab, UMI UMMISCO 209, IRD, FCSE, Thuyloi University, Hanoi, Vietnam

M. Visani
Laboratory L3i, University of La Rochelle, La Rochelle, France

© Springer International Publishing AG, part of Springer Nature 2018
A. Sieminski et al. (eds.), *Modern Approaches for Intelligent Information and Database Systems*, Studies in Computational Intelligence 769,
https://doi.org/10.1007/978-3-319-76081-0_30

1 Introduction

This study was led in the framework of the SWARMS project (Say and Watch: Auto-mated image/sound Recognition for Mobile monitoring Systems), implemented in Ha Noï, Vietnam. This project, led with epidemiologists and computer scientists, aims at better understanding (and eventually preventing) dengue fever epidemics. It is based on monitoring urban sites in order to detect potential risk zones for dengue fever transmission: mainly sites where rain water is collected or kept. The epidemi-ologists are especially interested in four types of zones: construction sites, pagodas, ponds and rubbish, which must be distinguished from "negative" images of Ha Noi streets (see Fig. 1). In a country like Vietnam where epidemiologists do not have access to any official record of such sites (except maybe pagodas), the SWARMS project relies on geo-localized pictures uploaded in OpenStreetMap. The dataset we consider is of very limited size: 1000 images distributed among 5 classes (including a negative class) to detect that were collected during field trips led in the framework of the project.

Recently, deep learning methods have spectacularly improved the state-of-the-art in visual object recognition and scene understanding. Following the 2012 ImageNet competition where deep Convolutional Neural Networks (CNNs) have almost halved the classification error rates of the other best competing approaches [1] on a data set of about a million images from 1,000 different classes, deep learning has been quickly adopted by the computer vision community. This success has been made possible thanks to the technological advances and increases in the amount of avail-able data [2], in particular the efficient use of GPUs, simple activation functions (e.g. ReLUs), regularization methods such as dropout, and data augmentation tech-niques that generate more training examples, generally by deforming the existing ones. Many new learning algorithms and architectures are currently being developed for deep neural networks, in order to further enhance their performances and/or effec-tiveness. Since the AlexNet architecture that achieved the ImageNet 2012 milestone [1], current developments for object classification include the development of very deep networks such as VGG16/VGG19 [3] and MSRANET [4], the development of wider deep networks, e.g. by adding Inception modules in the architecture [5, 6] and the development of wider and deeper, yet efficient, networks such as RsNet [7], Inception V4 + RsNet [8] and Identity RsNet [9]. Because of their very numerous parameters to adjust, deep learning models generally require massive amounts of annotated data to be trained, which might be very costly and even almost impossible

Fig. 1 Examples of the four classes to detect (construction sites, rubbish, ponds and pagoda) and the negative class

in some applications. To circumvent this problem, researchers can use tricks such as data augmentation, and transfer learning [10, 11].

In this paper, we present a framework to adapt the CNN to a very small and dedicated dataset. Our experiments show that by combining different techniques, we obtain more than 90% average precision for our classification task. We present our framework in Sect. 2, follow by experiments and discussion in Sect. 3. Section 4 concludes our research.

2 Proposed Framework

As mentioned in previous section, our research work is dedicated to a specific dataset which presents several restrictions: (i) small number of annotated image data; (ii) images may belong to multiple classes and (iii) our classes are non-mutually exclusives. Our framework, illustrated in Fig. 2, employs different techniques to overcome those issues. First, we propose some data augmentation techniques to enrich the training data. Then, a multi-scale and region-based classification extracts the most discriminative information from the data. A classification weighting scheme is learned to help increasing the classification performance in general. We detail each stage of our global framework in the following sub-sections.

2.1 Data Augmentation

Many applicative contexts have to manage small and very specific datasets, distributed into few categories. Such small amount of data is far from enough to properly train a CNN. For this reason, we apply several image augmentation techniques to enrich the training data:

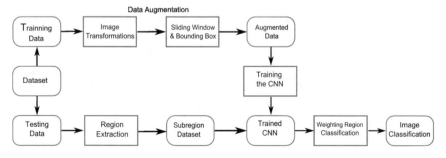

Fig. 2 The proposed framework

- Rescaling: original images are resized into different resolutions based on the smaller size (height or width) of them;
- Rotating: images are rotated of 5 °C to the left and 5 °C to the right;
- Flipping: images are vertically and horizontally flipped;
- Cropping: images are cropped, maintaining undistorted information.

By applying the above techniques, we multiply our image dataset by the number of selected image transformations. Following this stage, other augmentation techniques called sliding window and bounding box are applied to our dataset of transformed images including original ones to extract sub-regions of the images. We illustrate these techniques in Fig. 3. The sliding window technique scans the whole image shifting little by little by a small predefined step. Each position of the window returns a small image region that can be used to train the CNN. To do so, we need to extract an annotation that is linked to the image region. Figure 3 illustrates this methodology: the yellow windows present different regions extracted in an image containing a construction site. Although the original image is classified as a specific class (in the example "construction site"), the window-regions extracted from this image may not all belong to the same class. For example, in Fig. 3, the window-region number 1 should not be labeled as construction site as it contains mostly garbage and trees. Thus, we need to manually define a bounding box for each original image in the training set. This bounding box, the red one in Fig. 3, contains the region in the image that corresponds exactly to the specific class it belongs to. Then, a window-region is considered to be from the same class as the original image if it covers a proportion of the bounding box above a user fixed threshold. We call those window-regions, the positive ones. Note that, our exhaustive experiment show that a threshold of 70% is suitable for labeling the window-regions. In Fig. 3, window-region 2 is then labeled as a positive region as it covers more than 70% of the red bounding box, however window-region 1 is a negative region and it will not be used in the training of our proposed CNN framework. Note that, there is a difference between negative regions of any class image, which are discarded regions, i.e. not used to train the CNN and positive regions of the negative class that will be used to train it.

As output of the whole data augmentation step, we obtain a training dataset which contains much more data to train the CNN. For our specific dataset, 112511 positive window-regions are extracted from 1000 original images (200 images per class) as detailed in Table 1. Those positive regions are used for training our CNN in the next step. Note that, we have a lot of positive window-regions for the negative class because there is no bounding box in the negative class, so all possible regions extracted from "negative" class images are consider as positive to train the CNN.

2.2   Training a CNN

Our next step is to train a CNN for image classification. Three approaches have been developed and tested. The first approach, we build a whole new CNN from scratch and train it with the augmented dataset we obtained. Creating a new network comes

Fig. 3 Sliding windows and bounding box

Table 1 Number of window-regions after data augmentation

Class	Window-regions
Construction site	2685
Pagoda	3457
Pond	2611
Rubbish	1979
Negative	101779

with the problem of setting the numerous possibilities that such a network offer: setting number of layers, number of filters each layer, size of the filters, etc. Tuning those parameters to best fit the specific dataset and its application is still a headache to most researchers.

The second approach, we adopt an existing pre-trained CNN. This methodology has the advantage to use the features resulting from deep learning on massive image datasets. Then, these "generic" features are used to described our image dataset, and a classification algorithm as Support Vector Machine (SVM) is applied on the image features. This transfer learning approach can be used for different purposes such as Content Based Image Retrieval (CBIR) [12, 13], object detection [14, 15] or classification with other datasets of different natures [13]. We adopt the popular Inception v3 model [16] to extract features from our image data and use those features to build a classifier with SVM method. Inception v3 is a CNN which has been trained using ImageNet dataset [17] in order to build a classifier of 1000 categories.

Finally, the third approach is fine-tuning an existing CNN using the augmented dataset as training input. As the features extracted by CNN are task oriented, they work well on the data of similar characteristics than the training data, but may not be very effective for specific dataset or task. This problem can be solved by fine tuning the CNN [13, 18, 19]. The advantages of fine tuning a CNN are two folds: it is computationally less expensive than training a whole new CNN, and it helps to adapt the CNN features for a specific task. Similarly to the previous method, we use the Inception v3 architecture. However, we keep the first 19 layers untouched and train the last 3 layers of the network.

2.3 Multiscale and Region-Based Image Classification

As already mentioned before, image classification is not bound to one class per image. In fact, one image may belong to multiple classes. We deal with this issue by classifying regions of image using the trained CNN. Indeed, the classifier works with multiscale and region images to capture both global as well as local details of an image. We transform the image to be classified into different scales as we did to the training images. The images are then divided into sub-regions of the same dimension as the window-region. All the regions of an image are given as input to the trained CNN individually. The CNN returns a classification score for each of them. Finally, the classification scores of all regions are combine together to have the single classification score for the whole images. Figure 4 demonstrates this approach. The colors of each window represent an example of the result of the image classification by CNN, one color per class. In this example, we can see the black color for negative class, blue for pagoda, yellow for pond, green for rubbish and red for construction site. This figure also presents the multi-class problem linked to a specific dataset as one single image can be classified into several classes.

Fig. 4 Multiscale, region-based classification (black color for negative class, blue for pagoda, yellow for pond, green for rubbish and red for construction site)

2.4 Classification Weighting

Another issue of our dataset is that we have non-mutually exclusive classes as there are some regions or details that may appear in multiple classes. For example, a region that contains a wall may belong to both construction site or pagoda class, the only part that separates the image of a pagoda from construction site might be the roof of the pagoda. This phenomenon is very frequent in any dataset. Moreover, the object of interest in an image may cover a small area of the image while the framework classifies every window-regions of the image. Thus, many regions are misclassified (or classified as part of the negative class). Thus, the discriminative power of each window-region classification score for an image is different, they should not contribute equally in the final image classification decision. For instance, the negative windows mostly appear in every images, even in the image of other classes. On the contrary, a pagoda window is very discriminative since it appears only in the pagoda-class images. That is why we propose to include in our framework one last stage which is a weighting scheme specifically built with respect to the applicative context. For example, a bonus weighting on the few but discriminative image window-regions classified as "pagoda" may improve the classifier accuracy. As well as a penalty weighting is possible for "negative" regions as they outnumbers other classes. Our weighting scheme could be seen as a final layer of our neural network. As input, we take all window-region classification scores, and as output, we have all whole image classification scores. Thus, we propose a simple logistic regression algorithm to learn the weighting coefficient matrix. Our experiments show the importance of such a scheme in our framework to improve the classifier accuracy for a specific applicative dataset.

3 Experiments and Discussions

This section presents our specific experimental setup and the technical choices made. Then, we present a summary of the exhaustive experiments evaluating the performance of our proposal.

3.1 Setup and First Results

3.1.1 SWARMS Dataset

As mentioned in the Introduction Section, in the SWARMS project, we consider a specific dataset of 800 images collected during field trips. This dataset is divided into 4 classes "Pagoda", "Pond", "Construction site" and "Rubbish". To complete these classes, we have added a "Negative" class with 200 images from the Internet that are different from other classes. From this very small set of images, we apply the global framework. After the augmentation stage, we obtain more than 100 000 regions as detailed in Table 1.

3.1.2   Building a Whole New CNN

First, we implement a whole new CNN from scratch with lots of different setups. For example, we evaluate the performance using 5, 7 and 9 layers for the CNN with different filters, parameters, etc. For most of the setups we have tried, the average results was around 60% classification accuracy. Another important observation is the CNN results were hard to converge, with lots of up and down in the classification results with respect to the different parameter setups. This phenomenon may come from the fact that we do not have enough data to properly train a complete CNN. Moreover, even if we have an augmented dataset of 112511 regions, half of them are belonging to the negative class.

3.1.3   Adopting an Existing CNN and Directly Using its Features With a Support Vector Machine (SVM) Classification

For this framework, we use the transfer learning from the CNN given by Szegedy et al. [16]. Each image is considered as input of the CNN, returning a 1000-dimension vector and we use this vector to classify the image. Using this transfer learning methodology, we skip all the feature training of the CNN, all we need to train is the SVM classifier with respect to our 5 classes and the features extracted from the CNN. To do so, 80% of our dataset is used for training and the rest 20% for testing. Different SVM kernels were tested: linear, polynomial, RBF. The best performing SVM kernel for our application has been the third-order polynomial with an average of 82% accuracy. The performance of this approach is better than training a whole new CNN because we use a CNN that has been trained on the massive ImageNet dataset which contains millions of generic images. Even with a non dedicated dataset, the fact the CNN is learned on millions of image data returns more robust and discriminative features than the previous approach which was trained on only 1000 dedicated images.

3.2   Fine Tuning an Existing CNN Using Our Specific Dataset

For this framework, we keep the convolution layers of inception-v3 [16] and retrain the fully-connected layers. The number of outputs of the network is adjusted into 5 classes to adapt to our task. We use 80% of our dataset for training and 20% for testing with several tries. For this framework, we apply different methods to

Table 2 Multiscale, region based classification

Class	Construction site	Pagoda	Pond	Rubbish	Negative	Average
Recall (%)	90.4	83.2	85.5	72.7	82.5	85.1
Precision (%)	75	48.3	69	96	69.2	71.4

classify the image: (i) classify the whole image directly, (ii) multiscale, region based classification and multiscale, and (iii) region based classification with the additional weighting.

3.2.1 Classify the Whole Image

In this section, we directly use the original images to test our retrain CNN. This first section gives us a baseline performance of the framework without any data augmentation or weighting scheme techniques. The average accuracy obtained on all test images for several runs is 76.6%.

3.2.2 Multiscale, Region Based Classification

For this set of experiments, we use the complete data augmentation techniques detailed in Sect. 2. We rescale the image into 3 different dimensions: 1024, 512 and 256 as the minimum width or height from the original images. Then we use the sliding window (size 224 * 224 see [5]) with a shifting step of 32 pixels based method to extract different regions in the image. Each sub region is then classified with our retrained CNN and finally, we combine the classification scores of all regions to have a single score for the whole image. However, using this methodology, the average accuracy drops down to 71.4% as presented in Table 2. We can observe good results for the "Rubbish" class at 96% but the result for pagoda class is low, i.e. 48%. One may conclude that using multiscale and region based classification is useless as the result is worse than classifying directly the whole image. However, as we explained before, this issue comes from the numerous and less discriminative negative regions appearing in most regions of any image. It is also due to the non-mutually exclusive classes (like pagoda and construction site). This observation inspires us to treat the classification score of each class differently by adding a weight coefficient based on each window-region based classification.

Table 3 Multiscale, region based classification using the weighting scheme

Class	Construction site	Pagoda	Pond	Rubbish	Negative	Average
Recall (%)	95.4	80	93.5	66.7	87.5	90.1
Precision (%)	94.5	90.9	87.8	80	77.8	90.1

3.2.3   Multiscale, Region Based Classification With an Additional Weighting Scheme

In this approach, we use 20% of training dataset as a validation dataset to learn the additional weighting scheme using a simple logistic regression algorithm. As already mentioned, we use the window-region classification scores as input to learn a weighting coefficient matrix that improves the overall accuracy of the framework. For instance, to improve the "pagoda" classification, a weight coefficient of 7 has been determined for "pagoda" window-regions by the logistic regression as well as negative coefficients for all other classified window regions (close to 0 for the negative class). First experiment results show promising good results using such a weighting scheme with more than 90% accuracy. The obtained performance is shown in Table 3.

These promising results show the interest of our global framework for small and very specific applicative datasets where little apriori knowledge is given by expert users. Nevertheless, there is still room for improvement. In order to achieve automatic generalization to any specific class, we plan in a near-future to combine our framework with saliency maps obtained through the prediction of visual attention with deep CNNs [20]. Very schematically, the idea is to focus the CNN on regions of the images attracting human's eyes when we show images and ask them: "is there a pagoda?". Way beyond pagodas, this kind of strategy can be used with any very specific visual class where only few parts of the image are discriminative from other classes.

4   Conclusion

In this paper, we present a framework to adapt the CNN approach to a dedicated dataset containing very little information. With such dataset, data augmentation is a crucial step to have a correct training of a CNN. Multi-scale and region-based classification helps to exploit the locally discriminative information in the image. Finally, by learning an adaptive weighting scheme to each window-region, we enhance the performance of the framework significantly. The classification results, with more than 90% accuracy on a very small and specific dataset of (about 200 images per class) show the potential of CNN on small-scale image collection.

References

1. Krizhevsky, A., Sutskever, I., Hinton, G.E.: Imagenet classification with deep convolutional neural networks. In: Pereira, F., Burges, C.J.C., Bottou, L., Weinberger, K.Q. (eds.), Advances in Neural Information Processing Systems, vol. 25, pp. 1097–1105. Curran Associates, Inc. (2012). http://papers.nips.cc/paper/4824-imagenet-classification-with-deep-convolutional-neural-networks.pdf
2. Lecun, Y., Bengio, Y., Hinton, G.: Deep learning. Nature **521**(7553), 436–444, 5 (2015)
3. Simonyan, K., Zisserman, A.: Very deep convolutional networks for large-scale image recognition. CoRR, vol. abs/1409.1556. http://arxiv.org/abs/1409.1556 (2014)
4. He, K., Zhang, X., Ren, S., Sun, J.: Delving deep into rectifiers: surpassing human-level performance on imagenet classification. CoRR, vol. abs/1502.01852. http://arxiv.org/abs/1502.01852 (2015)
5. Szegedy, C., Liu, W., Jia, Y., Sermanet, P., Reed, S.E., Anguelov, D., Erhan, D., Vanhoucke, V., Rabinovich, A.: Going deeper with convolutions. CoRR, vol. abs/1409.4842. http://arxiv.org/abs/1409.4842 (2014)
6. Ioffe, S., Szegedy, C.: Batch normalization: accelerating deep network training by reducing internal covariate shift. CoRR, vol. abs/1502.03167. http://arxiv.org/abs/1502.03167 (2015)
7. He, K., Zhang, X., Ren, S., Sun, J.: Deep residual learning for image recognition. CoRR, vol. abs/1512.03385. http://arxiv.org/abs/1512.03385 (2015)
8. Szegedy, C., Ioffe, S., Vanhoucke, V.: Inception-v4, inception-resnet and the impact of residual connections on learning. CoRR, vol. abs/1602.07261. http://arxiv.org/abs/1602.07261 (2016)
9. He, K., Zhang, X., Ren, S., Sun, J.: Identity mappings in deep residual networks. CoRR, vol. abs/1603.05027. http://arxiv.org/abs/1603.05027 (2016)
10. Razavian, A.S., Azizpour, H., Sullivan, J., Carlsson, S.: CNN features off-the-shelf: an astounding baseline for recognition. CoRR, vol. abs/1403.6382. http://arxiv.org/abs/1403.6382 (2014)
11. Long, M., Wang, J.: Learning transferable features with deep adaptation networks. CoRR, vol. abs/1502.02791. http://arxiv.org/abs/1502.02791 (2015)
12. Zheng, L., Yang, Y., Tian, Q.: SIFT meets CNN: a decade survey of instance retrieval. CoRR, vol. abs/1608.01807. http://arxiv.org/abs/1608.01807 (2016)
13. Babenko, A., Slesarev, A., Chigorin, A., Lempitsky, V.S.: Neural codes for image retrieval. CoRR, vol. abs/1404.1777. http://arxiv.org/abs/1404.1777 (2014)
14. Ren, S., He, K., Girshick, R.B., Sun, J.: Faster R-CNN: towards real-time object detection with region proposal networks. CoRR, vol. abs/1506.01497. http://arxiv.org/abs/1506.01497 (2015)
15. Erhan, D., Szegedy, C., Toshev, A., Anguelov, D.: Scalable object detection using deep neural networks. In: IEEE Conference on Computer Vision and Pattern Recognition, ser. CVPR '14, pp. 2155–2162. IEEE Computer Society, Washington, DC, USA. http://dx.doi.org/10.1109/CVPR.2014.276 (2014)
16. Szegedy, C., Vanhoucke, V., Ioffe, S., Shlens, J., Wojna, Z.: Rethinking the inception architecture for computer vision. CoRR, vol. abs/1512.00567. http://arxiv.org/abs/1512.00567 (2015)
17. Russakovsky, O., Deng, J., Su, H., Krause, J., Satheesh, S., Ma, S., Huang, Z., Karpathy, A., Khosla, A., Bernstein, M., Berg, A.C., Fei-Fei, L.: ImageNet large scale visual recognition challenge. International Journal of Computer Vision (IJCV) **115**(3), 211–252 (2015)
18. Lin, S., Zhao, Z., Su, F.: Homemade ts-net for automatic face recognition. In: Proceedings of the 2016 ACM on International Conference on Multimedia Retrieval, ser. ICMR '16, pp. 135–142. ACM, New York, NY, USA. http://doi.acm.org/10.1145/2911996.2911999 (2016)
19. Margeta, J., Criminisi, A., Cabrera Lozoya, R., Lee, D.C., Ayache, N.: Fine-tuned convolutional neural nets for cardiac MRI acquisition plane recognition. Comput. Methods Biomech. Biomed. Eng. Im. Visual. https://hal.inria.fr/hal-01162880 Aug (2015)
20. 14th International Workshop on Content-Based Multimedia Indexing, CBMI 2016. Bucharest, Romania, June 15–17, 2016. IEEE. http://ieeexplore.ieee.org/xpl/mostRecentIssue.jsp?punumber=7496233 (2016)

How to Choose Deep Face Models for Surveillance System?

Vy Nguyen, Tien Do, Vinh-Tiep Nguyen, Thanh Duc Ngo and Duc Anh Duong

Abstract Face recognition suits well in situations that subjects may not cooperate, such as surveillance system, which can be deployed to track movements of a newly detected thief. In this retrieval task, the choice of face representation is highly important. The rise of Deep Learning in Computer Vision has led to the rise of deep models in face recognition, such as FaceNet, DeepFace, VGG Face B, CenterLoss C, VIPLFaceNet, ... However, when it comes to applications, which model should be chosen to ensure the balance amongst accuracy, computational cost and memory resource is still an open problem. In this work, evaluations some of state-of-the-art deep models (VGG Face B, CenterLoss C, VIPLFaceNet) were conducted under different settings and benchmark protocols to illustrate the trade-offs and draw conclusions not clearly indicated in the original works.

Keywords Face recognition · Deep learning · Evaluation · Identification · Face feature extraction · Face representation

V. Nguyen (✉) · T. Do · V.-T. Nguyen · T. D. Ngo · D. A. Duong
Multimedia Communications Laboratory, University of Information
Technology, Vietnam National University, Ho Chi Minh City, Vietnam
e-mail: vynt@uit.edu.vn

T. Do
e-mail: tiendv@uit.edu.vn

V.-T. Nguyen
e-mail: tiepnv@uit.edu.vn

T. D. Ngo
e-mail: thanhnd@uit.edu.vn

D. A. Duong
e-mail: duongda@uit.edu.vn

© Springer International Publishing AG, part of Springer Nature 2018
A. Sieminski et al. (eds.), *Modern Approaches for Intelligent Information and Database Systems*, Studies in Computational Intelligence 769,
https://doi.org/10.1007/978-3-319-76081-0_31

1 Introduction

Face recognition holds many advantages over other reliable biometrical methods (e.g. DNA, iris, and fingerprint) for being inexpensive and not annoying to people. Especially in surveillance system, subjects may not even cooperate [1]. Surveillance system can be deployed to detect VIP guests, to identify wanted fugitives, or to track movements of a newly detected thief. In the first two cases, gallery samples can be nicely stored beforehand (high-quality frontal face images), while gallery samples are possibly in an arbitrary pose and quality in the last case. So, the choice of face representation is critical for the application of tracking movements of a newly detected thief: A good choice may make the retrieval tasks trivial, a bad choice may puzzle the systems, or even worse, it may make retrieval tasks impossible.

Approaches determining face representation from 2D face images are diverse. We believe that there are two major categories: shallow methods and deep methods. These approaches have their own pros and cons. Still, deep methods are currently dominating as they have shown their superiority in practice. Jason Yosinski et al. went deeper in the transferability and generalization of CNN, and found some surprising results [2]. Erik Learned-Miller et al. acknowledged that most of the current dominant methods under LFW benchmark protocols employ CNN architecture with large-scale training sets [3].

We evaluated three state-of-the-art deep models, namely VGG Face B [4], CenterLoss C [5], and VIPLFaceNet [6], under benchmark protocols BLUFR [7] and TIFS_SI-2014 [8]. We also measured their execution times and memory consumptions. By studying their characteristics, which were not fully studied and reported in their original works, we hope this work will be able to illustrate the trade-offs between models and help the readers make a better choice.

2 Related Works

Since the publication of [9] in 2005, there has been an explosion of highly successful face recognition systems using CNN. Overall, most existing works focus on reporting the CNN architecture adapted from object recognition with some modifications that work for face recognition.

In 2014, Facebook researchers introduced a face recognition system called Deep-Face [10] and then later improved it into Fusion [11] in 2015. The human-level accuracy performance in face verification on the database LFW for face-cropped images [12] was also surpassed in 2015 by GuassianFace [13]. At the same time, many versions of DeepId were introduced using more and more complex architectures [14–17]. FaceNet [18], the current champion of various benchmark protocols, was also introduced in 2015 by Google researchers. Following the trend, VGG Face was introduced in 2015 as a model trained on fewer data. In 2016, VIPLFaceNet was introduced as a light model, and CenterLoss introduced new loss function for face representation.

Hu et al. [19] and Mehdipour Ghazi et al. [20] conducted evaluations on some deep models for face recognition. Grgic et al. [21] and Hong et al. [22] offered benchmark protocols for surveillance system. Peng et al. [23] evaluated how super-resolution methods improve face recognition on low-resolution images.

We only share VGG Face B with [19] (but under different benchmark protocols). The chosen deep models in our work are more state-of-the-art. Due to the fact that retrieval tasks possibly involve non-frontal face images, we chose to evaluate under benchmark protocols BLUFR and TIFS_SI-2014 instead of the benchmark protocols in [21] and [22] as they only comprises of frontal face images.

3 Experimental Settings

In this section, we introduce our chosen deep models and benchmark protocols. These models are state-of-the-art and publicy available.

3.1 Deep Models

3.1.1 VGG Face B

The model inspired by AlexNet [24] and VGGNet [25]. The authors of VGG Face developed three configurations of network A, B, D, which were later trained on the large-scale data collected and built by themselves. The one we employed in this work is the configuration B, which outperforms the other twos.

3.1.2 VIPLFaceNet

On the aim of building an open source deep face recognition SDK, Liu et al. [6] have carefully modified the AlexNet [24] so that the training time only takes 20% while the testing time only takes 60% for their VIPLFaceNet model. However, the model published on GitHub was trained using a different dataset. Though, our focus is not on the academic potential of each model, but rather their off-the-shelf performances.

3.1.3 CenterLoss C

Wen et al. [5] proposed a new loss function called center loss. They claimed that when center loss is combined with softmax loss, it addresses the problem of intra-class compactness more directly, while its training requirement for resources is as economic as softmax loss.

3.2 Benchmark Protocols

Many methods specialized for face recognition in surveillance system deal with low-resolution and recognition with one face image per person. However, in this work, we assume that the resolution of the camera is acceptable: face images captured with the width and height are expected to be 100 ± 25 pixels.

3.2.1 Benchmark of Large-Scale Unconstrained Face Recognition

Benchmark of Large-scale Unconstrained Face Recognition (BLUFR) [7] involves two benchmark protocols based on LFW database [26]: Open-set identification protocol and verification protocol.

3.2.2 TIFS_SI-2014

TIFS_SI-2014 was proposed in [8] based on the LFW database and YTF database. Until now these benchmark protocols still have no official name[1] although they has been used to evaluate several popular face recognition systems. They are usually referred as 1:N benchmark of LFW, however we do not prefer this informal name as it is confusing. Therefore, we refer the benchmark protocols by their folder name.

4 Evaluation Results and Analyses

4.1 Number of Parameters and Feature Extraction Time

The accuracy is indeed very important. However, in practice, people may prefer models which are less accurate but faster. For each of the three models, Table 1 shows our estimations of feature extraction times under two conditions: with and without the employment of MTCNN [27], which is a facial landmarks detector and also a face detector. The number of parameters in the testing phase is also presented. The original authors of the four models do not provide these pieces of information.

The number of parameters of each model in the testing phase is related to memory consumption. For the case of CenterLoss C, the authors extract the feature vector of one face image by doing forward propagation twice: once for the original image after normalization and once for its upside-down version, then they concatenate two

---

[1]And the benchmark protocols have no official website either. However, you can download the benchmark protocols here: http://biometrics.cse.msu.edu/Publications/Databases/TIFS_SI-2014_protocols.zip.

Table 1 Number of parameters and feature extraction time of the four models, estimated by us. The feature extraction time is measured on a single core of i7-4600U CPU or on a GPU Tesla C2050

Model	Number of parameters in testing phase	Feature extraction time		Feature extraction time with MTCNN	
		CPU (ms)	GPU (ms)	CPU (ms)	GPU (ms)
VGG Face B	134M	735.7	86.8	833.6	129.4
CenterLoss C	27.5M	582.6	47.8	680.5	90.4
VIPLFaceNet	18M	132.9	–	230.8	–
MTCNN	0.5M	97.9	42.6	–	–

feature vectors into one to represent the face image.[2] So, in Table 1, its feature extraction time was counted twice as 582.6 ms, however, its number of parameters was only counted once as 27.5M.

The information of feature extraction time of the three models in Table 1 should give the readers an idea about their relative speeds in CPU and GPU (except for VIPLFaceNet, the authors do not offer an implementation specialized for GPU). It should be noted if a model has a lower number of parameters, it does not necessarily have a faster feature extraction speed, e.g. CenterLoss C.

From Table 1, we see that VIPLFaceNet has the least number of parameters and fastest feature extraction time in CPU mode, CenterLoss C comes next and then VGG Face B comes last. In GPU mode, CenterLoss C is twice as fast as VGG Face B.

4.2 The Employment of Face Registration

As the readers can see from Table 1, the phase facial landmarks detection may consume quite an amount of time. Do we really need facial landmarks detector? What if facial landmarks detector fails to do its job? This section addresses those questions. As illustrated in the Fig. 1:

- By *tight-cropped face image*, we imply the face image includes only the face and not any or very limited background information.
- By *loose-cropped face image*, we imply the face image may include a small but non-trivial amount of background information, such as full hair style. In this work, a loose-cropped face image is obtained by extending the original crop region returned by the face detector more 16 pixels to the left, to the right, to the top, and to the bottom.

---

[2]Be vigilant that this does not imply CenterLoss C is invariant to upside down rotation.

(a) Tight cropped   (b) Loose cropped
    face image   face image

Fig. 1 Two kinds of face image cropping

BLUFR: VR @ FAR = 0.1%

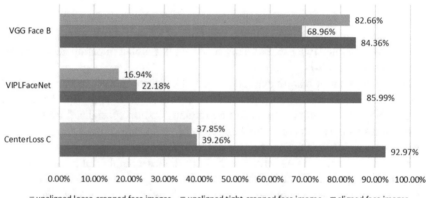

■ unaligned loose-cropped face images   ■ unaligned tight-cropped face images   ■ aligned face images

Fig. 2 Performance records of CenterLoss C, VGG Face B and VIPLFaceNet under BLUFR verification benchmark protocol

From Figs. 2, 3 and 4, we see that the performance of VGG Face B is quite stable under different conditions of input face images. Though, the employment of a facial landmarks detector does improve its performance a little bit. In the case inputs are unaligned face images, the model prefers loose-cropped ones over tight-cropped ones. In fact, VGG Face B was trained on unaligned face images directly cropped by DPM Cascade [28], which is a mix of tight-cropped and loose-cropped face images (see Fig. 5). A similar result is also reported by Mehdipour Ghazi et al. [20], however, we do not entirely agree with their comments: We are not sure if background information does help or DPM Cascade does produce many loose-cropped face images over tight-cropped face images.

When inputs are aligned face images, the overall performance of CenterLoss C comes first and VIPLFaceNet comes second. Under BLUFR open-set identification benchmark protocol and TIFS_SI-2014 closed-set identification benchmark protocol, the quality of face representation extracted by CenterLoss C is highest in terms of inter-class separability and intra-class compactness.

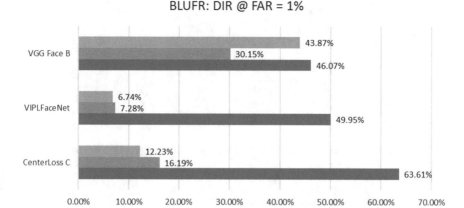

BLUFR: DIR @ FAR = 1%

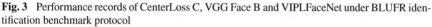

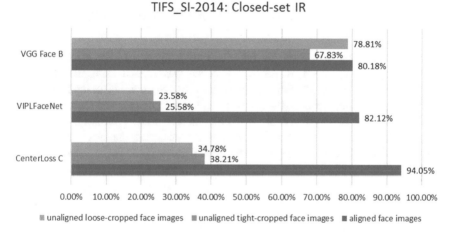

Fig. 3 Performance records of CenterLoss C, VGG Face B and VIPLFaceNet under BLUFR identification benchmark protocol

Fig. 4 Performance records of CenterLoss C, VGG Face B and VIPLFaceNet under TIFS_SI-2014 closed-set identification benchmark protocol

When inputs are unaligned face images, performances of CenterLoss C and VIPLFaceNet decrease dramatically, which show their dependence on facial landmarks detector. In this case, performances of CenterLoss C come second and performances of VIPLFaceNet come third. There is a large gap of performance between CenterLoss C and VGG Face B. CenterLoss C and VIPLFaceNet were both trained using aligned face images.

CenterLoss C and VIPLFaceNet also have better performances when inputs are unaligned tight-cropped face images compared to when inputs are unaligned loose-cropped face images. It is worth noting that CenterLoss C was trained using

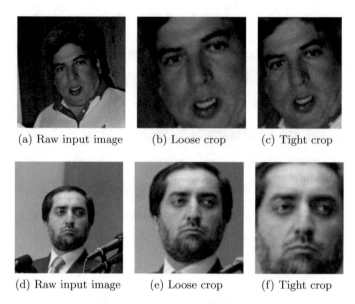

(a) Raw input image   (b) Loose crop   (c) Tight crop

(d) Raw input image   (e) Loose crop   (f) Tight crop

Fig. 5 A comparison on DPM cascade (**a, b, c**) and MTCNN (**d, e, f**) face detectors

aligned loose-cropped face images (as implicitly suggested by the extraction code provided by the authors[3]), while VIPLFaceNet was trained using aligned tight-cropped face images (as implicitly suggested in [6]). So, there is an interesting correlation between models being trained on aligned face images and their preference for tight-cropped face images over loose-cropped face images. We suspect the tight-cropped face images offer less variance in pixel values than the loose-cropped face images do. Therefore, when the facial landmarks detector fails, if the chosen model is trained on aligned face images, unaligned tight-cropped face images can be a better choice.

5   Conclusion and Future Work

We evaluated three state-of-the-art deep models (VGG Face B, CenterLoss C and VIPLFaceNet) on different benchmark protocols (BLUFR and TIFS_SI-2014), and measured the number of parameters and the feature extraction time of each model to point out the trade-offs between models.

If facial landmarks detector works well in a specific application, VIPLFaceNet or CenterLoss C may be a good choice depending on the application: VIPLFaceNet is the fastest model in CPU mode, CenterLoss C is the fastest model in GPU mode. If not, VGG Face B should be the choice due to its robustness. In the case of

---

[3]https://github.com/ydwen/caffe-face.

unaligned face images, VGG Face B prefers loose-cropped ones over tight-cropped ones.

A correlation is determined that those models trained using aligned face images tend to work better on unaligned tight cropped face images than unaligned loose cropped face images. We suspect unaligned tight cropped face images offer less variance than unaligned loose cropped face images do.

Our results help the readers to choose a deep model suitable for their needs in case gallery samples may not be frontal. However, resolution of face images in LFW database is not low enough to fit in situations where high-quality cameras are not available. Therefore, future works should assess these deep models in case frontal gallery samples are unavailable and of low-resolution.

References

1. Zhao, W., Chellappa, R., Phillips, P.J., Rosenfeld, A.: Face recognition: a literature survey. ACM Comput. Surv. (CSUR) 35(4), 399–458 (2003)
2. Yosinski, J., Clune, J., Bengio, Y., Lipson, H.: How transferable are features in deep neural networks? Adv. Neural Inf. Process. Syst. 3320–3328 (2014)
3. Learned-Miller, E., Huang, G.B., RoyChowdhury, A., Li, H., Hua, G.: Labeled faces in the wild: a survey. In: Advances in Face Detection and Facial Image Analysis, pp. 189–248. Springer (2016)
4. Parkhi, O.M., Vedaldi, A., Zisserman, A.: Deep face recognition. BMVC 1(3), 6 (2015)
5. Wen, Y., Zhang, K., Li, Z., Qiao, Y.: A discriminative feature learning approach for deep face recognition. In: European Conference on Computer Vision, pp. 499–515. Springer (2016)
6. Liu, X., Kan, M., Wu, W., Shan, S., Chen, X.: Viplfacenet: an open source deep face recognition sdk. arXiv:1609.03892 (2016)
7. Liao, S., Lei, Z., Yi, D., Li, S.Z.: A benchmark study of large-scale unconstrained face recognition. In: 2014 IEEE International Joint Conference on Biometrics (IJCB), pp. 1–8. IEEE (2014)
8. Best-Rowden, L., Han, H., Otto, C., Klare, B.F., Jain, A.K.: Unconstrained face recognition: identifying a person of interest from a media collection. IEEE Trans. Inf. Forensics Secur. 9(12), 2144–2157 (2014)
9. Chopra, S., Hadsell, R., LeCun, Y.: Learning a similarity metric discriminatively, with application to face verification. In: IEEE Computer Society Conference on Computer Vision and Pattern Recognition. CVPR 2005, vol. 1, pp. 539–546. IEEE (2005)
10. Taigman, Y., Yang, M., Ranzato, M., Wolf, L.: Deepface: closing the gap to human-level performance in face verification. In: Proceedings of the IEEE Conference on Computer Vision and Pattern Recognition, pp. 1701–1708 (2014)
11. Taigman, Y., Yang, M., Ranzato, M., Wolf, L.: Web-scale training for face identification. In: Proceedings of the IEEE Conference on Computer Vision and Pattern Recognition, pp. 2746–2754 (2015)
12. Kumar, N., Berg, A.C., Belhumeur, P.N., Nayar, S.K.: Attribute and simile classifiers for face verification. In: 2009 IEEE 12th International Conference on Computer Vision, pp. 365–372. IEEE (2009)
13. Lu, C., Tang, X.: Surpassing human-level face verification performance on LFW with gaussianface. arXiv:1404.3840 (2014)
14. Sun, Y., Chen, Y., Wang, X., Tang, X.: Deep learning face representation by joint identification-verification. In: Advances in Neural Information Processing Systems, pp. 1988–1996 (2014)

15. Sun, Y., Wang, X., Tang, X.: Deep learning face representation from predicting 10,000 classes. In: Proceedings of the IEEE Conference on Computer Vision and Pattern Recognition, pp. 1891–1898 (2014)
16. Sun, Y., Wang, X., Tang, X.: Deeply learned face representations are sparse, selective, and robust. In: Proceedings of the IEEE Conference on Computer Vision and Pattern Recognition, pp. 2892–2900 (2015)
17. Sun, Y., Liang, D., Wang, X., Tang, X.: Deepid3: face recognition with very deep neural networks. arXiv:1502.00873 (2015)
18. Schrofi, F., Kalenichenko, D., Philbin, J.: Facenet: a unified embedding for face recognition and clustering. In: Proceedings of the IEEE Conference on Computer Vision and Pattern Recognition, pp. 815–823 (2015)
19. Hu, G., Yang, Y., Yi, D., Kittler, J., Christmas, W., Li, S.Z., Hospedales, T.: When face recognition meets with deep learning: an evaluation of convolutional neural networks for face recognition. In: Proceedings of the IEEE International Conference on Computer Vision Workshops, pp. 142–150 (2015)
20. Mehdipour Ghazi, M., Kemal Ekenel, H.: A comprehensive analysis of deep learning based representation for face recognition. In: Proceedings of the IEEE Conference on Computer Vision and Pattern Recognition Workshops, pp. 34–41 (2016)
21. Grgic, M., Delac, K., Grgic, S.: Scface-surveillance cameras face database. Multimedia Tools Appl. **51**(3), 863–879 (2011)
22. Hong, S., Im, W., Ryu, J., Yang, H.S.: Sspp-dan: deep domain adaptation network for face recognition with single sample per person. arXiv:1702.04069 (2017)
23. Peng, Y., Gökberk, B., Spreeuwers, L., Veldhuis, R.: An evaluation of super-resolution for face recognition (2012)
24. Krizhevsky, A., Sutskever, I., Hinton, G.E.: Imagenet classification with deep convolutional neural networks. In: Advances in Neural Information Processing Systems, pp. 1097–1105 (2012)
25. Simonyan, K., Zisserman, A.: Very deep convolutional networks for large-scale image recognition. arXiv:1409.1556 (2014)
26. Huang, G.B., Ramesh, M., Berg, T., Learned-Miller, E.: Labeled faces in the wild: A database for studying face recognition in unconstrained environments. Technical Report, 07-49. University of Massachusetts, Amherst (2007)
27. Zhang, K., Zhang, Z., Li, Z., Qiao, Y.: Joint face detection and alignment using multitask cascaded convolutional networks. IEEE Sig. Process. Lett. **23**(10), 1499–1503. ISSN: 1070-9908. https://doi.org/10.1109/LSP.2016.2603342 Oct (2016)
28. Mathias, M., Benenson, R., Pedersoli, M., Van Gool, L.: Face detection without bells and whistles. In: European Conference on Computer Vision, pp. 720–735. Springer (2014)

GPU Video Stream Magnification as a Tool for Touchless Object Vibration Measurement

Dawid Sobel, Karol Jędrasiak and Aleksander Nawrat

Abstract Video stream magnification allows people to see otherwise not visible subtle changes in surrounding world. In the article we present a method for GPU-based touchless object vibration measurement using IR video camera in order to process video stream in real time. We compare acquired results by the proposed method with the results from accelerometers. The conducted experiments confirmed that the results obtained using the Video Magnification technique are not only visual effects, but also allow measurement of object vibration. The results of the proposed touchless vibration measurement approach are similar to classic accelerometer-based solutions, which we consider to be a promising starting point for further research with goal of vibration measurements for unmanned aerial vehicles (UAVs) as additional source of information for control algorithms.

Keywords Computer vision · Video magnification · GPU · Vibration measurement

1 Introduction

The world in which we live is constantly changing, so it is important to understand the processes taking place in the ever-changing environment that surrounds us. One of the ways to capture and visualize these changes is the use of vision systems. Whether using cameras, cameras, microscopes, satellites, etc., images and video streams are an invaluable source of information about process variables in the environment around us. Thanks to the advances in electronics that translate into the development of digital photography, the acquisition of images and video streams is now significantly simplified and more accessible to everyone. However, in spite of

D. Sobel · K. Jędrasiak (✉) · A. Nawrat
Institute of Automatic Control, Silesian University of Technology, Gliwice, Poland
e-mail: karol.jedrasiak@polsl.pl

A. Nawrat
e-mail: aleksander.nawrat@polsl.pl

© Springer International Publishing AG, part of Springer Nature 2018
A. Sieminski et al. (eds.), *Modern Approaches for Intelligent Information and Database Systems*, Studies in Computational Intelligence 769,
https://doi.org/10.1007/978-3-319-76081-0_32

the widespread use of image acquisition devices, its methods of computer vision analysis, in particular slow-moving processes invisible to the human eye, are still evolving. The article presents an imaging method that enables visualization of small amplitude changes that are difficult or impossible to be observed without human computer assistance [1], such as vibration of structural elements of technical infrastructure. The proposed approach can be applied when analyzing low-amplitude movements occurring in measuring infrastructure facilities that are currently using costly precision-based accelerometer installations. The main advantage of the video stream enhancement approach is the non-contact nature of data capture for analysis by video cameras. Presently used methods of measuring e.g. bridges or vehicle constructions often require the use of the facility for the duration of measurements, which entails huge costs.

2 GPU-Based Video Motion Magnification Algorithm

Image processing and computer vision algorithms require remarkable computational power. Currently fastest CPU Intel Core i7-5960X peaks at 384 GFLOPS while GPU nVidia GeForce GTX 980 Ti offers over 5.5 TFLOPS and remains cheaper. The performance ratio remains true also for cheaper products. The main reason that GPUs did not replace CPU is their architecture. They consist of hundreds or even thousands cores therefore GPUs are efficient only when algorithm can be split into massive amounts of parallel operations. Fortunately, often algorithms allow to treat image as a set of independent pixels and all of them can be processed simultaneously, which is utilized in [2–6].

Right now, there are two main technologies that allow GPU computations: openCL and CUDA. OpenCL permit implementation of applications for most CPU and GPU in alike manner, but there is the price-worse performance. This is the reason why proposed solution is based on CUDA technology, which is faster and allows lower level access to device although work only with nVidia products. Computations are done by threads (kernels) launched in a grid of blocks, which consist of chosen number of threads and often performance is limited by memory speed (even though bandwidth can exceed 200 GB/s), so manufacturer gave programmers access to few different kinds of memory.

2.1 Implementation

Proposed implementation involves openCV library in order to acquire and display frames from the video sequence. CUDA allows to perform all computations on GPU. The following steps of the VMM algorithm can be distinguished [7–9]: downsampling, upsampling, convolution, phase magnification, elements subtraction and addition. Because convolution was used in each part of processing it had to be efficient. Fortunately, it was used for Gaussian blurring and gradient in horizontal and vertical direction, so it was possible to take advantage of separability. The implementation involved each thread processing one pixel. The thread was responsible for reading values of pixels of video frame and process it directly from global memory. It was assumed that convolution mask had N elements, then each thread performed N reads of values of image and further N for values of mask. First suggested optimization was to use constant memory for the mask. Each thread accessed the same value of mask at the same part of thread and constant memory broadcast it to all threads at cost of 1 operation. Therefore, using GPU the number of operations k for M pixels is reduced from:

$$k = 2NM \tag{1}$$

to

$$k = NM + N \tag{2}$$

Further speedup was obtained by reading pixel values indirectly with use of shared memory, which is a lot faster than global memory. Shared memory is faster because all threads in block share it, so each thread reads one or two pixels (two because of block surroundings) from global memory. Reduction in that case is dependent on block dimension and length of mask, number of reads for X × Y image is:

$$k = \frac{X}{X_B}(X_B + N - 1) + N \tag{3}$$

for row convolution and:

$$k = \frac{Y}{Y_B}(Y_B + N - 1) + N \tag{4}$$

for column, where index B stands for block. Additionally CUDA technology supports launching threads in format of 3D grids which is natural for color image and speedup memory access because nearby threads require data from nearby addresses.

In case of down and upsampling each thread is connected with a group of pixels from smaller sub-image. In case of upsampling it allows to spare memory reads. If threads would be connected with a group of pixels from a larger image then they would have needed firstly read pixel in downsampled image and then write it to one pixel. In result for 2 × 2 region would require 8 read/write operations while proposed approach relies on reading values from pixel to register and then storing it in 4 pixels, so there is only 5 read/write operations.

(a)

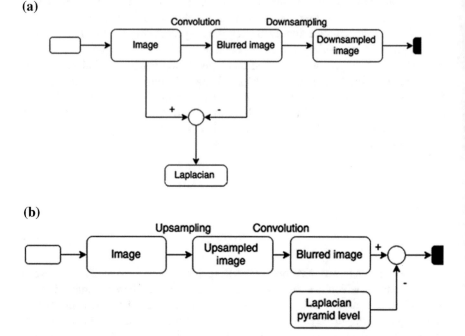

(b)

Fig. 1 **a** Laplacian pyramid construction scheme, **b** Laplacian pyramid collapsing scheme

Operations mentioned above let construct and collapse Laplacian pyramid. Unfortunately, pyramid construction requires to be performed in strict order (lower level of pyramid can not be constructed, before higher is at least partially ready) and because of that in lower levels part of GPU cores can idle (Fig. 1).

After first decomposition follows phase amplification which consists of two parts. That division is forced by blur operation during processing, because there is no way to synchronize all threads from inside. Therefore appears necessity to store data between kernel calls and that slow down processing. On the other side after decomposition pyramid levels can be processed in parallel and that is done by using CUDA streams mechanism.

In 2D Riesz transform that can be approximated with pair of filters: [0.5 0 −0.5] and $[0.5\ 0\ -0.5]^{\mathrm{T}}$. Their response R_1 and R_2 together with input band I form monogenic signal, which can be converted to spherical coordinates using Eqs. (5)–(7):

$$I = A cos(\phi) \tag{5}$$

$$R_1 = A sin(\phi) cos(\theta) \tag{6}$$

$$R_2 = A sin(\phi) sin(\theta) \tag{7}$$

Above equations allow to obtain local amplitude A, local phase φ and local orientation θ. In next step phase is temporally filtered, but because of sign ambiguity (pair (φ, θ) is equivalent $(-\varphi, \theta + \pi))$ filtered are quantities:

$$\phi sin(\theta), \phi cos(\theta) \tag{8}$$

Next, spatial filtering is performed on obtained values and they are recombined to get filtered local phase φ f:

$$\phi_f = cos(\theta) \frac{\phi cos(\theta) * K_\rho}{A * K_\rho} + sin(\theta) \frac{\phi sin(\theta) * K_\rho}{A * K_\rho} \tag{9}$$

$K\rho$ stands for Gaussian convolution kernel with standard deviation ρ and $*$ for convolution. After that follows computation of image with magnified phase.

$$O = I cos(\alpha \phi_f) + A sin(\phi) sin(\alpha \phi_f) \tag{10}$$

where α is phase amplification factor. The last part is VMM algorithm is image reconstruction, which was described earlier.

3   Tests

The proposed method was used for experimental verification of the ability to observe and measure the vibration of steel beams in a non-contact manner using image-processing techniques to enhance subtle changes in the image. During the tests a solid steel beam C140 and the steel beam LR25 × 25 × 3 were used. Based on the results of the tests, the strengths and weaknesses of the proposed method were determined in the context of non-contact vibration measurement of objects. Presented results were obtained on laptop supplied with Intel Core i7-3537U with 2.9 GHz clock during processing and nVidia GeForce GT 740 m. Video streams were acquired using a near-infrared video camera with a resolution of 2080 × 1552 pixels and a 60 Hz recording rate.

3.1   Experiment 1

The goal of the experiment was to observe the response to excitation of a solid steel beam C140 placed on two stable supports (Fig. 2) and weighted with two weights. During the experiment the beam was excited by a hammer hit. The beam vibrations

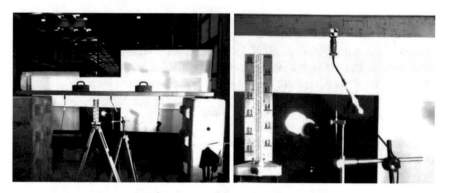

Fig. 2 (Left) Test bench scheme for experiment #1. (Right) Zoom of one of the measurement points

were recorded by three accelerometers pinned to the beam and by an infrared camera placed on a tripod at the distance of 312 cm from the beam. The camera recorded the movement of the tag affixed to the beam. The camera was fixed in manual acquisition and the video stream was undistorted.

3.2 Results of the Experiment 1

The video stream of the marker recorded by the camera was amplified using the present method. Marker offset was measured in pixels (Fig. 3, left). On this basis, the spectrum was calculated. There is a noticeable peak near 10 Hz (Fig. 3, right).

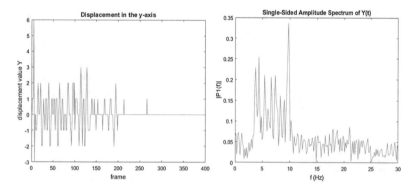

Fig. 3 (Left) Chart of position change in time of the marker observed by the camera, (right) plot of frequency spectrum calculated for the data of Fig. 3 (left)

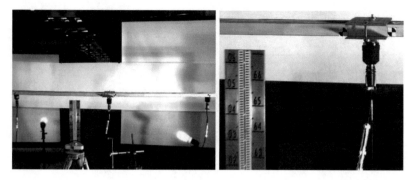

Fig. 4 (Left) Test bench scheme for experiment #2. (Right) Zoom of one of the measurement points

3.3 Experiment 2

The goal of the experiment was to observe the response to excitation of a steel beam LR25 × 25 × 3 placed on two stable supports (Fig. 4). During the experiment the beam was excited by a hammer hit. The beam vibrations were recorded by three accelerometers pinned to the beam and by an infrared camera placed on a tripod at the distance of 149 cm from the beam. The camera recorded the movement of the tag affixed to the beam. The camera was fixed in manual acquisition and the video stream was undistorted.

3.4 Results of the Experiment 2

The video stream of the marker recorded by the camera was amplified using the present method. Marker offset was measured in pixels (Fig. 5, left). On this basis, the spectrum was calculated. There is a noticeable peak near 5 Hz (Fig. 5, right).

3.5 Comparison

The non-contact results were compared with measurements from precision accelerometers considered standard in the construction industry. The obtained results for beam #1 are shown in Fig. 6. The peak measured by accelerometers was 10.78 Hz, the peak calculated using the proposed method of contactless vibration measurement was 9.699 Hz. The error of the non-contact method was 1.081 Hz.

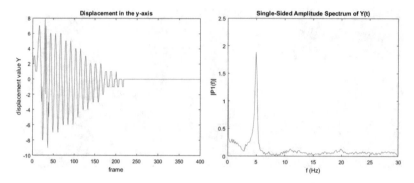

Fig. 5 (Left) Chart of position change in time of the marker observed by the camera, (right) plot of frequency spectrum calculated for the data of Fig. 5 (left)

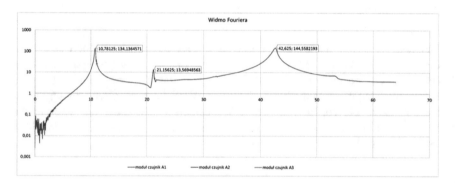

Fig. 6 Spectrum chart calculated on the basis of accelerometer readings during experiment #1

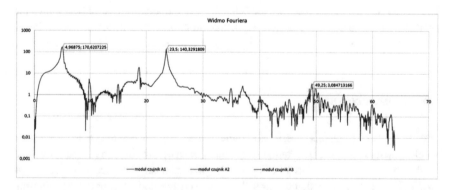

Fig. 7 Spectrum chart calculated on the basis of accelerometer readings during experiment #2

The obtained results for beam #1 are shown in Fig. 7. The peak measured by accelerometers was 4.968 Hz, the peak calculated using the proposed method of contactless vibration measurement was 4.95 Hz. The error of the non-contact method was 0.018 Hz.

4 Conclusions

The obtained results of the experiments have confirmed that there is a possibility to make contactless object vibration measurement based on image processing from the camera observing the measuring points. It has been confirmed that the results obtained using image magnification techniques are not only visual effects, but also allow the measurement of vibration. The results of the proposed non-contact vibration measurement approach are similar to classic accelerometer-based solutions, which we consider a promising starting point for further research. The quality of the measurement depends on the recording that is affected by: camera resolution, camera objects, matrix noise, video compression, measurement environment stability, distance from the measuring point. Further research will be focused on using the proposed method as additional/backup source of information for UAV control algorithms. Further work will be conducted using simulation systems [10].

Acknowledgements This work has been supported by National Centre for Research and Development as a project ID: DOB-BIO7/13/05/2015, WIMA—virtual mast as a platform for surveillance sensors for the Border Guard.

References

1. Sokabe, M., Sachs, F., Jing, Z.Q.: Quantitative video microscopy of patch clamped membranes stress, strain, capacitance, and stretch channel activation. Biophys. J. **59**(3), 722–728 (1991)
2. Sobel, D., Jędrasiak, K., Daniec, K., Wrona, J., Jurgaś, P., Nawrat, A.M.: Camera calibration for tracked vehicles augmented reality applications. In: Innovative Control Systems for Tracked Vehicle Platforms, pp. 147–162. Springer, Cham (2014)
3. Babiarz, A., Bieda, R., Jędrasiak, K., Nawrat, A.: Machine vision in autonomous systems of detection and location of objects in digital images. In: Vision Based Systems for UAV Applications, pp. 3–25. Springer International Publishing (2013)
4. Bereska, D., Daniec, K., Fraś, S., Jędrasiak, K., Malinowski, M., Nawrat, A.: System for multi-axial mechanical stabilization of digital camera. In: Vision Based Systems for UAV Applications, pp. 177–189. Springer International Publishing (2013)
5. Jędrasiak, K., Nawrat, A., Wydmańska, K.: SETh-link the distributed management system for unmanned mobile vehicles. In: Advanced Technologies for Intelligent Systems of National Border Security, pp. 247–256 (2013)
6. Jedrasiak, K., Andrzejczak, M., Nawrat, A.: SETh: the method for long-term object tracking. In: International Conference on Computer Vision and Graphics, pp. 302–315. Springer (2014)

7. Domżał, M., Jędrasiak, K., Sobel, D., Ryt, A., Nawrat, A.: GPU-based video motion magnification. In: Simos, T., Tsitouras, C. (eds.), AIP Conference Proceedings, vol. 1738, no. 1, p. 180017. AIP Publishing (2016)
8. Domżał, M., Sobel, D., Kwiatkowski, J., Jędrasiak, K., Nawrat, A.: Efficient Motion Magnification. In: Asian Conference on Intelligent Information and Database Systems, pp. 477–486. Springer, Berlin, Heidelberg (2016)
9. Wadhwa, N., Rubinstein, M., Durand, F., Freeman W.T.: Riesz pyramids for fast phase-based video magnification. ICCP, IEEE International Conference on Computational Photography. pp. 1–10 (2014)
10. Daniec, K., Iwaneczko, P., Jędrasiak, K., Nawrat, A.: Prototyping the autonomous flight algorithms using the Prepar3D® simulator. In: Vision Based Systems for UAV Applications, pp. 219–232. Springer International Publishing (2013)

Viewpoint Invariant Person Re-identification with Pose and Weighted Local Features

Chun-Huei Chen, Ju-Chin Chen and Kawuu W. Lin

Abstract In this study, we propose a viewpoint-invariant person re-identification scheme with pose priors and weighted local features. We divide the pose angle into three classes: $(0°, 180°)$, $(45°, 135°)$, and $90°$. Each of the classes has a weighted map. In addition, the texture-based feature, histogram of oriented gradients, is extracted to predict pose angle using support vector machine. Moreover, two additional features, salient color names and local binary patterns (LBP), are extracted. The former feature is computed using a weighted map with Gaussian distribution. The latter feature is computed using a weighted map based on the predicted pose angle. Then, the image representation is concatenated with salient color names and LBP. Finally, we adopt cross-view quadratic discriminant analysis for person re-identification.

Keyword Person re-identification · Pose angle estimation · Color names

1 Introduction

In recent years, surveillance systems are applied more and more. They are used in airport, train station, school, office, etc. The ranges of surveillance equipment are usually huge, and their field-of-view is different to cover more area to lower the obvious that large-scale surveillance equipment will have tons of video data. The benefit of saving these video data is what if something like threat or crime happen, we will need those video data to collect evidence. Online artificial surveillance system is usually used in train station, airport, etc. But it may cost lots of time and human resource. All these reasons lower the quality and effectiveness of surveillance system, therefore automatically analyzing a large amount of data could not

C.-H. Chen · J.-C. Chen (✉) · K. W. Lin
National Kaohsiung University of Applied Sciences, Kaohsiung, Taiwan, ROC
e-mail: jc.chen@cc.kuas.edu.tw

K. W. Lin
e-mail: linwc@cc.kuas.edu.tw

© Springer International Publishing AG, part of Springer Nature 2018
A. Sieminski et al. (eds.), *Modern Approaches for Intelligent Information and Database Systems*, Studies in Computational Intelligence 769,
https://doi.org/10.1007/978-3-319-76081-0_33

only enhance data operating speed but also significantly raise the quality of surveillance system [1]. The analysis of video can also be applied in field, person long term activity, person behavior, or even higher-level application. For example: strange behavior detection or strange person tracking.

However, using computer vision to realize high-level surveillance system application needs the ability of detecting person from multi cameras. With this ability we can analyze human behavior and do behavior detection. Expect tracking, person identification can also be used to build the relation between different cameras. That is person identification which can recognize the person image from different camera. Give a stable ID to captured image. We can use this ID build the moving locus, and continue tracking the locus. Person identification can be applied to not only multi cameras but also unique camera. For example, if someone keeps walking in and out through the camera or someone pick up the package or bag, at this time, we can use the ID to recognize is man the same or not. So person identification has a variety of uses.

The major difficulty of person identification is the appearance differences from person figures between disjoint cameras [2]. In Fig. 1, those images include same person in different situations. Figure (a) and (b) are captured from different camera in same day, and figure (c) and (d) are captured from different day. We can find out even in same day, there is a big difference in person appearance. So the challenge is viewpoint, pose, cluttered background, partial occlusions, low resolution, etc. To overcome these challenges, many research of person re-identification were proposed [3–11] and the solutions can be categorized into feature extraction [12] and metric learning [13–19]. In our work, a viewpoint invariant person re-identification

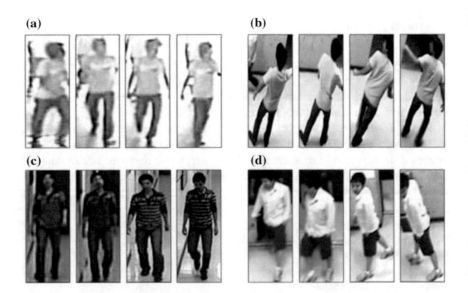

Fig. 1 Appearance differences captured by surveillance cameras

system is proposed. In the situation of knowing person pose angle, the corresponding weighted graph is applied to weight the feature vectors. Make the significant data keep higher importance. In the end, use the relation between similar and dissimilar pair to learn metric and build person re-identification system.

This paper is divided into five parts. In Sect. 1 we introduce the topic and difficulty of person re-identification, and simply relate research motivation. We review the system including the training and test process in Sect. 2, and the methods of pose angle prediction and feature extraction are presented in Sect. 3 in detail. In Sect. 4, experiment results are shown and discussed. Conclusion is made in Sect. 5.

2 System Overview

The proposed system consists of the training and test process and the flowchart is shown in Fig. 2. Section 2.1 is training process including the usage of SVM [20] to evaluate angle and how to use XQDA to train metric [18]. Section 2.2 is the test process. After input a test pair, the pose angles are estimated by SVM. Then the metric trained by XQDA is used to estimate the similarity of the test pair.

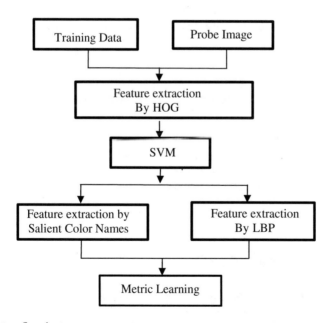

Fig. 2 System flowchart

2.1 Training Process

In the training process, HOG feature is extracted for the training data firstly, and then the extracted feature is weighted according to the labeled pose class. Note that three pose classes are designed in the proposed method and a specific feature weighted graph is designed for each pose class. And then use the model evaluated by SVM training angle divide each image into six parallel blocks. Use each block capture Salient Color Names feature and LBP feature to be the person re-identification metric training sample. Salient color name is a feature weighted graph with Gaussian distribution. After extracting features for each block, the metric for the similarity of the input pair is trained by XQDA in the end.

2.2 Test Process

In the test process, the HOG feature is extracted for the test data firstly, and then the SVM model is used to estimate the pose angle for the test data. Slice the testing images into six parallel blocks. The features of salient color name and LBP are extracted for each block. Salient color takes Gaussian distribution feature weighted graph. LBP use the angle class and the corresponding feature weighted graph to make eigenvector. At last, use the metric trained by XQDA to do person re-identification.

3 Viewpoint Invariant Person Re-identification

In this section, pose classification and metric learning are introduced in detail. Section 3.1 presents the feature weighted graph and HOG feature [21] is introduced in Sect. 3.2. Then the pose angle estimation is presented in Sect. 3.3. The extraction of weighted color and texture features are presented in Sects. 3.4 and 3.5, respectively.

3.1 Feature Weighted Graph

In the proposed method, three pose classes are defined. The first class includes 0 and 180°, the second class includes 45 and 135° and the third class is 90°. A specific weight graph with the size of 48 × 128 is designed for each pose class as shown in Fig. 3. The weight for the pixels in green color is set to 2 and the weight is set to 1 for the remaining ones.

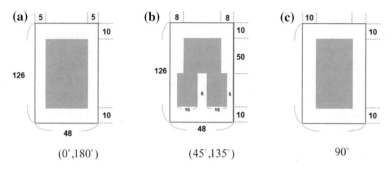

Fig. 3 Three classes of feature weighted graph

3.2 Histogram of Oriented Gradients

For evaluating angle, we will explain the process of HOG [21]. At first, set the image size to 48×128, and then turn the figure into gray level. Apply edge detection algorithm to calculate the oriented gradients and magnitude, and then slice the image into numbers of 16×16 blocks. Each block slices into 2×2 cells, so each cell is 8×8. Each cell is represented by a 9-dimensional histogram of gradient orientation, and hence each block is a 36-dimensional feature vector. Each block moves 8 pixels with overlap so each image is represented by seventy five 36-dimensional HOG feature vectors. Finally, a 2700-dimensional feature vector is extracted for each image by concatenating all feature vectors in one row.

3.3 Pose Angle Classification

The follow is about the method of pose angle classification. According to the VIPeR database [22], it provides the true angle in five types $(0°, 45°, 90°, 135°, 180°)$. After our work, we find out the angle of figure can divide into three classes by their texture $((0°, 180°), (45°, 135°), 90°)$. We use HOG to capture the texture feature. In training data, give corresponding label to each image with the ground truth angle, and then use the training data to train the angle model by SVM with RBF kernel. In the test process, the well-trained SVM model is used for the pose angle classification. Each test image is classified into one of three classes.

3.4 Feature Extraction: Salient Color Names

At first, we will introduce the part about capturing features. For person re-identification, color is the most important feature, and the features of color names have been proposed [23–25] and applied to image retrieval [26] and person re-identification [25]. This paper followed the structure of SCNCD [25] to find salient color, and use LBP calculation to find texture feature. In RGB color space, we can see three channels as a cube. Color range of each channel is from 0 to 255. Divide each channel into 32 parts. Overall, there are $32 \times 32 \times 32 = 32768$ cubs. Each little cube represents an index. Each index contains $8 \times 8 \times 8 = 512$ color. Calculate a histogram of 16 kinds of salient color for each index. There will be 32768 histograms. Slice each figure to six stripes. Take the corresponding salient histogram with each index of color. Furthermore, according to the classified angle the histogram is multiplying with a weight function. The following is feature weighted function

$$w_{ik} = \exp\left(\frac{(y_{ik} - \mu)^2}{2\sigma^2}\right) \tag{1}$$

where $\mu = L/2$ and $\sigma = L/4$, After obtaining the histogram of 16 kinds of colors, the normalization is applied and hence a salient color feature in RGB color space is extracted. Besides, we also transform RGB color space to HSV, Normalize RBG, and L1L2L3 color spaces. After transforming, normalize the value from 0 to 255. Divide the figure into six parts. Use the color index of each part, the histogram of salient color in various color space can be extracted as well.

According to the classification result of angle each feature weight of colors multiply to the histogram of the silent color. After that, cumulate all result to 16 kinds of salient color histogram. Normalize the histogram and concatenate all histogram obtained in various color space. Therefore, the color feature for a stripe is a 64-dimensional vector. By concatenating the features for 6 stripes, a 384-dimensional color feature vector is used to represent an image.

3.5 Feature Extraction: Local Binary Pattern

LBP is a texture-based feature [26, 27]. the image "is transformed" into gray level and sliced into six parts. The basic operation for LBP is 3×3 image pixels and the gray value of the center pixel is set as the threshold. Compare the other 8 gray level values with threshold. If the gray level is smaller than threshold, label the the binary code as 0. If the threshold is bigger than others, label the binary code as 1. After calculation, eight binary codes can be obtained.

According to the class of feature weighted graph, the weighted LBP feature with a 59-dimensional histogram can be obtained for each image stripe. Then all the LBP features of six stripes are normalized and concatenated into one 354 feature vector as the texture-based feature for the image.

3.6 Cross-View Quadratic Discriminant Analysis (XQDA)

Many metric learning algorithms have been proposed and applied to person re-identification [13–18]. Among them, XQDA [18] is the state-of-the art. In the proposed system, XQDA is applied to measure the distance between two images. Since the color and texture features are extracted for the image, two kinds of features are concatenated into a 738-dimensional vector before the training of the metric. 316 training pairs from the VIPeR database are used in the training process. The covariance matrices for the similar and dissimilar pairs are constructed. Then the projecting matrix can be obtained and used to measure the distance of the image pair.

In the test process, a testing data is projected into the same feature space with the training data by the projecting matrix. Then the distance between the probe image and other gallery image can be calculated via the metric. After ranking, we will get the result of person re-identification.

4 Experimental Results

We trained and tested 632 images in VIPeR database [22]. When the angle is 0° and 180°, the accuracy rate is 85.76%. The accuracy rate for 90° data is 84.21%, but the results for 45° and 135° is only 40.78% (Table 1).

In order to analyze the performance of various features, four feature types are used in the experiments. Type 1 is Salient Color Names with Gaussian feature weighted graph and LBP feature angle class with the result of feature weighted graph. Type 2 is Salient Color Names with Gaussian distribution weighted graph and LBP feature with the result of corresponding feature weighted graph. Type 3 is the angle class estimated by Salient Color Names with corresponding feature weighted graph and the LBP feature according to the estimated angle class with the result of corresponding feature weighted graph. Type 4 is the result of Salient Color Names and LBP feature without weight. Table 2 summarized the accuracy rate by various features. We can find that Type 4 without weight in Rank 1 is 28.92%, and the result of Type 2 with pose classification and weight in Rank one is 31.13%. We can notice that the accuracy is higher (Table 2).

Figure 4 shows the results by the proposed method. The image captured by cam_a is Probe. Its similarity with the other image captured from camera cam_b is ranked.

Table 1 The accuracy rate for angle estimation	Ground truth	Predict		
		0° and 180°	45° and 135°	90°
	0° and 180°	85.76	13.59	0.64
	45° and 135°	47.36	40.78	11.84
	90°	2.33	13.45	84.21

Table 2 Experimental results for various features

Feature data	Rank 1	Rank 5	Rank 10	Rank 20
Type 1	31.20	61.61	75.79	87.05
Type 2	31.13	61.58	75.75	87.02
Type 3	29.93	60.91	74.96	87.59
Type 4	28.92	59.46	73.35	86.74

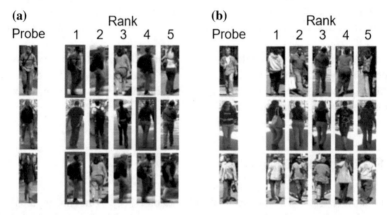

Fig. 4 The result of person re-identification

When the similarity between the test image and one gallery image is high (that is the distance between them is small), the gallery image would be ranking in the front, and vice versa. Figure 4(a) shows the success examples and the correct matching is marked in a red rectangle. The failure examples are shown in Fig. 4(b). It is observed that color features would be dominated the similarity in the person re-identification.

5   Conclusion

In this study, we propose a viewpoint-invariant person re-identification system with pose priors. One of three angle classes is estimated for each detected window before matching. According to the estimated angle class, the feature weight is given. Then two features, salient color names and local binary patterns, are extracted for each window and concatenated. The matching algorithm, Cross-view quadratic discriminant analysis for person re-identification, is applied for person re-identification. The proposed method can provide about 70% accuracy rate for angle estimation and 30% accuracy rate on average for person re-identification in rank 1.

References

1. Tu, P., Doretto, G., Krahnstoever, N., Perera, A.G.A., Wheeler, F., Liu, X., Rittscher, J., Sebastian, T., Yu, T., Harding, K.: An intelligent video framework for homeland protection. In: SPIE Defense and Security Symposium (2007)
2. Wang, S., Lewandowski, M., Annesley, J., Orwell, J.: Re-identification of pedestrians with variable occlusion and scale. In: ICCV, pp. 1876–1882 (2011)
3. Gray, D., Tao, H.: Viewpoint invariant pedestrian recognition with an ensemble of localized features. In: ECCV, vol. 5302, pp. 262–275 (2008)
4. Farenzena, M., Bazzani, L., Perina, A., Murino, V., Cristani, M.: Person re-identification by symmetry-driven accumulation of local features. In: CVPR, pp. 2360–2367 (2010)
5. Liu, C., Gong, S., Loy, C.C., Lin, X.: Person re-identification: what features are important? In: ECCV, vol. 7583, pp. 391–401 (2012)
6. Zha, R., Ouyang, W., Wang, X.: Unsupervised salience learning for person re-identification. In: CVPR, pp. 3586–3593 (2013)
7. Kuo, C.H., Khamis, S., Shet, V.: Person re-identification using semantic color names and RankBoost. In: WACV, pp. 281–287 (2013)
8. Simonnet, D., Lewandowski, M., Velastin, S.A., Orwell, J., Turkbeyler, E.: Re-identification of pedestrians in crowds using dynamic time warping. In: ECCV, vol. 7583, pp. 423–432 (2012)
9. Prosser, B., Zheng, W., Gong, S., Xiang, T.: Person re-identification by support vector ranking. In: BMVC, pp. 1–11 (2010)
10. Cheng, D.S., Cristani, M., Stoppa, M., Bazzani, L., Murino, V.: Custom pictorial structures for re-identification. In: BMVC (2011)
11. Hirzer, M., Beleznai, C., Roth, P.M., Bischof, H.: Person re-identification by descriptive and discriminative classification. In: SCIA, pp. 91–102 (2011)
12. Kviatkovsky, I., Adam, A., Rivlin, E.: Color invariants for person re-identification. TPAMI 35, 1622–1634 (2013)
13. Zheng, W.S., Gong, S., Xiang, T.: Re-identification by relative distance comparison. TPAMI 35, 653–668 (2013)
14. Weinberger, K.Q., Blitzer, J., Saul, L.K.: Distance metric learning for large margin nearest neighbor classification. In: Advances NIPS (2005)
15. Weinberger, K.Q., Saul, L.K.: Fast solvers and efficient implementations for distance metric learning. In: ICML, pp. 1160–1167 (2008)
16. Davis, J.V., Kulis, B., Jain, P., Sra, S., Dhillon, I.S.: Information-theoretic metric learning. In: ICML, pp. 209–216 (2007)
17. Koestinger, M., Hirzer, M., Wohlhart, P., Roth, P.M., Bischof, H.: Large scale metric learning from equivalence constraints. In: CVPR, pp. 2288–2295 (2012)
18. Liao, S., Hu, Y., Zhu, X., Li, S.Z.: Person re-identification by local maximal occurrence representation and metric learning. In: CVPR, pp. 2197–2206 (2015)
19. Roth, P.M., Hirzer, M., Koestinger, M., Beleznai, C., Bischof, H.: Mahalanobis distance learning for person re-identification. In: Person Re-identification, pp. 247–267 (2014)
20. Chang, C.C., Lin, C.J.: LIBSVM: a library for support vector machines. TIST 2(27) (2011). Software available at http://www.csie.ntu.edu.tw/~cjlin/libsvm/
21. Navneet, D., Triggs, B.: Histograms of oriented gradients for human detection. In: CVPR, vol. 1, pp. 886–893 (2005)
22. Gray, D., Brennan, S., Tao, H.: Evaluating appearance models for recognition, reacquisition, and tracking. In: PETS (2007)
23. van de Weijer, J., Schmid, C., Verbeek, J., Larlus, D.: Learning color names for real-world applications. Image Process. 1512–1523 (2009)
24. van de Weijer, J., Schmid, C.: Applying color names to image description. In: ICIP, vol. 3, pp. III-493–III-496 (2007)

25. Yang, Y., Yang, J., Yan, J., Liao, S., Yi, D., Li, S.Z.: Salient color names for person re-identification. In: ECCV, vol. 8689, pp. 536–551 (2014)
26. Liu, Y., Zhang, D., Lu, G., Ma, W.Y.: Region-based image retrieval with high-level semantic color names. In: Proceedings of the 11th International Multimedia Modelling Conference, pp. 180–187 (2005)
27. Ojala, T., Pietikäinen, M., Harwood, D.: Performance evaluation of texture measures with classification based on Kullback discrimination of distributions. In: ICPR, vol. 1, pp. 582–585 (1994)
28. Mäenpää, T., Pietikäinen, M.: Texture analysis with local binary patterns. In: Handbook of Pattern Recognition and Computer Vision, pp. 197–216 (2005)

Breast Cancer Detection Using Modern Visual IT Techniques

Sebastien Mambou, Petra Maresova, Ondrej Krejcar,
Ali Selamat and Kamil Kuca

Abstract Nowadays, cancer is a major cause of women death, especially breast cancer which is most seen on ladies older than 40 years. As we know, several techniques have been developed to fight breast cancer, like a mammography, which is the preferred screening examination for breast cancer. However, despite mammography test showing negative result, there are still patients with breast cancer diagnostic, found by other tests like ultrasound test. It can be explained by potential side effect of using mammography, which can push patients and doctors to look for other diagnostic technique. In this literature review, we will explore the digital infrared imaging which is based on the principle that metabolic activity and vascular circulation, in both pre-cancerous tissue and the area surrounding a developing breast cancer, is almost always higher than in normal breast tissue. In the same way, an automated infrared image processing of patient cannot be done without a model like the hemispheric model, which is very well known. As novelty,

S. Mambou · O. Krejcar (✉) · A. Selamat · K. Kuca
Faculty of Informatics and Management, Center for Basic and Applied Research,
University of Hradec Kralove, Rokitanskeho 62, 500 03 Hradec Kralove
Czech Republic
e-mail: ondrej.krejcar@uhk.cz; ondrej.krejcar@remoteworld.net

S. Mambou
e-mail: sebastien.mambou@uhk.cz

A. Selamat
e-mail: aselamat@utm.my

K. Kuca
e-mail: kamil.kuca@uhk.cz

P. Maresova
Faculty of Informatics and Management, Department of Economy,
University of Hradec Kralove, Rokitanskeho 62, 500 03
Hradec Kralove, Czech Republic
e-mail: Petra.Maresova@uhk.cz

A. Selamat
Faculty of Computing, Universiti Teknologi Malaysia, 81310, Johor Bahru, Johor, Malaysia

© Springer International Publishing AG, part of Springer Nature 2018   397
A. Sieminski et al. (eds.), *Modern Approaches for Intelligent Information
and Database Systems*, Studies in Computational Intelligence 769,
https://doi.org/10.1007/978-3-319-76081-0_34

we will give a comparative study of breast cancer detection using modern visual IT techniques view by the perspective of computer scientist.

Keywords Breast · Cancer · Detection · Visual techniques
Neural network · SVM

1 Introduction

In a healthy body, natural systems control the creation, growth and death of cells. We can see the apparition of cancer cells when these systems don't work right and cells don't die at the normal rate, this results to more cell growth than cell death. In addition, breast cancer occurs when cells in the breast divide and grow without their normal control. It is a world problem which is present in every country, for example, in USA one in eight women will be diagnosed with breast cancer in her lifetime and more than 40,000 will die [1]. A reduction of this rate can be achieved with a better screening and early detection, also with an increased in awareness, and continuous improvement in treatment options. In this literature review, we will explore the Digital Infrared Imaging which is based on the principle that metabolic activity and vascular circulation, in both pre-cancerous tissue and the area surrounding a developing breast cancer, is almost always higher than in normal breast tissue. As we know, several techniques have been developed to fight breast cancer, like a mammography, which is the preferred screening examination for breast cancer [2]. Our duty is to give a clear picture of what has been done so far and what we esteemed to be the most convenient way among the state-of-the-art techniques. Later on, we will give the elements that will enhance our ideas.

2 Previous Techniques and Comments

Authors in [3] did a review, based on several articles from 2002 to 2010, which reveals Screen-film mammography (SFM) limitations, which are illustrated by its false-negative rate ranges from 4 to 34%. Even if, mammography has been considered for long time as the gold standard for breast cancer screening and detection, the need of new techniques to overcome those limitations came in evidence. Also publication [4] has presented near-infrared fluorescent (NIRF) as key roles in clinical diagnosis, as well as evaluation of disease status and treatment of tumour. However, it is very important for the image processing to have a Clair strong NIRF signal, so that the image taken, can content a lot of information very near to the real state of the breast. As we know, as early is the detection of the tumor, as better and successful is the treatment. In addition, authors [5] talked about the difficulty to obtain tumor parameters such as: metabolic heat, tumor depth and diameter from a thermogram, Furthermore, another paper [6] mentioned the limitations of computed tomography (CT) and magnetic resonance imaging (MRI), which have a low sensitivity for sub-centimetre lesions due to their limited spatial resolution. In addition, [7] mentioned another bad point of a successive mammography test for a period of 10 years, according to his study paper, the rate of false positive diagnosis for women after doing a mammogram every year for 10 years is 49.1%. Other aspect has been explored in [8], where he advised to do (Sentinel Lymph Node) SLN biopsy, in order to reduce the a higher risk of disease (breast cancer) progression, in addition authors [9] has advised, to not take breast cancer's thermography result like sufficient information for decision taking.

2.1 Related Works

2.1.1 A Doubt for Infrared Base Techniques

Article [9] illustrates the limitation of the thermography image as important tool in decision taking technique due to its high number of false positive observe by conducting, for example a test on 126 breasts of 63 patients (58 females and five males). An assumption has been taken so that the mean age of the patients was 47.6 years (range 26–82 years). After processing the thermal image, cancerous lesions were finally diagnosed in 20 breasts and there were no bilateral cancers. The outcome of the study was that the thermal diagnosis as non-sufficient for the primary evaluation of symptomatic patients.

2.1.2 Infrared Thermal Imaging

Paper [3] deal with pre-digital mammography (FFDM) and digital infrared thermal imaging (DITI) as imaging modalities that would overcome mammography limitations. Due to its ability to selectively optimize contrast in areas of dense parenchyma, Digital mammography is superior to screen-film mammography in younger women with dense breasts, however we observe due to the density of their breast tissue, a lower sensitivity of mammography. In Other hand, the utilization of Digital Infrared Imaging is based on the principle that, metabolic activity and vascular circulation in both pre-cancerous tissue and the area surrounding a developing cancer, is almost always higher than in normal tissue, also it is reported that the results of thermography can be correct 8–10 years before mammography can detect a mass in the patient's body. In addition of that, a shortage of qualified radiologists, also causes an urgent demand for the development of computer technologies as computerized thermal imaging (CTI).

Furthermore, paper [6] discussed about thermal diagnosis, which is still improving every year, furthermore, several thermal diagnosis show that breast cancer patients with abnormal thermograms have fast growing tumours. For achieved a better test results, a protocol must be follow before a patient undergoes

(a)

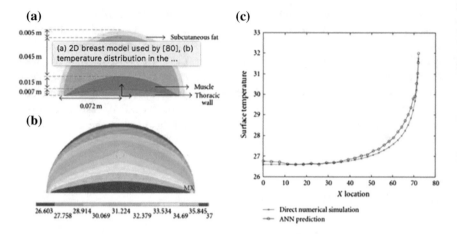

(c)

(b)

Fig. 1 Thermal diffusion in a conventional breast representation [10]

an examination, for the testing procedure and environment during the examination and for the post processing of the obtained thermograms. In addition, thermal diagnosis can be further optimize by the help of Artificial Neural Network (ANNs) where many images both with cancer and without breast cancer must be provided, to feed the input layer and then processed in the hidden layers. The output from the last hidden layer serves as input to the neurons in the output layer and a decision is taken. A better improvement can be to combine ANNs, genetic algorithms, and computer simulations, so that they relate the skin surface temperature with the tumor depth, diameter, and heat generation, all of these, by considering the computational domain as layered semi-spherical breast (Fig. 1a). Furthermore, it is good to notice that ANN showed good agreement with the numerical simulation and surface temperatures (Fig. 1c). For better understanding of the breast cancer several geometric models have been propose so far, among them, we have: Rectangular domain (Fig. 2) which provided a first insight on predictive models that relate the surface temperature with the tumor size and location, however it does not represent

Fig. 2 Rectangular domain which provided a first insight on predictive models that relate the surface temperature with the tumor size and location [11]

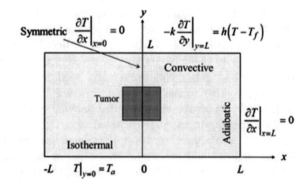

Fig. 3 Hemispherical domain with non-concentric layers, which is a very popular model due to its ability of reproducing surface temperatures [12]

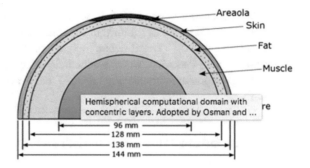

Areaola
Skin
Fat
Muscle

Hemispherical computational domain with concentric layers. Adopted by Osman and ...

96 mm
128 mm
138 mm
144 mm

the real shape of the breast; hemispherical computational domains (Fig. 3) gave results in better agreement with experimental data. Also, the temperature distribution illustrated the effect of the different layers in the surface temperature.

In our context of breast tumor detection, the surface temperature of the breast is the solution the bio-heat transfer equation, so we just need to determine the value of the parameters of this equation. An application of the inverse modelling method requires however a model for the breast. Furthermore, in our context the bio-heat transfer equation is solved for a set of initial values of the thermos physical properties of the breast tissues.

We need also optimization techniques such as the Gradient Descent Method, the Levenberg-Marquardt algorithm, or Genetic Algorithms so to estimate the value of the Thermal physical properties. However, as temperature can only be measured at the surface of the breast and the temperature profile inside the breast remains unknown, the inverse modelling problem remain an ill posed with no unique solution. In the other side, paper [13] elaborate on the IR radiation of human body which is most often in the range of 2–20 μm wavelength, it can be detected very precisely with our nowadays infrared camera [14, 15]. Two factor must be keep for achieved a success rate of computer diagnosis: higher quality of the sensing element (infrared camera) and Improvement of the image processing algorithms throughout the whole process. In the same way, Independent component analysis (ICA) is a subspace projection technique that projects data from a high-dimensional space to a lower-dimensional space. The ICA's tumor analysis method is composed of 3 concatenated phases: separation of the original image in two chrominance (Cb, Cr) and luminance components (Y), obtaining the independent components (ICA) of the image and post-processing for segmentation of the tumorous areas. This analysis brings out features that are inappreciable in the original image which are associated with areas of high risk tumor, because of their high body temperature link to an extreme tone within the original image. However, ICA is to solve problem which consists in considering the observation data matrix X (digital image) as a linear combination of independent components, i.e. $X = A \cdot S$ where S contains the independent components and the mixing matrix, and its coefficients describe uniquely the mixed source regions and can be used as extracted features. In short, ICA attempts to 'un-mix' the data by estimating a Demixing matrix W where

$Y = W \cdot X$, so the goal of ICA is to recover A and Y using the information contained in X. One possible algorithm for obtaining the matrix $W = A$ power (-1) (the demixing matrix). The different outputs of ICA will be given as input of **automated post processed**, which by the help of several internal computation will proceed each input and compare the result so that the output is with a best accuracy as possible.

2.1.3   Near-Infrared Fluorescent (NIRF) and Agent

Article [2] introduced the concept of agent which represents in this context, a composition of molecule which when uptake, can envelop the tumor and produce a stronger near-infrared fluorescent (NIRF) signal suitable for a good localization in the brain. The IR780-phospholipid micelle (Fig. 1) is suitable for this purpose, as it has the ability to pass through the natural barrier of brain, localize and attach itself to tumour cells (Fig. 2).

2.1.4   The Diameter and Depth of the Tumor

Paper [3] shows the diameter and depth of the tumor as a problem to be solve using thermal techniques, the full width at half maximum (FWHM), to estimate the depth of a small heat source from the isothermal distribution of hot spots at the surface. Furthermore, due to the complex structure of the breast heat transfer, the application of a vascular model requires a detailed knowledge of the microvasculature network.

In order to determine the relationship between depth and heat transfer during cooling, it is helpful to define the thermal penetration depth which is not yet precise due to the complex structure of the breast. However an assumption can be made by considering hemispherical domaine. Furthermore, it appears that after cooling of the breast, the response time (the considerable among of heat detect at the surface the breast) increases with tumor depth. In addition, we have observe that the tumor diameter have no significant effect on the response time for shallow tumors. We have also observe that the smaller the tumor is, the longer is the response time. During our experience, we have observe certain principle in order to measure a correct values or make our thermal sensor more accurate. One of the important procedure was to keep the room's temperature stable within the range of 18–22 °C.

2.2   Economical Aspects of Breast Cancer

Breast cancer is the most common cancer in women, making up 23% of all newly diagnosed oncology cases. 85% of families in which one of the parents will suffer from cancer without the help of family, friends, nonprofit organizations or even debt. Caring for patients with breast cancer has a major impact on the budget, so it is a priority of public health to support effective preventive programs that could be cost-saving in the long run by reducing the incidence of cancer or offering therapeutic interventions that would increase patients' survival chances, thus reducing the indirect costs of morbidity and mortality [16]. The reality of the cost burden is evidenced by many studies, for example authors [10] show that loss of productivity associated with breast cancer in Poland were €583.7 million in 2010 and €699.7 million in 2014. During this period, costs accrued for 0.162–0.171% of GDP. Public finance expenditure for social insurance benefits to BC sufferers ranged from €50.2 million (2010) to €56.6 million (2014), an equivalent of 0.72–0.79% of expenditures for all diseases. Lost opportunities costs in public finance revenues accounted for €173.9 million in 2010 and €211.0 million in 2014 [10].

Authors at [12] estimated that $245 million USD in cohort of Mexican women in medical costs and income losses owing to breast cancer could be saved over a cohort's lifetime. Medical costs account for 80% of the economic burden; income losses and opportunity costs for caregivers account for 15 and 5%, respectively [12].

Paper [17] evaluated cost utility of Breast Cancer Index (BCI) from a US third-party payer perspective. Use of BCI was projected to be cost saving. In the newly diagnosed population, net cost savings were $3803 per patient tested. In the 5 years post diagnosis population, BCI was projected to yield a net cost savings of $1803 per patient tested.

Authors at [18] estimated the socioeconomic cost and burden for breast cancer patients in Korea between 2007 and 2010. The prevalence of treated breast cancer increased from 7.9 to 20.4%. The total socioeconomic costs incurred by breast cancer increased by approximately 40.7% from US $668.49 million in 2007 to US $940.75 million in 2010. Medical care costs for 2010 were 1.4 times greater (US $399.22 million) than for 2007 (US $278.71 million. The direct non-medical costs rose from US $50.69 million in 2007 to US $75.83 million in 2010, a 49.6% increase. Regarding the economic burden of breast cancer, the total indirect costs were US $339.09 million in 2007 and increased by 37.3% to US $465.70 million in 2010.

However, cost growth may not always be linked to rising prevalence but also to the development of new healing methods. For example, standard chemotherapy in the Czech Republic for breast cancer costs around 200 EUR. Biological treatment, which is suitable for one fifth of patients, costs roughly 30,000 EUR. Hormonal standard treatment comes to 200 EUR, while other hormonal preparations are worth 9,000 EUR. The cost of radiotherapy [14, 15, 19] is between 1,700 and 2,600 EUR. At present, emerging gene or epigenetic therapy that directly influences genetic information responsible for tumor growth will be even more expensive [11]. Oncology cost growth is expected in the 7.5–10.5% range annually through 2020, when global oncology costs will exceed $150 billion (Fig. 4).

Based on all mentioned facts, the reason to find any new detection method or even small improvement of current ones is more than evident.

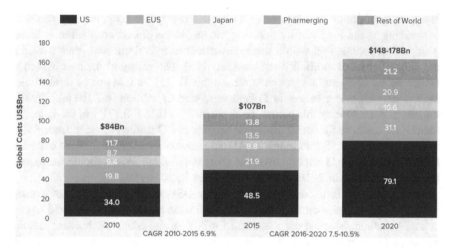

Fig. 4 Global oncology costs and growth, 2010–2020 [20]

3 Discussion of Findings

3.1 Differences Between no Touch BreastScan and Sentinel BreastScan Techniques

The study of the breast cancer base on modern techniques gives us a better view of the issue relate to the accurate detection of the tumor. Base on the previous paper discussed in "related works section", we have considered some features that we judge relevant enough to show us, a general difference between the sentinel BreastScan and the no Touch BreastScan (Table 1). It appears that, most of techniques nowadays use Neural network to reduce the number of false positive.

Table 1 Difference between the sentinel BreastScan and the no touch BreastScan

Feature	No touch BreastScan	Sentinel BreastScan
Temperature sensitivity (°C)	0.05	0.08
IR Camera resolution (pixel)	640 * 512	320 * 240
Number of IR cameras	2	1
Wavelength range	3.5–10.5	7–12
Transient IR	Yes	Yes
Cooling method	Cold air	Cold air
Cooling time (min)	5–6	3–6
Analyst time (min)	Immediate	4–5
Neural network (nn)	Yes	Yes
Age < 50 sensitivity without nn [17]	78%	67%
Age < 50 sensitivity with nn [17]	89%	78%

We have also observe an important improvement in breast cancer detection for the women with age < 50, with a sensibility ≥ 78% using No Touch software.

3.2 Proposed Model

Through the reviews of the different papers discussed in this Literature review, we have noticed the importance of the image processing which is well done by a human been but still not well achieve by Artificial intelligent. We attend in this discussion to emphasize on the need of Computer Aid Device (CAD) which will help us to better understand the thermal images, capture by our different thermal camera. CAD will be a new neural network SVM model (Fig. 5) (assuming already trained) which will take Thermal images as input and as output will classifier images as cancer or not.

It is good to mention that each neuron will apply Super vector Machine classification so that for any incoming image having a close similarity with an image of a breast cancer, the model will be powerful enough to differentiate both images.

Our propose architecture (Fig. 5) can be assimilate to a flow chart where each module represent a specific component. We can explant each as follow:

- **Pre-processing of breast Thermal images**: here, each images are converted to a matrix of features which will be submit as input to the next module.

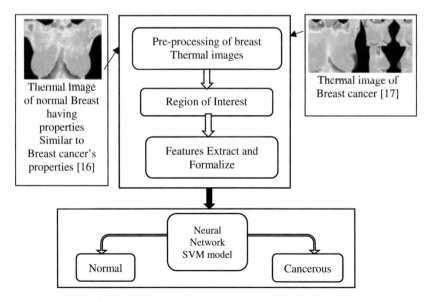

Fig. 5 Propose architecture for breast cancer detection

- **Region of Interest**: the areas with more gradient colour will be localize as they are more susceptible to have a Cancer. They will be transfer as input to the next component.
- **Features Extract and Formalize**: the localize area will be extract and convert to the appropriate format suitable for the input of the **Neural Network SVM model**.
- **Neural Network SVM model**: Through several computations base on SVM at each layer of the neural network, the system will classify the input and will output the given status of the thermal image (Normal or Cancerous).

4 Conclusion

During presented literature review, it was question to present some works done on breast cancer in the point of view of a computer scientist [21]. In that thought, we have presented the different most common techniques used to detect the breast Cancer, their strengths and dear weaknesses. We came out however, with one among them which present a good future, due to its non-immersive property and its big number of data which needs to be process with more efficient techniques. Infrared imaging couple with an agent prior administrate to a patient can lead to a very precise detector of tumor. We will use in our future study a camera with 0.5 thermal sensitivity and we will propose a model of breast which can help us to achieve a more precise diagnostic.

Acknowledgements This work was supported by internal students project at FIM, University of Hradec Kralove, Czech Republic (under ID: UHK-FIM-SP-2018).

References

1. Breast Cancer Facts, National Breast Cancer Foundation (2016)
2. Dongola, N.: Mammography in Breast Cancer. Medscape Logo (2016)
3. Köşüş, N., Köşüş, A., Duran, M., Simavlı, S., Turhan, N.: Comparison of standard mammography with digital mammography and digital infrared thermal imaging for breast cancer screening. J. Turk. Ger. Gynecol. Assoc. (2010)
4. Li, S., Johnson, J., Peck, A., Xie, Q.: Near infrared fluorescent imaging of brain tumor with IR780 dye incorporated phospholipid nanoparticles. J. Trans. Med. (2017)
5. Amria, A., Pulko, S.H., Wilk, A.J.: Potentialities of steady-state and transient thermography in breast tumour depth detection: a numerical study. Comput. Methods Programs Biomed. (2016)
6. Boogerd, L.S.F., Handgraaf, H.J.M., Lam, H.-D., Huurman, V.A.L., Farina-Sarasqueta, A., Frangioni, J.V., van de Velde, C.J.H., Braat, A.E., Vahrmeijer, A.L.: Laparoscopic detection and resection of occult liver tumors of multiple cancer types using real-time near-infrared fluorescence guidance. Surg. Endosc. (2017)

7. Kandlikar, S.G., Perez-Raya, I., Raghupathi, P.A., Gonzalez-Hernandez, J.L., Dabydeen, D., Medeiros, L., Phatak, P.: Infrared imaging technology for breast cancer detection—Current status, protocols and new directions. Int. J. Heat Mass Trans. (2017)
8. Tsutomu Namikawa, T.S.: Recent advances in near-infrared fluorescence-guided imaging surgery using indocyanine green. Surg. Today (2015)
9. Kontos, M., Wilson, R., Fentiman, I.: Digital infrared thermal imaging (DITI) of breast lesions: sensitivity and specificity of detection of primary breast cancers. Clin. Radiol. (2011)
10. Łyszczarz, B., Nojszewska, E.: Productivity losses and public finance burden attributable to breast cancer in Poland, 2010–2014. BMC Cancer 17(1), 676 (2017)
11. National Oncology Program. Czech Oncological Society (2011)
12. Unar-Munguía, M., Meza, R., Colchero, M.A., et al.: Economic and disease burden of breast cancer associated with suboptimal breastfeeding practices in Mexico. Cancer Causes Control (2017)
13. Boquete, L., Ortega, S., Miguel-Jiménez, J.M., Rodríguez-Ascariz, J.M.: Automated detection of breast cancer in thermal infrared images, based on independent component analysis. J. Med. Syst. (2012)
14. Kubicek, J., Bryjova, I., Faltynova, K., Penhaker, M., Augustynek, M., Maresova, P.: Evaluation of gama analysis results significance within verification of radiation IMRT plans in radiotherapy. Lecture Notes in Computer Science, vol. 10449, pp. 541–548 (2017). https://doi.org/10.1007/978-3-319-67077-5_52
15. Augustynek, M., Korpas, D., Penhaker, M., Cvek, J., Binarova, A.: Monitoring of CRT-D devices during radiation therapy in vitro. BioMedical Engineering Online, 15 (1), article no. 29 (2016). https://doi.org/10.1186/s12938-016-0144-7
16. Smidova, I.: Alcohol and breast cancer—economic costs. Hygiena 51(1), 17–21 (2012)
17. Gustavsen, G., Schroeder, B., Kennedy, P., et al.: Health economic analysis of breast cancer index in patients with ER+, LN− breast cancer. Am. J. Manag. Care 20(8), 1 (2014)
18. Kim, Y.A., Oh, I.H., Yoon, S.J., et al.: The economic burden of breast cancer in Korea from 2007–2010. Cancer Res. Treat. 47(4), 583–590 (2015)
19. Bryjova, I., Kubicek, J., Molnarova, K., Peter, L., Penhaker, M., Kuca, K.: Multiregional segmentation modeling in medical ultrasonography: extraction, modeling and quantification of skin layers and hypertrophic scars. Lecture Notes in Computer Science, vol. 10449, LNAI, pp. 182–192 (2017). https://doi.org/10.1007/978-3-319-67077-5_18
20. IMS Health, MIDAS, Dec 2015; Market Prognosis, Mar 2016. IMS Institute for Healthcare Informatics, May 2016
21. Cardoso, F., Harbeck, N., Bergh, J., Cortés, J.: Research needs in breast cancer. Ann. Oncol. (2016)

Contactless Identification System Based on Visual Analysis of Examined Element

**Lukas Kolda, Ondrej Krejcar, Ali Selamat, Peter Brida
and Kamil Kuca**

Abstract Identification methods based on biometry are going through great expansion lately. They are being applied in common electronics and home computers. As an example can be used implementation of biometry in Windows 10 as a function Windows Hello. This article is applying to project and realization of experimental multibiometric system for laboratory verification of theoretical knowledge. The system identifies the user using biometric characteristics of hand contour and bloodstream on the dorsum of the hand. This article deals with project and realization of hardware and software for experimental biometric system. The software part is written in C# using OpenCV library and Emgu CV wrapper.

Keywords Biometry · Identification · Bloodstream · Visual recognition
Multibiometry

L. Kolda · O. Krejcar (✉) · A. Selamat · K. Kuca
Faculty of Informatics and Management, Center for Basic and Applied Research,
University of Hradec Kralove, Rokitanskeho 62,
500 03 Hradec Králové, Czech Republic
e-mail: ondrej.krejcar@uhk.cz; ondrej.krejcar@remoteworld.net

L. Kolda
e-mail: lukas.kolda@uhk.cz

A. Selamat
e-mail: aselamat@utm.my

K. Kuca
e-mail: kamil.kuca@uhk.cz

A. Selamat
Faculty of Computing, Universiti Teknologi Malaysia, 81310
Johor Baharu, Johor, Malaysia

P. Brida
FEE, Department of Multimedia and Information-Communication Technologies,
University of Zilina, Univerzitna 1, 010 26 Zilina, Slovakia
e-mail: peter.brida@fel.uniza.sk

© Springer International Publishing AG, part of Springer Nature 2018
A. Sieminski et al. (eds.), *Modern Approaches for Intelligent Information
and Database Systems*, Studies in Computational Intelligence 769,
https://doi.org/10.1007/978-3-319-76081-0_35

1 Introduction

With rapid progress in IT area the use of biometry for verification and identification purposes has become real. The interconnection of identity with particular individual is called personal identification. Verification (authentication) means the problem of confirmation or denial of personal identification of person who proves given identity Identification means the problem of finding the identity for an unknown person. The use of biometric identification systems is beginning to be more widespread in today's world. Reasons for applying biometry in applications that needs the user's identification instead of present cards, keys or log-in data are clear: the problem with the data loss or forgetting are falling off, the danger of copying or falsification is decreasing.

Unfortunately, the biometry is not perfect solution. Analysis and sometimes even the scanning of biometric characteristics is complicated process, besides the outcome is not clear. The result is only a probability of success (that is more than 99% at the best systems). Among various biometric characteristics that can be recognized at humans, is recognizing by hands the eldest [1]. At human hand can be recognized for example: the hand's geometry, fingerprints, palm grooves, grooves on knucklebones, the bloodstream image etc [2, 3]. When the person reach adulthood, these characteristics become relatively steady for the rest of the life, so they can be used for person's identification or verification [4].

Several methods how to increase the safety of identification exists. The system can work with characteristics that are difficult to imitate, or that are hidden. As an example can be used fingerprints scanning. However the disadvantage is high acquisition cost.

The second option how to increase the system's safety is scanning of more biometric characteristics in one system (at least two, but also more). For example, the combination of both hand shape and scanning of the bloodstream. That system has relatively acceptable price, but his safety increases [5].

2 Goals

The goal of our work is to create experimental multibiometric identification system that will combine biometric characteristics gained from hand geometry and bloodstream on the dorsum of the hand. With this combination of two types of biometric characteristics we will try to reach better results, than if only one type of biometric characteristic was used. The combination of results will be performed on level of partial appraisal of correspondence. This process was used with consideration on very various biometric characteristics, where is not possible to combine the appraisal on the level of gross data (Sensor-Level Fusion) neither on the level of gained biometric characteristics (Feature-Level Fusion).

3 System Description

Function of the whole software system of user identification consists from steps that are depicted in diagram on Fig. 1.

While defining biometric characteristics for identification using bloodstream we came out from work of Lingyu Wang and collective [6]. While defining biometric characteristics we used the similarity of bloodstream and fingerprints. While personal identification using fingerprints is instead of comparison of whole picture (based on the pattern) used comparison of critical points position, so called markants. Markant in case of appraising fingerprints can be the beginning and the end of papillary line, bifurcation (split), hook, eyelet, etc. For needs of designed experimental software we defined two types of markants: branching of veins and ending of veins at the edge of the picture [7–9]. The number of found markants may vary for every user, but for one user should be the same at every scanning. Because the cut-out from original picture on which is performed extraction of biometric characteristics of bloodstream is correctly oriented and focused considering the examined hand of the user the positions of markants should remain at the same position at multiple repetition of scanning of one user. Because of that the experimental software works with coordinates of markants related directly to the bloodstream image.

(1) *Extraction of biometric characteristics*

For extraction of markants from the picture we used again the principle of pixel interconnection number. The algorithm scans every pixel of the bloodstream skeleton picture. For each pixel that belongs to the object of interest is counted the

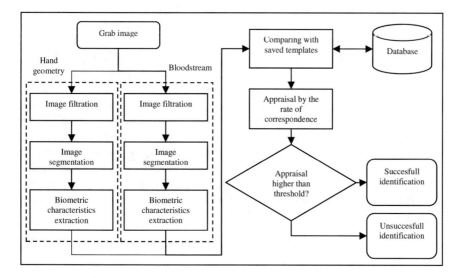

Fig. 1 Diagram of function of projected identification system

number of pixel interconnection. Every pixel which has the number of connections bigger than 3 is the point of branching and is labelled as markant. Every pixel which has the number of connections smaller than 1 and is at the edge of the picture is the ending point of vein and is also labelled as markant.

3.1 Appraisal by the Rate of Correspondence

(1) *Calculation of appraisal of partial metrics*

In devised experimental project we used following methods for calculating the rate of correspondence:

1. Euclidean distance: It is a distance of two points in n-dimension space. It is counted according to the equation:

$$d = \sqrt{\sum_{i=1}^{N} (x_i - t_i)^2} \tag{1}$$

   where n is number of dimensions, which is in case of comparing the template and testing data the number changed biometric characteristics, x_i is i-th element of tested data and t_i is i-th element of template. Number of template dimensions and tested data is equal. Final value is the sum of all differences between the template and tested data.

2. Hamming's distance: The term Hamming's distance comes from the theory of information. While comparing two series of the same length Hamming's distance tags the smallest number of positions in which they differ. Or the number of exchanges that needs to be performed for change of one of the series to the other. Sanchez-Reillo and collective [10] in his work generalized Hamming's distance to the suitable form for appraising the similarity of biometric data. He suggested use the comparison based on the number of disapproving biometric characteristics. The result is metric that don't measures the size of difference, as the Euclidean distance do, but gives the number of single biometric characteristics, at which is while comparing the tested data with template the difference bigger then standard deviation for biometric characteristic. The standard deviation is defined for each characteristic while generating the template. Determined difference was chosen based on the condition that characteristics for one user at multiple photographing will never be completely the same. The expected distribution of values for given characteristic matches to the normal distribution. Hamming's distance is counted according to equation:

$$d(x_i, \bar{x}_i) = \#\{i \in \{1 \dots N\}/|x_i - \bar{x}_i| > \sigma_i\} \tag{2}$$

where x_i is biometric characteristic of tested data with order number i, $\bar{x}_i$ is average of biometric characteristic (from template) with order number i, N is the total number of biometric characteristics for given template, σ_i is standard deviation (from template) with order number i.

3. Hausdorff's distance: It is a method which measures how far from each other are two sets of points in metric space [11]. In other words, two sets are close to each other if in surroundings of every point from one set occurs any point from the second set. Hausdorff's distance (or just HD) is the longest distance of all these distances between the pair of points that are created from the points of the first and second set of points. The more are the two sets of points similar, the smaller is HD. Considering that the biometric characteristic of bloodstream is consists of set of points, at which is important the position in the picture (it is actually some kind of pattern) [12], we tried to use HD in devised experimental software for the calculation of correspondence rate, because HD naturally compares the similarity of shapes. The disadvantage of HD is high sensitivity on outer values. Oriented HD tagged as $\bar{H}$ between sets of points A and B matches the maximal distance from all pairs $x \in A$ and $y \in B$. Oriented HD is expressed in equation:

$$\bar{H}(A,B) = \max_{x \in A}\{\min_{y \in B}\{||x,y||\}\} \tag{3}$$

where $||.,.||$ is any appraisal function, mostly Euclidean distance. Oriented HD is not symmetric, so stands $\bar{H}(A,B) \neq \bar{H}(B,A)$.

(2) *Normalization of partial metrics*

Before the result unification of single metrics is necessary to normalize those results. Single metrics provide results in various "dimensions". For appraisal normalization in experimental software we used the min-max method. Normalized appraisal *no* is counted according to equation:

$$no = \frac{o - \min_{i=1}^{N} o_t^i}{\max_{i=1}^{N} o_t^i - \min_{i=1}^{N} o_t^i} \tag{4}$$

where o is rough appraisal, N is number of elements in set of tested data, o_t^i the element of tested data.

3.2 Image Scanning

(1) *Proposal of scanning device*

Image scanning is the first step while processing the scene and has a big influence on gained results of image appraisal. For gaining required image which would be suitable for subsequent processing could be used various configurations of camera position and lighting. Basic configurations are: direct lighting, side lighting and back lighting. Those basic configurations are depicted on Fig. 4. Always depends on requirements of actual assignment. For example for measuring the object's shape is the most suitable the back lighting, where the object contour is highlighted. While planning the scanning device we used the direct lighting as a suitable configuration based on the performed tests.

For application of image scanning always try to suppress external impacts of surroundings, that could negatively influence processing and then appraising of image. One of options how to prevent those negative impacts is the use of additional scene illumination using special industrial lights, which intensively illuminates and "drown out" surrounding basic lighting. Especially appropriate is to supplement the lens with filter that lets through only the radiance with the same wave length as used light, through which is the effect notably strengthened. Requirements for those lightings are primarily:

- sufficient intensity,
- homogeneous scene illumination,
- constant intensity of illumination in time (as a result of wearing).

Because of requirements on scene illumination we used industrial types of lighting on projecting hardware components of scanning device. The main reason is the requirement of homogeneity, which is essential for subsequent image processing.

Because of the same reason we used a special camera in scanning device, that don't perform any automatic corrections of image, unlike the common cameras. Table 1 summarizes parameters of used components.

(2) *Realization of scanning device*

For image scanning we used the digital camera with resolution 640 × 480 pixels and bit depth 8 bits on pixel. That means that the final picture from camera is grey-toned with 255 shades of grey. Camera uses for data transfer FireWire interface. We used lens with fixed focal distance from producing series VCN with focal distance 4.5 mm. The important part of lens was IR filter. Using the filter the

Table 1 Approximate estimation of prices of used components

CCD camera GuppyPRO F-031B	Allied Vision Technologies GmbH	600 EUR
Lens VCN 1,4/4,5 f = 4.5 mm	Vision & Control GmbH	150 EUR
IR filter	Heliopan Lichtfilter-Technik Summer GmbH & Co KG	40 EUR
Lighting SFD 42/12 IR	Vision & Control GmbH	700 EUR
Total		1490 EUR

Fig. 2 Set of camera and
lighting

camera scanned only IR part of spectrum. This was important for suppress of daylight on scanned image.

Another part of scanning device was the source if IR lighting. We performed tests with various types of IR lighting sources. We tested the following types of lighting: direct circular lighting, direct linear lighting, diffusion DOM lighting and back lighting. Based on tests results we used the DOM type as the most suitable lighting type that uses LED with wave length 850 nm as lighting source. With this type of lighting we get pictures with homogeneous lighting, well-recognizable vein structure on the dorsum of the hand and at the same time with adequate hand contrast against surroundings for appraising hand contour.

The whole set of light and camera was placed on construction from aluminium sections which was installed on adjustable stand in a way that the height of the set above the pad could be alternatively adjusted. As a background was used black matte surface so it will be gained the highest possible contrast between the scanned user's hand and the background. The whole set of camera, optical system and lighting is depicted on pictures Figs. 2 and 3 (Fig. 4).

Fig. 3 Set of camera and
lighting

Fig. 4 Possible types of scene illumination

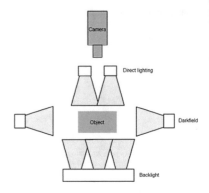

4 Testing of Developed Application

4.1 Methodology of Testing

For testing of projected biometric identification system we had the use of pictures of hands from ten people. From each person were gotten four samples (four various images) on which were performed the system testing. In examined group were both men and women, their age was between 25–60 years. Qualities of examined persons are depicted in following Tables 2 and 3.

For testing we chose that small group because the set of hardware components for image scanning was too big for transport: ca. 0.8 × 0.8 × 0.6 m and lent only for limited time. Because of these reasons we were not able to provide more data for testing.

We performed the testing system on personal computer with following parameters (given are only the parameters that have an effect on running of application):

- OS—Windows 10, 64 bits
- CPU—Intel Core i5-5257U with frequency 2.7 GHz (maximal turbo frequency 3.1 GHz)
- RAM—8 GB

Table 2 Division of tested persons according to gender

Men	Women	Total
8	2	10

Table 3 Division of tested persons according to age

20–30	30–40	40–50	50–60	Total
2	5	2	1	10

Table 4 Times needed for doing single steps of image performance and appraisal

Programm step	Hand geometry		Bloodstream	
	Average time (ms)	Maximal deviation (ms)	Average time (ms)	Maximal deviation (ms)
Image preprocessing	1.25	14.75	3.5	8.5
Segmentation	1.08	8.9	152.7	70.3
Extraction of characteristics	667.5	168.5	8.83	3.26
Calculating rate of correspondence and evaluation	0.97	3	1	1.9

4.2 Tests of Experimental System's Speed

Appraisal of one tested image lasted in interval from 0.6 to 0.9 s. Computation most demanding was the algorithm of extraction of biometric characteristics. The processing speed of single program steps and image appraisal are depicted in Table 4.

4.3 Tests of Reliability of Experimental System

For evaluating the biometric system we chose indicators FMR, FNMR and EER, that gives the frequency of system mistakes. At first we performed the comparison of single metrics on above mentioned data sample. That means that the total rate of correspondence comprised only one metric. As a threshold value we chose 0.5 and the interval of minimal distance from 0 to 0.5. For metrics normalization we used min-max method.

In next part of testing we chose the best metric for hand geometry (Hamming's metric) and best metric for bloodstream (MHD) [13–15]. With these metrics we created a multi-biometric system and again tested the performance of this system on above described data sample. Results are depicted in graph on Fig. 5.

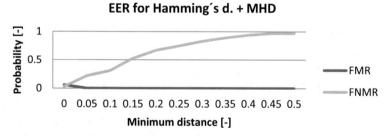

Fig. 5 EER for combined Hamming's and MHD metrics

Results confirmed the expected performance increase of multi-biometric system compared to biometric system with one biometry. EER value decreased at multi-biometric system a half compared to single bloodstream biometry. Also the course of FNMR mistake (frequency of not-finding sameness) has smoothened and mildly decreased.

4.4 Discussion of Results

Used algorithms of single metrics show various accuracy and are weighted by certain factors with whose improvement would be possible to reach better identification results.

Negative factor was low image contrast for bloodstream checking. While filtering the image came about the vein interruption in the image or the other way around was as a vein considered a shadow on hand's skin.

Evaluation of single characteristics was performed using comparing frequencies of two types of mistakes (false identification and not-finding sameness) and EER value. The lower is EER value, the higher is precision of given metric. The lowest value had the metric Modified Hausdorff's distance (MHD), for which had the EER value 11%.

Afterwards we chose the best metrics for given biometric characteristics (Hamming's distance and MHD) and combined them into final appraisal. Resulting multi-biometric system had the EER value 5%, thereby was confirmed the precondition that during combination of metrics increases the precision of system.

For putting the experimental biometric system in conditions of real use must the whole system be further improved. The biggest problem would probably be unwanted surrounding influences of surrounding. For example direct daylight, which would influence the intensity of scanned image. Another problem should be the change of base qualities at continuous wearing, on which would be hands placed while scanning. Also should be improved the user affability of hardware for hand scanning, in order to faster and more comfort use [16].

5   Conclusions

In the article we described multi-biometric system, which recognizes the user based on hand geometry and bloodstream on the dorsum of the hand. System was tested on the sample of users, on which was able to identify the user with high precision. Projected biometric characteristics proved good distinguishing ability. EER value of the system on tested data reached 5% value. For using the system in real condition [17, 18] would be proper to optimize the algorithm of biometric characteristics extraction and with this noticeably shorten the time needed for user identification.

Acknowledgements This work was partially supported by the project "Smart Solutions for Ubiquitous Computing Environments" FIM, University of Hradec Kralove, Czech Republic (under ID: UHK-FIM-SP-2018) and by project of Slovak VEGA grant agency, Project No. 1/ 0263/16.

References

1. Hand-based biometrics. Biom. Technol Today. **11**(7), 9–11 (2003). https://doi.org/10.1016/s0969-4765(03)07018-8. ISSN 09694765
2. van Tilborg, H.C.A., Jajodia, S.: Encyclopedia of Cryptography and Security, 2nd ed., pp. 1353–1354. Springer, New York (2011). ISBN 978-1-4419-5906-5
3. Saxena, N., et al.: Hand geometry: a new method for biometric recognition. Int. J. Soft Comput. Eng. (IJSCE) **2**(6), 2231–2307 (2013)
4. Yörük, E., Dutağaci, H., Sankur, B.: Hand biometrics. Image Vis. Comput. **24**(5), 483–497 (2006). https://doi.org/10.1016/j.imavis.2006.01.020. ISSN 02628856
5. Jain, A.K., Hong, L., Kulkarni, Y.: A multimodal biometric system using fingerprint, face and speech. In: Proceedings of 2nd International Conference on Audio- and Video-Based Biometric Person Authentication, Washington DC, pp. 182–187 (1999)
6. Wang, L., Leedham, G., Siu-Yeung Cho, D.: Minutiae feature analysis for infrared hand vein pattern biometrics. Pattern Recognit. **41**(3), 920–929 (2008). https://doi.org/10.1016/j.patcog.2007.07.012. ISSN 00313203
7. Alpar, O., Krejcar, O.: Quantization and equalization of pseudocolor images in hand thermography. Lect. Notes Comput. Sci. **10208**, 397–407 (2017). https://doi.org/10.1007/978-3-319-56148-6_35
8. Assefa, D., Krejcar, O.: Novel edge detection scheme in the trinion space for use in medical images with multiple components. Lect. Notes Comput. Sci. **9876**, 231–241 (2016). https://doi.org/10.1007/978-3-319-45246-3_22
9. Kolda, L., Krejcar, O.: Biometrie hand vein estimation using bloodstream filtration and fuzzy e-means. In: IEEE International Conference on Fuzzy Systems, art. no. 8015736 (2017). https://doi.org/10.1109/fuzz-ieee.2017.8015736
10. Sanchez-Reillo, R., Sanchez-Avila, C., Gonzalez-Marcos, A.: Biometric identification through hand geometry measurements. IEEE Trans. Pattern Anal Mach. Intell. **22**(10), 1168–1171. https://doi.org/10.1109/34.879796. ISSN 01628828
11. Dubuisson, M.-P., Jain, A.K.: A modified Hausdorff distance for object matching. In: Proceedings of 12th International Conference on Pattern Recognition, pp. 566–568. IEEE Computer Society Press (1994). https://doi.org/10.1109/icpr.1994.576361. ISBN 0-8186-6265-4
12. Zhang, T.Y., Suen, C.Y.: A fast parallel algorithm for thinning digital patterns. Commun. ACM **27**(3) (1984)
13. Jian, A.K., Duta, N.: Deformable matching of hand shapes for user verification. In: Proceedings 1999 International Conference on Image Processing (Cat. 99CH36348), pp. 857–861. IEEE (1999). https://doi.org/10.1109/icip.1999.823019. ISBN 0-7803-5467-2
14. Wong, A.L.N., Shi, P.: Peg-free hand geometry recognition using hierarchical geometry and shape matching. In: Proceedings of the IAPR Workshop on Machine Vision Applications, pp. 281–284 (2002)
15. Haeger, S.: Feature Based Palm Recognition. University of South Florida (2003)
16. Alpar, O., Krejcar, O.: Dorsal hand recognition through adaptive YCbCr imaging technique. Lect. Notes Comput. Sci. **9876**, 262–270 (2016). https://doi.org/10.1007/978-3-319-45246-3_25

17. Bobkowska, K., Inglot, A., Mikusova, M., et al.: Implementation of spatial information for monitoring and analysis of the area around the port using laser scanning techniques. Pol. Marit. Res. **24**(1), 10–15 (2017)
18. Maresova, P., Penhaker, M., Selamat, A., Kuca, K.: The potential of medical device industry in technological and economical context. Ther. Clin. Risk Manag. **11**, 1505–1514 (2015). https://doi.org/10.2147/TCRM.S88574

Part VI
Sensor Networks and Internet of Things

Integrated Data Access to Heterogeneous Data Stores for IoT Cloud

Shodai Watanabe and Akihito Nakamura

Abstract Recently, Internet of Things (IoT) attract attention. The authors are developing a cloud platform for IoT applications. The IoT cloud needs to deal with various types of data and large data sets depending on applications and purpose of use. That is, the IoT cloud necessarily includes heterogeneous data stores in a mixed manner. For example, relational databases and NoSQL databases have different connection methods and query languages. This configuration complicates the system design and increases the development cost. This paper presents a configuration method of data access component (DAC) that absorbs the connection method and the query language differences among data stores. This allows us to develop IoT applications without worrying about data store differences and later replacements. In the implementation, we used specific DACs optimized for specific data stores and a multi-purpose DAC Apache MetaModel. With a large scale data set of more than one million records under most configurations, the response time for various kinds of queries are less than 1 second.

1 Introduction

Recently, Internet of Things (Internet of Things) attract attention. The authors are developing a cloud platform for IoT applications. The IoT cloud need to deal with various types of data and large data sets depending on applications and purpose of use. That is, the IoT cloud necessarily includes heterogeneous data stores in a mixed manner. For example, relational databases and NoSQL databases have different connection methods and query languages. This configuration complicates the system design and increases the development cost. When replacing the current working data

S. Watanabe
Kubota Systems Inc., 1-2-47, Shikitsuhigashi, Naniwa-ku, Osaka 556-8601, Japan
e-mail: shodai.watanabe@kubota.com

A. Nakamura (✉)
University of Aizu, Aizu-Wakamatsu, Fukushima 965-8580, Japan
e-mail: nakamura@u-aizu.ac.jp

© Springer International Publishing AG, part of Springer Nature 2018
A. Sieminski et al. (eds.), *Modern Approaches for Intelligent Information and Database Systems*, Studies in Computational Intelligence 769,
https://doi.org/10.1007/978-3-319-76081-0_36

423

store with another type of data store, the application programs need to be modified at the data access component (DAC).

There are some techniques for transparent access to different types of data stores. Object-relational mapping (ORM) is a technique to wrap the implementation-specific details of data store drivers in an API [1–4]. ORM tools transparently operate data manipulation and query functions via programming language-specific APIs. Java Database Connectivity (JDBC) [5, 6] is an API which defines how applications may access a data store. It provides standard methods to query and manipulate data. Both ORM and JDBC are oriented towards only relational databases. For IoT data stores, we need more types of emerging data stores so called NoSQL; document, key-value, graph, etc.

In this paper, we discuss a configuration method of DAC that absorbs the connection method and the query language differences among different types of data stores including relational and NoSQL databases. This allows us to develop IoT applications without worrying about data store differences and later replacements, i.e. *heterogeneous transparency*. As a practical approach to the problem, we develop a hybrid DAC with a built-in multi-purpose DAC and pluggable DAC mechanism. Various kinds of specific DACs for the specific data stores can be integrated. In the implementation, Apache MetaModel is utilized for multi-purpose DAC.

The remainder of this paper is organized as follows. In Sect. 2, we introduce the architecture of IoT cloud and the sequence of query processing. In Sect. 3, we discuss our approach to accessing heterogeneous data stores. Based on the proposed scheme, we show the performance evaluation result in Sect. 4. Section 5 concludes the paper.

2   IoT Cloud

We are developing and operating a cloud computing platform for IoT applications especially [7]. Upon the platform, we are building a disaster prevention system utilizing renewable energy.

In this section, we briefly decribe this IoT cloud platform.

2.1   Overall Architecture

Figure 1 shows the overall architecture of the IoT cloud. The IoT cloud uses a standard IEEE 1888 protocol [8] for communication between the edge and server. The protocol is built upon SOAP [9]. We use Java to implement the system and applications.

These is a possibility of addition, upgrade/replacement, and removal of sensors and devices, software tools, and applications. For example, a new type of high-precision sensor is introduced and the application is upgraded with change of data store schema.

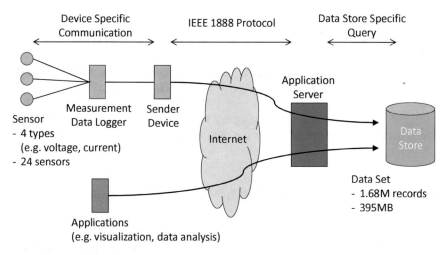

Fig. 1 IoT cloud architecture

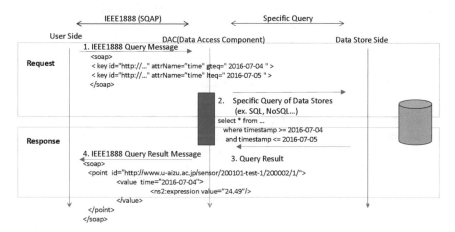

Fig. 2 Query processing sequence

2.2 Query Processing

Figure 2 shows the sequence of query processing. The data access component (DAC) is located in the application server.

- *Step 1*: A user specifies sensor information, attribute information and query condition. Then, the user creates an IEEE 1888 query as a Java object.
- *Step 2*: The user creates a stub of IEEE 1888 and marshal a Java object to SOAP/XML. The user sends the query message to the DAC.
- *Step 3*: The DAC receiving the query message unmarshals the query message from XML to a Java object.

- *Step 4*: The DAC creates a specific query for the data store based on sensor infor-
  mation, attribute information and query condition for arbitrary sensor.
- *Step 5*: The DAC issues the query to the data store.
- *Step 6*: The DAC receives the query result from the data store, and marshals the
  Java object to an IEEE 1888 message. Then, the DAC sends this message to the
  user.
- *Step 7*: The User unmarshals the query result message to the Java object and
  obtains the query result of the sensor and the query condition.

3   Proposed Hybrid Data Store Access Scheme

In this section, we present our hybrid data store access scheme. There are three types
of DACs: specific, multi-purpose common, and hybrid. A specific DAC is dedicated
for one type of data store. On the other hand, a multi-purpose common DAC can be
used for various types of data stores. We create a best mix of the advantages of these
two types of DACs; *hybrid* DAC.

3.1   Specific Data Access Component

A specific DAC is dedicated for one type of data store. Figure 3 shows the concept.
The advantage is optimization. That is, the DAC can be optimized at the performance
based on the knowledge of the data store. However, if the current working data store
is replaced to others, it requires rebuilding of the DAC.

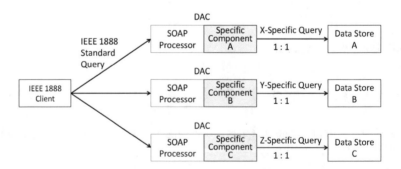

Fig. 3   Specific data access component

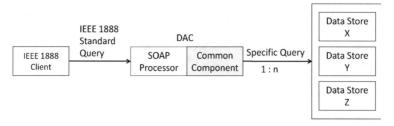

Fig. 4 Common data access component

3.2 Common Data Access Component

A common DAC provides a common interface to various types of data stores. That is, this component can be used as a general-purpose connector and query API. Figure 4 shows the concept. This DAC configuration has several advantages. First, it makes possible to access heterogeneous data stores by one component. Second, it does not require to rebuild a new DAC on replacement of data stores. Third, it reduces the cost of development. However, a potential disadvantage is that it has possibility resulted in performance degradation. It is difficult to optimize single component for many types of targets.

In the implementation, we used Apache MetaModel [10] as the common component of this type of DAC. MetaModel provides a connector and query API of general purpose for the data stores of relational and heterogeneous data model.

3.3 Hybrid Data Access Component

This DAC is a best mix of the advantages of the specific and common DACs. The common component can be used for various types of data stores while each specific component can be used for the specific data store with high performance. If the common component does not support a type of data store, it is complemented with a specific component. Therefore, it is possible to use different components as the situation demands, e.g. performance requirement and development cost. In the implemetation, we need to append a data access object to switch between components. thereby supporting various types of data stores by one DAC.

4 Performance Evaluation

In this section, we show the performance evaluation result of the proposed hybrid DAC for the large scale data set.

We use the voltage and current data set that was obtained from twnty six smart grid sensors deployed on our IoT cloud application for about 2 months from July 4 to September 2. The volume of the dataset is 396 MB and the number of the records is 1.68 million.

4.1 Configuration of Performance Evaluation

Table 1 shows the cases of the type of DAC and the data store system for performance evaluation. The performance is measured against the queries on electric power sensors and data shown in Table 2 (Fig. 5).

Table 1 Evaluation cases: DAC type and data stores

	DAC		Data store	
	Specific	Common	PostgreSQL	MongoDB
Case 1	x		x	
Case 2		x	x	
Case 3	x			x
Case 4		x		x

Table 2 Test queries

Query ID	Query condition	Num. of result records
Q1-1	timestamp $\geq$ 2016-07-04 15:00:00 and timestamp $\leq$ 2016-07-04 16:00:00	60
Q1-2	timestamp $\geq$ 2016-07-04 15:00:00 and timestamp $\leq$ 2016-07-04 16:00:00 and value $\geq$ 24 v and value $\leq$ 24.5 v	47
Q1-3	timestamp $\geq$ 2016-07-04 15:00:00 and timestamp $\leq$ 2016-07-04 16:00:00 and value > 24.5 v	13
Q2-1	value $\geq$ 23.40 v and value $\leq$ 23.42 v	59
Q2-2	value $\geq$ 23.40 v and value $\leq$ 23.42 v and timestamp $\geq$ 2016-07-07 07:00:00 and timestamp $\leq$ 2016-07-07 08:00:00	12

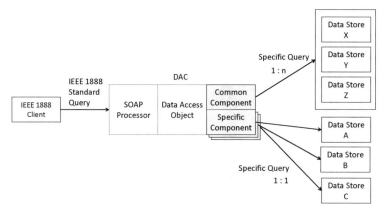

Fig. 5 Hybrid data access component

Table 3 Test environment

Infrastructure	MacBook Pro (2.9 GHz Intel Core i5, 16 GB RAM, 256 GB SSD)
Host OS	macOS Sierra
OS-level virtualization	Docker 1.12.6
Container OS	CentOS 7.3
Java	Oracle Java 1.7.0
Application server	Tomcat 7.0.73
SOAP processor	Apache Axis2 1.7.4
Data store 1	PostgreSQL 9.2.18 [11]
Data store 2	MongoDB 3.4.1 [12]

The system environment of the performance evaluation is shown in Table 3. We deployed the system on an OS-level virtualization environment; Docker container.

4.2 Query Complexity Versus Response Time

First, we examined how the response time changes, when a user changes the query condition to the data stores. The bar graph in Fig. 6 shows the result for the DAC configuration cases 1 and 2; PostgreSQL with specific and common DACs. In both cases, the response times of the queries are shorter than 0.7 s to 1.68 M records. Even if the query condition is complicated, there is no significant change in the response time. It can be saied that these cases are practically applicable to large scale data stores.

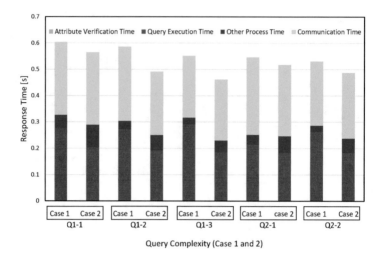

Fig. 6 Query complexity versus response time (DAC configuration Case 1 and 2, 1.68 M records)

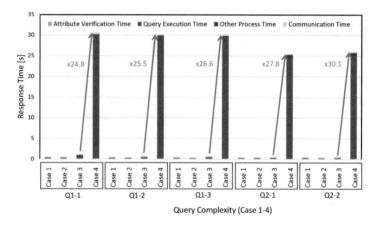

Fig. 7 Query complexity versus response time (DAC configuration Case 1–4, 1.6 M records)

Next, we examined all the 4 cases of the DAC configuration. Figure 7 shows the result. The case 3, i.e. the specific DAC for MongoDB, exhibits the same performance as the case 1 and 2 and the response time is less than or about 1 s. Therefore, these three cases are expected to be practically useable for large scale data stores. However, the case 4 has a performance degradation about from twenty five to thirty times compared with the case 3. This degradation is caused by query execution time, i.e. taking much time to exchange requests and responses between the DAC and the data store. It is predicted that the Apache MetaModel, used for the case 4 DAC, is not optimized well for MongoDB.

4.3 Data Set Size Versus Response Time

We examined how the response time changes, when the number of records in the data store changed (Fig. 8). This graph applied Case 1 and Case 2. And, the query conditions are Q1-1. The number of records are ten thousand to 1.68 million. Case 1 and Case 2 are found to be within 1 s from this graph. As the number of records increases, the query execution time also increases. Therefore, the response time increase in proportion to the execution time of the query. Therefore, Case 1 and Case 2 expected to be practically useable for large scale data stores. Also, Case 2 process higher speed compared with Case 1. Because Case 2 is predicted to be more optimized compared with Case 1 form this graph. Therefore, Case 1 is expected to leave room for further improvement.

Figure 9 shows the graph with Case 3 and Case 4 added. Case 4 is acceptable up to a hundred thousand records. However, Case 4 has a performance degradation about fifteen times in the case of more than five hundred thousand records. As the cause, the Case 4 is expected to be not optimized. As the solution, the Case 4 requires sharding and optimization of the data store in the case of more than five hundred thousand records.

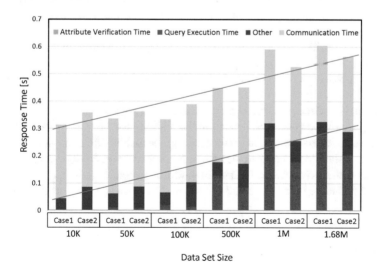

Fig. 8 Data set size versus response time (DAC configuration Case 1 and 2, Q1-1)

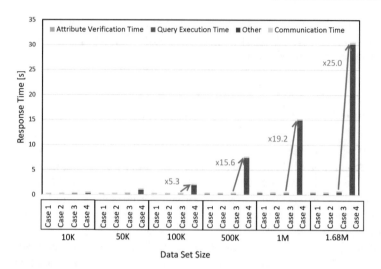

Fig. 9 Data set size versus response time (DAC configuration Case 1–4, Q1-1)

5 Concluding Remarks

Since IoT systems need to handle a varaiety of sensors and the data stores are demanded to store different types of data. In addition, number of sensors generate huge volume of data at high velocity. Therefore, in the IoT systems development and operation, scalability is not the only thing that matters data store subsystem but also heterogeneity of data. In this paper, we proposed a data access component (DAC) to provide transparent access to heterogeneous data stores. The DAC has a hybrid structure of specific DACs optimized for the specific data stores and multi-purpose common DAC. We utilized Apache MetaModel for the latter. It allows system developers to replace current data stores to different ones or even to coexist heterogeneous ones without modification of applications.

We performed the performance evaluation of the proposed scheme. Each specific DAC rightly shows good performance and scalability for the specific data store. The common DAC also shows good performance and scalability for relational data stores. However, a significant performance degradation occurs on NoSQL data stores. We determined the exact cause; Apache MetaModel generates slow queries for MongoDB. If this issue is fixed, we will see better performance under this configuration.

References

1. O'Neil, E.J.: Object/relational mapping 2008: hibernate and the entity data model (EDM). In: Proceedings of the ACM SIGMOD'08, pp. 1351–1356 (2008)

2. Cabibbo L., Carosi A.: Managing inheritance hierarchies in object/relational mapping tools. In: Advanced Information Systems Engineering, LNCS 3520, pp. 135–150. Springer (2005)
3. Adya, A., Blakeley, J.A., Melnik, S., Muralidhar, S.: Anatomy of the ADO.NET entity framework. In: Proceedings of the ACM SIGMOD'07, pp. 877–888 (2007)
4. JBoss Inc.: Hibernate. http://hibernate.org
5. Java Community Process: JSR 221: JDBC 4.0 API Specification. https://jcp.org/en/jsr/detail?id=221
6. Mason, T., Lawrence, R.: Dynamic database integration in a JDBC driver. In: ICEIS, no. 1, pp. 326–333 (2005)
7. Kikuchi, S., Nakamura, A., Yoshino, D.: Evaluation on information model about sensors featured by relationships to measured structural objects. Adv. Internet Things (AIT) $6(3)$, 31–53 (2016)
8. IEEE Standard for Ubiquitous Green Community Control Network Protocol. IEEE Std 2014(1888): 1888-2011 (2013)
9. W3C: SOAP Version 1.2 Part 1: Messaging Framework, 2nd Edn. W3C Recommendation (2007)
10. Apache MetaModel. http://metamodel.apache.org
11. PostgreSQL. https://www.postgresql.org
12. MongoDB. https://www.mongodb.com

Path Estimation from Smartphone Sensors

Jan Racko, Peter Brida, Juraj Machaj and Ondrej Krejcar

Abstract Nowadays the knowledge about position is very important for localization based services. Thanks to knowing the position many services can be provided, such as information about people in our surrounding, firemen can be navigated during movement while rescue action, or just simply tracking position of different things in buildings. Global Navigation Satellite System (GNSS) was commonly used in outdoor environment, but if we are in a building GNSSs are unusable. This is mainly because of multipath propagation which can cause huge localization errors. Therefore, many research teams have started to develop different systems for location estimation in indoor environment. In this work, we proposed position estimation system based on inertial sensors in smartphone with average accuracy below 0.6 m.

Keywords Accelerometer · Gyroscope · Positioning

1 Introduction

There are many technologies which can be used for location estimation in indoor environment, for example wireless technologies, inertial sensors, light sensors and cameras. The most widely used is wireless local area network—Wi-Fi. Position is

J. Racko · P. Brida (✉) · J. Machaj
FEE, Department of Multimedia and Information-Communication Technologies,
University of Zilina, Univerzitna 1, 010 26 Zilina, Slovakia
e-mail: peter.brida@fel.uniza.sk

J. Racko
e-mail: jan.racko@fel.uniza.sk

J. Machaj
e-mail: juraj.machaj@fel.uniza.sk

O. Krejcar
Faculty of Informatics and Management, University of Hradec Králové,
Rokitanského 62, 500 03 Hradec Králové, Czech Republic
e-mail: ondrej.krejcar@uhk.cz

© Springer International Publishing AG, part of Springer Nature 2018
A. Sieminski et al. (eds.), *Modern Approaches for Intelligent Information and Database Systems*, Studies in Computational Intelligence 769,
https://doi.org/10.1007/978-3-319-76081-0_37

estimated based on comparison of current received signal strength measurements and measurement included in radio map. Algorithms from Nearest Neighbour (NN) family are commonly used [1]. Comparison of K-Nearest Neighbour (KNN) with Weighted K-Nearest Neigbour (WKNN) can be found in [2]. Another wireless technologies used for localization are Ultra-Wide Band (UWB), Radio-Frequency Identification (RFID), Bluetooth. The biggest disadvantage of these approaches is that in many cases maximum localization error can be over 10 meters, especially in Wi-Fi case [3, 4].

Using inertial sensors can be also considered as appropriate solution for indoor localization. To calculate position from inertial sensors several steps have to be done. The first assumption for our proposal is that smartphone will be held in hand during walking. In such case Pedestrian Dead Reckoning (PDR) method can be applied. PDR is a sequential navigation system which calculates actual position from previous one. Basic concepts of PDR can be found in [5]. Output from accelerometer is commonly used for step detection, which provides information about acceleration in all three axes. To detect exact steps, which have been done we have used zero crossing method [6]. It is also very important to know the length of steps, which depends on age, gender, walker's height. Step length is also affected by current activity. During running we are doing longer steps, then during walking. The last step is estimation of heading angle. Commonly used sensor for this purpose is gyroscope [7, 8], thus magnetometer is affected by metal parts of the building causing errors in heading estimation.

In last few years, smartphones became very important part of life. Example of position estimation by using PDR on smartphone can be found in [9], where gyroscope, integrated in a smartphone, has been used. The main goal was to detect when the error in heading estimation occurred, if Global Positioning System (GPS) was available. Error between estimated and real heading was determined by Mahalanobis distance. In [10] SmartPDR system is described. For heading esti- mation, combination of accelerometer and gyroscope has been used and magne- tometer was used for correction. For step recognition, peak detection has been used. Maximum localization error of proposed system was 2 m, and maximum heading estimation error was 20°, what is insufficient for indoor navigation. What is more important, fact that heading error was less than 5° for more than 90% of results.

Combination of smartphone with other indoor localization techniques can improve accuracy. In [11] combination of security cameras and PDR is described. Final position was estimated using trilateration based on received signal strength from wireless network. Heading was estimated using security cameras, which captured frames of person, who moves along the corridor. Hybrid system based on combination of PDR and location fingerprinting is presented in [12, 13]. Path was calculated by accelerometer, gyroscope and magnetometer. In this case peak detection was used for step detection. Important assumption was that the radio map of area is available. Initial position was estimated by using KNN algorithm. During a walk, smartphone's inertial sensors have been used to estimate path and position was corrected by location fingerprinting. For position estimation only fingerprints in small area around estimated position could be used. Chosen area was represented with diameter of 1 m. This prevents shifting final position too far from actual position due to choosing remote fingerprints.

In this paper we describe indoor positioning system based on inertial sensors, accelerometer and gyroscope, implemented in smartphone. In our experiments, device was held in hand during the whole period of measurement. Proposed method was tested at the Department of Multimedia and Information-Communication Technologies at the University of Žilina. In the first case, path was calculated without correction, in the second case a particle filter has been chosen for correction.

The rest of the paper is organized as follows. Section 2 describes basics of PDR and particle filter. In Sect. 3 experimental setup can be found. Section 4 shows experimental results, and Sect. 5 will conclude the paper.

2 Theoretical Backround

2.1 Pedestrian Dead Reckoning

The first important calculation is step detection. For this purpose acceleration measurements are done. During walking, smartphone is hold in hand and there is a significant acceleration is in Z axis. But acceleration patterns can also be found in X and Y axes. To track attitude of device accurately norm of acceleration is used:

$$a(t) = \sqrt{a_x^2(t) + a_y^2(t) + a_z^2(t)} - g, \tag{1}$$

where g is gravity acceleration and a_x, a_y, a_z represents measured accelerations in all axes [8]. After calculation of norm of acceleration a detection method can be applied. In our approach we decided to use zero crossing method. Other options are peak detection or frequency analysis [6]. In case of zero crossing method the step is detected when signal is rising or decreasing. Therefore, following condition has to be used: next step can be detected only if previous step was detected more than 0.5 s earlier. This will reduce number of false detected steps. Moreover, two thresholds were set in positive and negative part of the signal. If signal crosses both thresholds, step is accepted, which indicate walking. In remaining cases standing was detected.

The second important calculation is step length. As mentioned, length of a step depends on gender, age or height and movement speed. In Fig. 1 dependency of step lengths in relation to frequency is shown.

However, in this paper constant step length of 0.75 m was used and only two motion states, walking and standing, were assumed.

The last step is heading angle estimation. The attitude estimation of smartphone relative to the global frame can be done by integrating angular velocity measured by gyroscope in body frame:

$$\omega_b(t) = \big(\omega_x(t),\ \omega_y(t),\ \omega_z(t)\big), \tag{2}$$

where ω_x, ω_y, and ω_z are angular rotations in body frame for all axis. One way to present the attitude of the IMU is to use Direction cosine matrix (DCM). DCM is 3-by-3 rotation matrix:

$$\mathbf{C} = \begin{bmatrix} \cos\theta\cos\psi & \cos\theta\sin\psi & -\sin\theta \\ \sin\varphi\sin\theta\cos\psi - \cos\varphi\sin\psi & \sin\varphi\sin\theta\sin\psi + \cos\varphi\cos\psi & \sin\varphi\cos\theta \\ \cos\varphi\sin\theta\cos\psi + \sin\varphi\sin\psi & \cos\varphi\sin\theta\sin\psi - \sin\varphi\sin\psi & \cos\varphi\cos\theta \end{bmatrix}, \tag{3}$$

where symbols ϕ, θ, ψ represent Euler angles roll, pitch and yaw. Rotation matrix must be updated all the time for tracking of IMU orientation. Updated matrix **C** (t + Δt) can be written as:

$$\mathbf{C}(t+\Delta t) = \mathbf{C}(t)\left(\mathbf{I} + \frac{\sin\sigma}{\sigma}\mathbf{B} + \frac{\cos\sigma}{\sigma^2}\mathbf{B}^2\right), \tag{4}$$

where Δt is sampling interval and:

$$\mathbf{B} = \begin{bmatrix} 0 & -\omega_z\Delta t & \omega_y\Delta t \\ \omega_z\Delta t & 0 & -\omega_x\Delta t \\ -\omega_y\Delta t & \omega_x\Delta t & 0 \end{bmatrix}, \tag{5}$$

$$\sigma = |\Delta t\omega_b|, \tag{6}$$

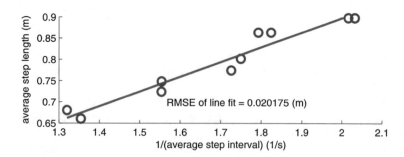

Fig. 1 Step length [12]

and $\mathbf{I}$ is 3-by-3 identity matrix. After that, we are able to calculate heading angle, i.e., yaw angle, Ψ from updated rotation matrix [14]:

$$\psi = \arctan_2(C_{2,1}, C_{1,1}). \tag{7}$$

After step detection and heading angle estimation, position can be calculated as:

$$\begin{bmatrix} P_{Ek} \\ P_{Nk} \end{bmatrix} = \begin{bmatrix} P_{E_{k-1}} + l_k \sin(\psi_k) \\ P_{N_{k-1}} + l_k \cos(\psi_k) \end{bmatrix}, \tag{8}$$

where P_{Ek} and P_{Nk} represent east and north position, l_k is step length and ψ_k is heading angle at moment k.

2.2 Particle Filter

To combine various sources of information optimally, Bayesian filters can be used. However, this method cannot be used if measurement model is nonlinear. A popular approximation method of the Bayesian filters is called Particle Filter (PF). This method approximates the posterior state distribution by using particles. Representation of particles as individual points gives advantage with map combination process, known as map-matching.

Similarly to Kalman Filter (KF), PF has both prediction and update steps. First, particles $x^{(k)}$, $k = 1, \ldots, N$, are drawn from proposal distribution in time moment t.

$$x_t^{(k)} \approx \pi\left(x_t^{(k)} \middle| x_{t-1}^{(k)}, y_{1\ldots t-1}\right), \tag{9}$$

where $y_{1\ldots t-1}$ represents measurements moment before t. We assume that states establish Markov model. This means that the current state x_t depends only on the previous state x_{t-1}. During the update phase, weights are recalculated according to the observation likelihood as:

$$w_i^k = w_{i-1}^k \frac{p\left(y_t \middle| x_t^{(k)}\right) p\left(x_t^{(k)} \middle| x_{t-1}^{(k)}\right)}{q\left(x_t^{(k)} \middle| x_{t-1}^{(k)}, y_i\right)}, \tag{10}$$

After update step weights are normalized. Over the time, there will be only few particles of the particle filter, which will have non-zero weight. Propagation particles with very low weight negatively affect posterior distribution. This problem is known as degradation, and can be avoided by resampling [15].

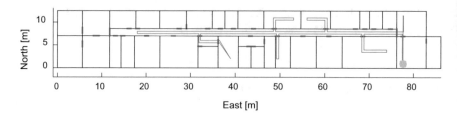

Fig. 2 Reference path with starting and ending point

3 Experimental Setup

Measurement has been done at the Department of Multi-media and Information-Communication Technologies at the University of Žilina by using smartphone Samsung Galaxy S6 Edge with IMU MPU 6500. For data collecting, Androsensors application was used and sampling frequency was set on 100 Hz. During whole measurement device was held in hand and walked path is shown in Fig. 2. For evaluation of measured data script in MATLAB was created. As was mentioned above, in the proposed script the step length was constant at 0.75 m.

Errors were quantified by cross track method. Walked path have been divided into smaller parts and has been compared with strait lines of reference path. Starting and ending points were in the same position marked as green circle shown in Fig. 2.

Path was split based on calculated heading angle. If significant changes in heading angle occurred it's marked starting or ending point of straight line of walked path. Segments consist of detects steps and each step has coordinates in 2D. Distance between point and line has been calculated by

$$D = \frac{|(y_2 - y_1)x_0 - (x_2 - x_1)y_0 + x_2y_1 - y_2x_1|}{\sqrt{(y_2 - y_1)^2 + (x_2 - x_1)^2}}, \tag{11}$$

where $P1[x_1; y_1]$, $P2[x_2; y_2]$ denote starting and ending point of the lane, and $[x_0; y_0]$ represent coordinates of the point.

4 Results

Reference path shown in Fig. 2 was walked with smartphone held in hand. In the algorithm it is assumed that we have information about initial heading angle. However, calculated path without correction is still absolutely useless due to high errors as can be seen from Fig. 3.

Correction using PF provides significant improvement. Calculation started with $N = 200$ particles. Each particle has weight 1/N. If particle has weight, it means that it is active. After first step, particles move in different directions and if a particle

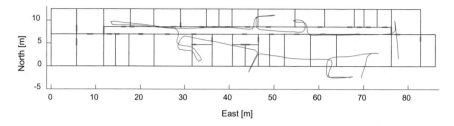

Fig. 3 Walked path calculated without correction

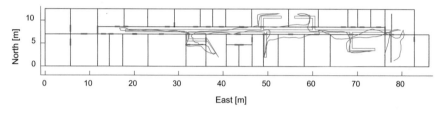

Fig. 4 Corrected path (red line) and reference path (blue line)

crosses a wall it was eliminated and weight of such particle was set to 0. If number of active particles drops below 40, resampling phase starts. During resampling new 200 particles were calculated around last estimated position with heading angle of previous particles. In Fig. 4 path corrected by PF can be seen as a red line and reference path as blue line.

In Table 1 statistical values of positioning error are shown for both corrected and uncorrected path.

From the table it can be seen that mean positioning error was decreased by more than 40%. On the other hand, the maximum error was not reduced so significantly. For deeper analysis of achieved results CDF of achieved errors can be seen in Fig. 5.

From Fig. 5 it is obvious that PF provided significant improvement of accuracy. In case of uncorrected path was maximum position error 6.8 m. For 50% of results error was lower than 1 m and for 80% of results error was app. 2 m. On the other hand, with use of PF correction the maximum error was 6.6 m what is comparable with previous case, however, for 50% of results the localization error was less than 0.5 m and for 80% of results the error was app. 1 m.

Table 1 Statistical results of position errors

Position error (m)	Uncorrected path	Corrected path
Mean	1.03	0.59
Minimum	0.001	0.001
Maximum	6.8	6.2

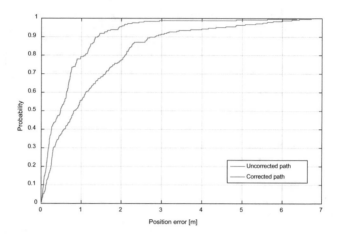

Fig. 5 CDF function for uncorrected and corrected path

5 Conclusion

In last couple of years quality of smartphone sensors rapidly increased. Devices primary designed for calling or texting have become appropriate for indoor navigation solutions.

In this paper, system based on PDR, that utilized data from smartphone sensors was proposed and tested. Data measured by application Androsensors were evaluated in MATLAB. From presented results it is clear that raw data smartphone's low-cost sensors alone can't provide accurate position estimation. Maximum error was 6.8 m and mean error was 1.03 m. Therefore PF was introduced to improve accuracy. Even though the maximum error was 6.2 m, mean error was reduced to 0.6 m and more than 80% position estimations were with error app. 1 m.

It has to be noted that even if we are talking about sensors implemented in mobile devices, Samsung Galaxy S6 is equipped with MPU6500, which can be considered as relatively accurate IMU sensor. There are still many smart devices in middle and low class, which are equipped with various types of less accurate or older IMU sensors. Therefore not all smartphones are suitable for indoor navigation yet.

Acknowledgements This work was partially supported by the Slovak VEGA grant agency, Project No. 1/0263/16 and by project "Smart Solutions for Ubiquitous Computing Environments" FIM, University of Hradec Kralove, Czech Republic (under ID: UHK-FIM-SP-2018).

References

1. Machaj, J., Brida, P.: Optimization of rank based fingerprinting localization algorithm. In: International Conference on Indoor Positioning and Indoor Navigation, January 2013

2. Ding, H., Zheng, Z., Zhang, Y.: AP weighted multiple matching nearest neighbors approach for fingerprint-based indoor localization. In: 4th International Conference on Ubiquitous Positioning, Indoor Navigation and Location Based Services, January 2017
3. Machaj, J., Brida, P., Piche, R.: Rank based fingerprinting algorithm for indoor positioning. In: International Conference on Indoor Positioning and Indoor Navigation, November 2011
4. Bobkowska, K., Inglot, A., Mikusova, M., et al.: Implementation of spatial information for monitoring and analysis of the area around the port using laser scanning techniques. Pol. Marit. Res. 24(1), 10–15 (2017)
5. Levi, R.W., Judd, T.: Dead reckoning navigational system using accelerometer to measure foot impacts. United States Patent No. 5583776 A (1996)
6. Davidson, P.: Algorithms for autonomous personal navigation system. Juvenes Print TTY, Tampere (2013). ISBN: 978-952-15-3174-3
7. Renaudin, V., Combettes, Ch., Peyret, F.: Quaternion based heading estimation with handheld MEMS in indoor environment. In: IEEE/ION Position, Location and Navigation Symposium, pp. 645–656, May 2014
8. Kang, W., Nam, S., Han, Y., Lee, S.: Improved heading estimation for smartphone-based indoor positioning systems. In: IEEE 23rd International Symposium on Personal, Indoor and Mobile Radio Communications, pp. 2249–2453, September 2012
9. Liew, L.S., Wong, W.S.H.: Improved pedestrian dead-reckoning based indoor positioning by RSSI based heading correction. Sens. J. 16(21) (2016)
10. Loh, D., Zihajehzadeh, S., Hoskinson, R., Abdollahi, H., Park, E.J.: Pedestrian dead reckoning with smartglasses and smartwatch. Sens. J. 16(22) (2017)
11. Lu, Q., Liao, X., Xu, S., Zhu, W.: A hybrid indoor positioning algorithm based on wifi fingerprinting and pedestrian dead reckoning. In: 27th Annual International Symposium on Personal, Indoor and Mobile Radio Communications, December 2016
12. Bojja, J., Parviainen, J., Collin, J., Hellevaara, R., Kappi, J., Alanen, K., Takala, J.: Robust misalignment handling in pedestrian dead reckoning. In: 84th Vehicular Technology Conference, March 2017
13. Leppakoski, H., Collin, J., Takala, J.: Pedestrian navigation based on inertial sensors, indoor map, and WLAN signals. In: IEEE International Conference on Acoustics, Speech and Signal Processing, pp. 1569–1572, March 2012
14. Elbes, M., Al-Fuqaha, A., Rayes, A.: Gyroscope drift correction based on TDoA technology in support of pedestrian dead reckoning. In: IEEE Globecom Workshop, pp. 314–319, December 2012
15. Wang, F., Lin, Y.: Improving particle filter with a new sampling strategy. In: 4th International Conference on Computer Science & Education, pp. 408–412, July 2009

A Multi-metric Routing Protocol to Improve the Achievable Performance of Mobile Ad Hoc Networks

Vu Khanh Quy, Nguyen Tien Ban and Nguyen Dinh Han

Abstract In recent years, with the rapid development of mobile technology, mobile ad hoc networks (MANETs) have been deployed in many areas such as rescue, military, medical applications and smart cities. Due to the characteristics of MANETs, routing protocols must be designed to be flexible, energy-efficient and highly performance achievable. In this study, we propose a multi-metric routing protocol with a cost function based on a set of three parameters: the length of the queue of packets at nodes, the link quality and the route hop count. Depending on the actual working condition of the network, one or more parameters may be involved in calculating the route cost. The simulation results show that our proposed routing protocol with the new cost function outperforms the conventional AODV protocol as it gains a better achievable performance.

Keywords MANET · Multi-metric routing protocol · High performance routing · AODV

1 Introduction

A Mobile Ad hoc Network (MANET) is a set of two or more nodes that are used for wireless communication and networking capacity. These nodes are formed by the wireless hosts without using preinstalled infrastructure and do not use a centralized administration [1]. In a MANET, a node can communicate with other nodes directly or indirectly via intermediate nodes.

V. K. Quy (✉) · N. D. Han (✉)
Hung Yen University of Technology and Education, Hung Yen, Vietnam
e-mail: quyvk0705@gmail.com

N. D. Han
e-mail: hannguyen@utehy.edu.vn; nguyendinhhan@gmail.com

N. T. Ban
Posts and Telecommunication Institute of Technology, Hanoi, Vietnam
e-mail: bannt@ptit.edu.vn

© Springer International Publishing AG, part of Springer Nature 2018
A. Sieminski et al. (eds.), *Modern Approaches for Intelligent Information and Database Systems*, Studies in Computational Intelligence 769,
https://doi.org/10.1007/978-3-319-76081-0_38

MANETs are widely used in many fields such as military, healthcare, disaster recovery and security [2–4]. However, recent experiment results show that the practical performance of MANETs is quite low [5, 6]. Note that the practical performance of a MANET is often evaluated by basic criteria such as *delay*, *packet delivery ratio* and *throughput* [7]. To improve the performance of a MANET, several methods have been proposed [8–12]. As pointed out in [4], an appropriate routing protocol will greatly affect the performance and power consumption of the entire network. For this reason, designing effective routing protocols has always been a hot topic that attracts special attention of a huge of researchers over the years. Indeed, due to the mobile ad hoc behavior of the nodes, a MANET has no fixed topology. Thus, routing in MANETs is a major problem and it is still open.

According to a hierarchy of routing protocols for MANETs [13], routing protocols can be classified by routing method or network topology. Based on the routing method, we have *proactive routing* and *reactive routing*, and based on the network topology, we have *flat routing*, *hierarchical routing*, and *location-based routing*. As claimed in [13], reactive routing protocols are energy-efficient and use less resource than proactive routing protocols. The two popular reactive routing protocols proposed for MANETs by the IETF (Internet Engineering Task Force) are AODV (Ad hoc On Demand Distance Vector) [8] and DSR (Dynamic Source Routing) [9]. However, the performance of a MANET depends on several criteria such as network structure, network size and traffic data [11]. There is currently no general solution or algorithm available for all cases. Each routing protocol has its own advantages and disadvantages, and only suitable for a certain situation. Fortunately, it is possible to combine existing solutions to provide a more efficient routing method. In this study, we propose an on-demand routing protocol to improve the achievable performance for MANETs. In reality, we modify the conventional AODV protocol to obtain a better one. Unlike AODV, our protocol exploits a combined costing function from three parameters: the queue length, link quality and hop number to select the appropriate route for data transmission. Hence, we named our new routing protocol MM-AODV (Multi-Metric AODV).

The rest of the paper is organized as follows. In the next section, we present the related work. The proposed routing protocol MM-AODV is presented in Sect. 3. Section 4 presents various simulation results to verify the achievable performance of the proposed protocol. Conclusions are given in Sect. 5.

2 Related Work

In this section, we review several approaches to enhancing the achievable performance of MANETs. Our main focus will be on multi-path routing protocols proposed recently for MANETs. It is surprised that modifying the traditional AODV to gain more efficient routing protocols is a popular approach. Here, we consider some AODV variants resulting from this approach such as MAR-AODV [10], A_WCETT [11] and AAODV [12].

Actually, to make use of mobile agents, the authors in [10] proposed a load distribution algorithm, namely MAR-AODV, as a modified AODV protocol which uses mobile agents to improve the performance of MANETs. The major contribution of this work is a method to select the route that can ensure the load balancing traffic in a network. The simulation results show that the probability congestion of MAR-AODV is lower than that one of AODV protocol. In [11], an on demand routing protocol for 5G MANET named A_WCETT, has been introduced. This protocol is also a modification of AODV. Like MAR-AODV, it adapts the mobile agent technology, but it uses a novel metric for routing. The metric is a function of the loss rate, the bandwidth and the end-to-end delay of the link. Beside this, A_WCETT exploits a new tunable parameter that always allows it to choose a high-throughput and low-delay route between a source and a destination. Hence, the achievable performance of the MANETs in 5G can be improved remarkably with the modified routing protocol.

In [12], an intelligent routing protocol for MANETs called ANFIS (Adaptive Neural Fuzzy Inference System) Aided AODV (AAODV for short) is first introduced. Here, the ANFIS is an advanced technique which adjusts itself to changes in system. The use of ANFIS in routing protocol will result in efficient guiding of packets from source to destination. As a consequence, the network performance obtained with the new protocol AAODV is much better than that of the original AODV.

A very recent approach to improving MANET performance is to use multi-parameter cost functions to calculate and make the decision to choose the route for data transmission [7]. Another approach used by many authors is to integrate MANET with cloud computing. For example, in [14] the authors introduce a solution to enhance the performance of a MANET using a mechanism that enables cloud integration with the MANET. The mechanism that the authors give in this work is very effective, greatly improves the performance of the MANET. In addition, approaches to prevent the performance degradation are used in [13, 15].

Note that all routing protocols mentioned above use a fixed set of routing metrics. This may prevent them to work in future mobile converged networks. For this reason, we suggest the use of various routing metrics in our work.

3   Proposed Routing Protocol

The main goal of the proposed routing protocol is to improve the overall performance of MANETs. Indeed, we modified the conventional AODV protocol to establish a better one. We use this approach to inherit AODV's advanced features as it always has high performance and stability in different network structures. To ensure that the proposed routing protocol has the best performance, we include three main parameters that most influence the network performance in the cost function for route selection. They are: the queue length, link quality and hop number. Then, we made the necessary changes in the AODV protocol to select the optimal routes. The detailed modifications are presented in the following subsections.

Fig. 1 The graph model of a
MANET

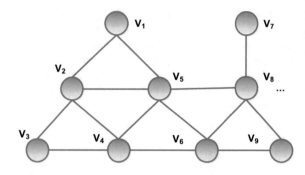

3.1 System Model

In this subsection, we define a MANET network model as shown in Fig. 1. From
the system model, we can achieve the interconnection network model. We define
$G = (V, E)$ is a communication graph of the MANET, where $V = \{V_1, \ldots, V_n\}$ is
the set of mobile nodes in network, E is the set of links. A mobile node can connect
directly with other nodes. It also can communicate with other nodes indirectly via
intermediate nodes.

3.2 Routing Metrics

The traditional AODV protocol selects the route with the smallest hop count.
However, the shortest route may not give the best throughput or latency. Note that,
the network performance is measured by three main criteria: throughput, transmission delay, and packet delivery ratio. To improve the network performance, we
propose a cost function with the following three parameters.

- $Path_{(i)}$ is the number of hops between the source node and node i.
- Link quality ($ETX(i)$): this parameter reflects the link quality of the route from
  the source node and node i. We use the technique in [11]. To increase the
  network performance, Couto et al. (see [11]) proposed a routing parameter to
  calculate the cost of the route, it is ETX (Expected Transmission Count). ETX is
  the expected number of transmissions at the link layer to successfully transmit a
  packet over a connection. To determine ETX, each node sends small probe
  packets to its neighboring nodes. Then, based on the number of probe packets
  sent and packets received, each node evaluates the packet forwarding capability.
  The probabilities that a packet is sent and received successfully, denoted by d_f
  and d_r, respectively. Then, the probability that a packet is transmitted successfully over a connection, is: $d_f \times d_r$. The expected number of transmissions
  required to successfully deliver a packet on a link (a direct link between two
  nodes) is calculated by equation:

$$\mathrm{ETX} = \frac{1}{d_f \times d_r} \tag{1}$$

The ETX of a route, is the sum of the ETX for each link in the route.

- $LengthQueue(LQ_{(i)})$: this parameter reflects the processing capability of the network node. It is clear that, in MANET environment, each node has a certain configuration and capacity (RAM, CPU, etc.). Therefore, the processing speeds and queue buffers are different from nodes to nodes. A node has an empty queue buffer is ready to receive and process packets rather than a network node with a full queue buffer. The result is that, packets will be processed faster and the latency will decrease. The routing technique based on mobile node queue buffer has been used extensively in [7].

$$LQ_{(i)} = \frac{L_i}{L_{max}} \tag{2}$$

Where L_i and L_{max}, respectively, are the number of packets in the MAC queue and the largest queue size of the node i. In order to choose an appropriate route, we propose a cost function of as follows:

$$CP_{(i)} = \begin{cases} \propto \times Path_{(i)} + \beta \times ETX_{(i)} + \gamma \times LQ_{(i)} \\ \propto + \beta + \gamma = 1 \end{cases} \tag{3}$$

Therefore, the cost of route P is:

$$CP_{(P)} = \sum_{l=1}^{P-1} CP_{(i)} \tag{4}$$

Based on the cost $CP_{(P)}$, the source node will select the lowest cost route.

In order to receive routing information, the cost will be calculated as in Eq. (3), and to make the decision to choose the route as in Eq. (4), we use the reserved field in the header of the RREQ (Router REQuest) packet to save the cost value CP. This method has been proposed in many works [10, 11]. The use of the reserved field in the header of the RREQ packet helps the mobile node determine the cost. This also helps to keep performance and power consumption unchanged.

4 Performance Evaluation

In this subsection, we setup a simulation to evaluate the performance of the proposed routing protocol MM-AODV. The parameters used for simulation, results and analysis are given in the following subsections.

4.1 The Metrics Used for Performance Evaluation

As we mentioned in the introduction section, performance of a MANET can be evaluated using various factors. The performance of MM-AODV protocol is evaluated by metrics as follows:

- *Packet Delivery Ratio (PDR)* (in %): the ratio of the number of packets delivered to the destination nodes over the number of packets sent by the source nodes.
- *Average End-to-End Delay (Delay)*: the time taken for a packet to be transmitted across a network from source to destination.

4.2 Simulation Parameter

In this subsection, we set up a simulation to evaluate the performance of MANETs operating with the proposed protocol MM-AODV (on NS2 simulation software version 2.34).

Our system consists of 200 mobile nodes, distributed randomly in an area 1.000 m × 1.000 m. We use the standard IEEE 802.11g with the numbers of end-to-end connections are: 15, 20, 25, 30, 35, 40, and 45 respectively. Simulation time for all experiments is set to be 300 s. The simulation parameters are given in Table 1.

In this simulation, we select and setup the set of performance factors $\{\alpha, \beta, \gamma\}$ as shown in Table 2. Depending on network conditions, values of performance factors can be adjusted.

Note that, the values $\{\alpha = 0, \beta = 0, \gamma = 1\}$ imply that the protocol MM-AODV uses the smallest hop count based routing method, or equivalently, MM-AODV can be considered as the common AODV protocol.

In the first experiment (see Fig. 2), we evaluate the performance of the MANET that operates with MM-AODV. The simulation results show that, when the number of end-to-end connections in the network is less than 25 pairs, MM-AODV (with $\alpha = 1$, or equivalently, it is identical with the common AODV protocol) gives better

Table 1 Simulation parameter

Parameter	Value
Simulation area	1.000 × 1.000 m
Number nodes	200
Mobile node speed	2 m/s
Mobility model	Random Waypoint
Traffic type	CBR/UDP
Packet size	512 byte
Wireless MAC interface	802.11g
Simulation time	300 s

Table 2 The set of performance factors for simulation

Performance factors $\{\alpha, \beta, \gamma\}$	α	β	γ
$(0.33, 0.33, 0.33)$	0.33	0.33	0.33
$(1, 0, 0)$	1	0	0
$(0, 1, 0)$	0	1	0
$(0, 0, 1)$	0	0	1

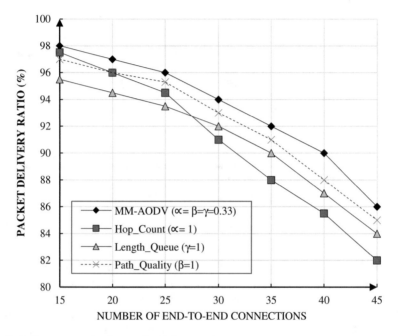

Fig. 2 Performance evaluation: packet delivery ratio

PDR. However, when the network traffic increases, both PDRs of MM-AODV (with *Link Quality* and *Queue Length*) are better. The MM-AODV (with $\alpha = \beta = \gamma = 0.33$) always provides the highest and stable PDRs. This result can be explained as follows. When the traffic in the network is low, congestion does not occur. Therefore, the smallest hop count based routing method is optimal. However, if the network traffic increases, congestion and conflict in the network will occur. In that case, routing based on *Link Quality* or *Queue Length* or a combination of the above parameters will produce higher PDR results.

In the second experiment, we evaluate the performance with the metric: Average End-to-End Delay (Delay). As shown in Fig. 3, in all cases, delay has an uptrend as the number of end-to-end connections increases. When the number of end-to-end connections in the network is less than 25 pairs, the delay of the routing method is low. Of course, the smallest hop count based routing method gives the lowest delay. However, when the number of end-to-end connections is greater than 25 pairs, the

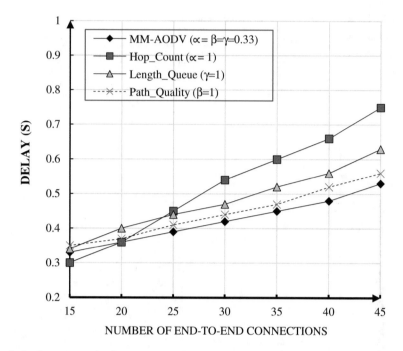

Fig. 3 Performance evaluation: average end-to-end delay

delay of this routing method increases rapidly and it will be the highest value. The routing method (with $\alpha = \beta = \gamma = 0.33$) always has the lowest delay.

As discussed so far, we can claim that MM-AODV protocol (with $\alpha = \beta = \gamma = 0.33$) always returns the best and stable PDRs as well as delay values. Also, when the network traffic is low, the smallest hop count based routing method is optimal. However, when the network traffic is medium and high, this routing method has some limitations and achieves lower performance than that of other routing methods. It is possible that one can set up different performance factors to adapt the actual conditions such as network structure, mobile node speed, traffic data, etc.

5   Conclusions

In this paper, we propose a routing protocol for MANETs, called MM-AODV which uses a multi-metric cost function in order to select the highest performance route. Moreover, we setup the set of performance factors $\{\alpha, \beta, \gamma\}$. Depending on the network condition, MM-AODV protocol will select an appropriate route, which can be the route with: smallest hop count ($\alpha = 1$), link quality ($\beta = 1$), queue length ($\gamma = 1$) or the route with the set of performance factors ($\alpha = \beta = \gamma = 0.33$). The simulation results show that MM-AODV protocol with the above set of performance factors obtains the best achievable performance. One limitation of this work is that we have not considered the amount of energy consumed when sending probe packets to neighboring nodes (in order to get information about link quality). However, we think that it is negligible and is the price to pay for high-performance routes.

In addition, the set of performance factors is fixed in the entire simulation process. Indeed, the performance may depend on the type of applications such as Video, VoIP, and common data tranfer. For further studies, we will propose a method to deal with these issues.

Acknowledgements This research was supported by Center for Research and Applications in Science and Technology, Hung Yen University of Technology and Education, under grant number UTEHY.T029.P1718.01.

References

1. Siddhant, D., Mane, P.B., Vanjale, M.S.: A survey on energy efficient routing protocol for MANET. In: 2nd IEEE International Conference (iCATccT), pp. 160–164. IEEE Press (2016)
2. Lav, G., Raj, J., Gabor, V.: Survey of important issues in UAV communication networks. IEEE Commun. Surv. Tutor. **18**(2), 1123–1152 (2016)
3. Babatunde, O., Naoki, S., Juntao, G.: Secure payment system utilizing MANET for disaster areas. IEEE Trans. Syst. Man Cybern. Syst. **PP**(99), 1–13 (2017)
4. Shivashankar, et al.: Designing energy routing protocol with power consumption optimization in MANET. IEEE Trans. Emerg. Top. Comput. **2**(2), 192–197 (2014)
5. Renisha, P.S., Rajesh, R.: A survey: optimal node routing strategies in MANET. In: IEEE International Conference (SAPIENCE), pp. 260–267. IEEE Press (2016)
6. Akansha, C., Vishnu, S.: Review of performance analysis of different routing protocols in MANETs. In: IEEE International Conference on Computing, Communication and Automation (ICCCA), pp. 541–545. IEEE Press (2016)
7. Evripidis, et al.: Multi-metric energy efficient routing in mobile ad-hoc networks In· IEEE Military Communications Conference, pp. 1147–1151. IEEE Press (2014)
8. RFC3561. https://www.ietf.org/rfc. Accessed 18 Aug 2017
9. RFC4728. https://www.ietf.org/rfc. Accessed 18 Aug 2017
10. Cuong, C.T., Tu, V.T., Hai, N.T.: MAR-AODV: innovative routing algorithm in MANET based on mobile agent. In: 27th IEEE International Conference on WAINA, pp. 62–66. IEEE Press (2013)
11. Quy, V.K., Han, N.D., Ban, N.T.: A_WCETT: a high-performance routing protocol based on mobile agent for mobile ad hoc networks in 5G (in Vietnamese). J. Inf. Technol. Commun. **17**(37), 14–21 (2017)
12. Vivek, S., Bashir, A., Doja, M.N.: ANFIS aided AODV routing protocol for mobile ad hoc networks. J. Comput. Sci. **13**(10), 514–523 (2017)
13. Lineo, M., Elisha, O.O.: Effect of varying node mobility in the analysis of black hole attack on MANET reactive routing protocols. In: 2016 Information Security for South Africa (ISSA), pp. 62–68 (2016)
14. Han, N.D., Younghwa, C., Minho, J.: Green data centers for cloud-assisted mobile ad-hoc network in 5G. IEEE Netw. **29**(2), 70–76 (2015)
15. Pimmy, G., Rakesh, J.K., Sanjeev, J.: Green communication in next generation cellular networks: a survey. IEEE Access J. **5**, 11727–11758 (2017)

Novel Aproach for Localization of Patients in Urgent Admission Department

Jan Kubicek, Libor Michalek, Tomas Urbanczyk, Jaromir Konecny, Martin Tomis, Filip Benes, Jiri Svub, Pavel Stasa and Leopold Pleva

Abstract The Real-time Locating Systems (RTLS) are also well known as real-time location systems have recently become significantly important part of many location aware systems. RTLS systems are primarily used for the object tracking placed indoor. This area relates to the healthcare environment where tracking and monitoring of patients are important. The paper deals with the overview of the possible applications of the RTLS systems in the healthcare facilities and their potential benefits. The paper also brings conceptual information about two currently developed systems with the cooperation of Trauma Center of University Hospital in Ostrava with a target of the patient database and mobile application development on the base of the Internet of Things allowing for fully digital archiving patient records. The second system focuses on the development of RTLS system based on the Infra Red (IR) system with anchors where the location of the patient is acquired using IR tags.

1 Introduction

Real-Time Location Systems (RTLS) are designed to identification and localization of the tagged equipment and patients while they are moving in the hospital. A possibility of the equipment tracking may have the significant potential to effectively

J. Kubicek (✉) · T. Urbanczyk · J. Konecny
Department of Cybernetics and Biomedical Engineering, VŠB–Technical University of Ostrava, 17. listopadu 15, Ostrava-Poruba, Czech Republic
e-mail: jan.kubicek@vsb.cz

L. Michalek · M. Tomis
Department of Telecommunications, VŠB–Technical University of Ostrava, 17. listopadu 15, Ostrava-Poruba, Czech Republic

F. Benes · J. Svub · P. Stasa
Institute of Economics and Control Systems, VŠB–Technical University of Ostrava, 17. listopadu 15, Ostrava-Poruba, Czech Republic

L. Pleva
Trauma Center, Ostrava University Hospital, 17. listopadu 1790, Ostrava-Poruba, Czech Republic

© Springer International Publishing AG, part of Springer Nature 2018
A. Sieminski et al. (eds.), *Modern Approaches for Intelligent Information and Database Systems*, Studies in Computational Intelligence 769, https://doi.org/10.1007/978-3-319-76081-0_39

455

manage inventory stored in the hospital environment including the process of routine preventive maintenance, thus improving the availability of needed items and in the same time reducing equipment rentals.

The second important issue is a tracking of the clinical staff. In this regard, the tracking procedure is focused on the clinical processes improvement, including a detection of the routine breakdowns in care provisions, and disciplining clinicians and other hospital staff underperforming in their responsibilities. The most important and challenging issue of the RTLS systems in the hospital environment is the patient tracking. The procedure primarily focuses at their localization and indication of clinical staff when a patient gets lost or leaves a particular environment which they should move in. Also, an important aspect of the patient tracking is a verification of the particular patient before applying medical procedures [1–7].

The following paper concerns with recent overview of research performed in this area, focuses on procedures related to Emergency Department of Ostrava hospital and describes our novel approach for localization of patients.

2   Overview of Location Systems

The **Location Based Services** (LBS) and **Real Time Location Systems** (RTLS) grows significantly over the past few years. The key areas for these systems comprise navigation, enterprise services, location-specific health information or user tracking [8]. The primary aim of LBS system is to determine what the location of the user is.

In general, localization systems can be divided into two categories

1. RF (Radio Frequency),
2. Non-RF (Ultrasound, Infrared Light, Laser, and others).

The main advantage of RF versus non-RF is the possibility of using it in an environment without direct visibility, which almost all non-RF localization systems require [9]. RF Indoor Positioning Systems (IPS) use wireless technologies such as Bluetooth, RFID, WiFi, and other. However, most of these technologies are not designed specifically for localization, so they need to be modified or expanded for localization purposes. One example is the Wireless Local Area Network (WLAN), which is rather expensive to implement and these devices are too large for fixing to small objects. These technologies provide only the estimated location of an object which position is calculated, for example, by the triangulation method [10, 11].

Using and deploying RTLS in a dynamic environment involves various issues. For example, GPS signal level is too weak to work inside buildings. Ultrasound can provide very high accuracy but requires a direct view configuration. Similarly, visual RTLS suffers from reflection. Also, the object can be lost if it moves very quickly or is shaded [12]. To choose the most appropriate technology, a comparison in Table 1 is presented [13, 14].

Table 1 Comparison among indoor localization technologies

Technology	Accuracy [m]	Coverage [m]	Typical Environment
Vision	0.001–0.1	1–10	Indoor
Infrared	0.01–1	1–5	Indoor
Ultrasound	0.01	2–10	Indoor
Wi-Fi	1–10	20–50	Indoor/Outdoor
RFID	0.1–1	1–10	Indoor
Bluetooth	1–10	1–30	Indoor/Outdoor

2.1 Overview of Location Systems in Healthcare Environment

Hospital real-time location systems, also known as indoor positioning systems standardly involve various types of hardware solutions with associated software interface. In fact, those location systems work on the principle of the hardware tags which can be placed on the respective property or person [15, 16]. Those tags are sharing its position via the sensor network that triangulates its position. The network data are consequently acquired by a software interface so that users can analyze the graphical representation of all tag locations or particular, the position of the tag can be obtained with a target of person or equipment localization [17, 18].

The most frequently used RTLS technologies involve the radiofrequency identification (RFID), Wi-Fi, Wireless Local Area Network (WLAN), ultra-wide band (UWB), infrared (IR), ZigBee, Bluetooth, or ultrasound. On the other hand some hospitals and commercial companies do experiments with the RTLS systems implementation by using these methods with the aim of the hybrid networks providing more accurate tracking of the tags.

The recent scientific literature dealing with the RTLS systems have tends to focus only on technological aspects of individual RTLS systems. Those studies are particularly focused on case studies of pilot projects. Many of those projects lack of information about specific problems related to particular hospital wards such as workflow or inventory management. Those studies have in common fact that the location systems have been tested and designed in the specific isolated conditions. Consequently, there is a lack of the information describing how those systems can be utilized in the holistic hospital environment, or how particular hospital environment might influence, and theoretically impede technical RTLS functionality. On the other hand the holistic environment may contain a lot of unexpected obstacles disallowing proper function of the RTLS systems [19, 20].

In [21] a UHF radio frequency identification system was proposed for patient identification through passive RFID tags, real-time location service of medical assets, and drug inventory control and monitoring. Experiments were performed to evaluate the system in a realistic hospital environment. The overall results were not ideal but satisfactory to a great extent. A study in [22] says that between 210.000 and

440.000 patients each year who go to the hospital for care suffer some preventable harm that contributes to their death. The results of a pilot study in [23] support proof of concept for continued experimentation with the application of RTLS technology to obtain time-use estimates for bedside nurses and particularly to quantify nurse-patient interaction time. RTLS time-use estimates demonstrate good concordance and reliability with direct observation. In [24] RFID technology presents a promising disruptive technology for automated tracking of high-value assets in the pathology laboratory.

In [25], a RFID is an emerging technology used for automated data capture of patient flow through the hospital and instant alert when patients turn up at the wrong operating room or other location. In [26], a Taiwanese hospital is using passive RFID to help correctly identify surgical patients and track their operations to ensure they get the correct procedures and the proper medications at the right time. The plastic bands are embedded with 13.56 MHz passive RFID. Each read-write tag has enough capacity to store a patient's name, medical record number, gender, age and doctor's name, and additional information can be stored on the tag if needed. The tag's ID number is then associated with patient records stored on the hospital's back-end information system. A comprehensive development framework for the implementation of an RFID-enabled system in a medical organization has been described in [27]. With the application of RFID system proposed in [28], emergency department based observation unit are becoming effectively evaluated for the assessment and treatment of patients whether who may require inpatient or intensive care unit management or monitoring.

In [29], a patient localization was performed using the Ekahau Real-Time Location Solution while questionnaire evaluated the results. In the study [30], an Android-based smartphone application that provides outpatients visiting the hospital with a personalized treatment schedule and indoor navigation service is developed. To provide indoor navigation service using RTLS, a total of 231 Bluetooth APs in the two basement-level floors were installed. Paper [31] presented an RFID-based system for mobile object positioning in hospitals. The approach is based on passive RFID tags to meet the requirements for compatibility and scalability. The purpose of the research in [32] was to assess real-time location systems (RTLS) that have been implemented in U.S. hospitals. The project was a 3-year qualitative study of 23 U.S. hospitals that had implemented.

In the prototype, presented in [33], authors examined the potential use of an existing RTLS system Ekahau to improve patient safety regarding having correct creation and deletion of patient-device associations, in real-time and with minimal effort. The final goal of the LAURA project [34] is the design and the implementation of a lightweight system based on Wireless Sensor Networks (WSNs) and Zigbee IEEE 802.15.4 for the automatic supervision of patients within the nursing institute. Interesting approach presents [35], where multiple indoor positioning systems in a healthcare environment were analyzed.

3   Procedures at Urgent Admission

The current trend of the University Hospital of Ostrava (UHO) is a utilization of the newest technologies which help to make the work more efficient and therefore achieve the higher patient standard for care. The Urgent Admission of UHO is one of the departments in the Central admission. In our research, we also focus on a development of the information system which can register patients data during its sojourn at Urgent admission. This idea relates to the Urgent admission upgrade while we develop the Real Time Location System for the patient localization in the area of Central Admission.

3.1   Issue Definition

For the purpose of domestic and international accreditation, and also for medical procedures improvement, it is necessary to analyze and evaluate the individual records. The spend time at the Urgent Admission depends on the particular diagnosis which a patient is admitted. A benefit is a presence of the consultants and ability of prompt diagnosis. Presently on the Urgent Admission, the paper form as documentation for the intensive care record is still used. This form is called as "ZIP". The records are written on the base of agreed symbols in the timeline such as medicaments, vital functions the blood pressure, pulse the saturation of blood with oxygen, but also performed procedures, took laboratory specimens, ordered and prepared products from the blood bank and, finite set of entities. This record is created as duplicate through the copy paper which is attached to the patients card and moving to another department. The original record is stored on the ward. On the other hand, substituting the paper form by the electronic ones carries certain aspects as follow.

Advantages

- The well-arranged A3 format structure.
- Intuitive for filling (template is stored on the ward).
- Quick response in reading and writing.
- It is marked by the identification patients label containing basic records: name, surname, birth, insurance company, place of living, a diagnosis which a patient was admitted.
- An exact record of the medicaments weight over the time.

Disadvantages

- The vast extent of the documents.
- Two sheets of form can be moved, and therefore an error in the timeline can appear.
- In the original form, the types such as opiates, antibiotics, transfusion preparations are marked by different color while the copy is monochrome.
- A risk of the douse the form with water or with the disinfection.

- Sometimes unreadable handwriting.
- In the case of the acute state, an inaccurate record to the paper form can occur.
- There is not the Electrocardiography (ECG) record over the time; only the pulse is indicated.

3.2 General Process Description on Urgent Admission

Three adjustable beds, two for quarantine and two childrens beds have the Urgent Admission at disposal. There is also the ultrasonic examination room and radio diagnostic part including X-ray and Computed Tomography (CT). It means that a patient can move up to eight rooms.

Fig. 1 The general process at urgent admission of UHO

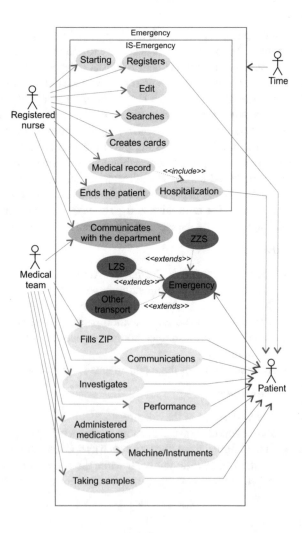

The patient admission is usually reported by the dispatching or ambulance service and consequently carried to the Urgent Admission. Treatment, patient registration, and management of medical records are running at same time. The whole process is shown on Fig. 1.

The 10 minutes physiological functions acquired by respective devices are indicated by the record. The time which the patient spends on the Urgent Admission depends on the duration of examinations, on the examination room workload or staff workload. It frequently occurs that a patient is examined, stabilized and finally ready to transfer to another ward but there is no free bed due to logistic reasons.

4 Development of RTLS for Urgent Admission

We currently develop the RTLS system based on the tag localization with the target of patient identification and localization on the hospital emergency department. After arrival, the patient is marked with the tag which is fastened for all time while staying in the hospital ward. Figure 2 shows the overall RTLS architecture of the system, which is in the process of development.

We also want to provide the tag by the buttons for predefined events calls. The RTLS system is activated by the tag, and in an appropriate time by pushing the

Fig. 2 Overall RTLS architecture

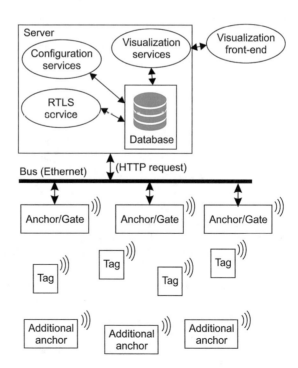

respective button, the respective event is recorded. The following time events we record: the patients arrival, diagnosis determination and, the patients leaving. This information helps us to form a time sequence for the patient and subsequently to determine the exact time for individual rooms in hospital emergency department. Also, we analyze individual time for which patient stayed in a certain ward. Moreover, we focus on determination of examinations length, mainly Computed Tomography (CT) examination.

We also observed that the staff is extremely loaded during preparation for patient reception and follow-up treatment. In this time, a nurse usually performs the necessary administrative procedures. The tag registration would induce next time demand, and therefore it is postponed until patient leaves the emergency department. In spite of the nurse has a spare time before patients reception, she can perform a registration to the RTLS system. Nevertheless, more likely the registration is done later.

5   Discussion and Conclusion

In the area of the healthcare facilities, RTLS systems may serve for localization of portable medical equipment, but the most challenging area is a localization of the patients and medical staff. The clinical staff would be as little load as possible. This should be the main contribution of the research. The efficiency of the hospital can be improved by ensuring that the appropriate medical staff and medical equipment are in the right place at the right time.

A development framework of RTLS system for an Urgent Admission in Ostrava University Hospital has been described in this paper. It is very tough to develop an automatic system which is able to record the whole process in Urgent Admission. Moreover, the current process in very conservative. Regardless, the idea represents the device which can be assigned to a patient, recording the values in the real-time to the database.

By complex analysis of RTLS system outputs (as the duration of examinations or other procedures), we can predict certain limitations of the whole process as well as optimize workload of the staff. The RTLS system helps in the logistic and the medical procedures timing. RTLS systems may also provide crucial information which deals with the movement of people and assets. Since the data of position from the RTLS systems are acquired and stored, real-time information can be utilized, while RTLS-based indicators can also be used to support operational aspects such as processes for nurses, physicians, or other staff members, as well as overall patient and resident care.

Information from RTLS especially helps to improve the care process with a target of understanding cycle time, required an amount of the time for which patient is receiving care, utilization rate, how often are equipment and rooms utilized, where patients are waiting and current patients state in the care process. All these aspects provide a level of intelligence in an organization which consequently helps to make decisions in the area of the healthcare.

An important feature of the healthcare is also a reduction or removal of wasteful or unnecessary steps which leads to providing better healthcare and additional time for further patients. Those facts improve the whole context of the healthcare without adding resources.

Regarding appropriate location equipment, the choice of RTLS technology should be very carefully considered. Utilized technology or hardware may not work well despite all its merits, if not properly matched to the intended application or the care facilitys (physical) environment, budget and future expansion plans (the latter will require an adequately scalable RTLS solution).

Acknowledgements The work and the contributions were supported by the project SV4507741/ 2101, 'Biomedicinske inzenyrske systemy XIII'. This contribution was supported by the Technology Agency of Czech Republic as a part of competence centres project titled Centre for Applied Cybernetics 3—identification number TE01020197. This work was supported by the project SP2017/158, Development of algorithms and systems for control, measurement and safety applications III of Student Grant System, VSB-TU Ostrava.

References

1. Davis, S.: Tagging along. rfid helps hospitals track assets and people. Health Facil. Manag. **17**(12), 20–24 (2004)
2. Vilamovska, A.M., Hatziandreu, E., Schindler, H.R., van Oranje-Nassau, C., de Vries, H., Krapels, J.: Study on the requirements and options for rfid application in healthcare. RAND Corporation (2009)
3. Glabman, M.: Room for tracking. RFID technology finds the way. Mater. Manag. Healthc. **13**(5), 26–38 (2004)
4. Revere, L., Black, K., Zalila, F.: RFIDs can improve the patient care supply chain. Hosp. Top. **88**(1), 26–31 (2010)
5. Frisch, P.H., Booth, P., Miodownik, S.: Beyond inventory control: understanding RFID and its applications. Biomed. Instrum. Technol. **44**(4), 39–48 (2010)
6. Thuemmler, C., Buchanan, W., Fekri, A.H., Lawson, A.: Radio frequency identification (RFID) in pervasive healthcare. Int. J. Healthc. Technol. Manag. **10**(1–2), 119–131 (2009)
7. Lee, I.: Encyclopedia of E-Commerce Development, Implementation, and Management. Number sv. 1 in 1. Business Science Reference (2016)
8. Liu, Y., Yang, Z.: Location, Localization, and Localizability—Location-Awareness Technology for Wireless Networks. Springer Science Business Media (2010)
9. Zhao, Y., Smith, J.R.: A battery-free RFID-based indoor acoustic localization platform. In: 2013 IEEE International Conference on RFID (RFID). IEEE (2013) 110–117
10. Zhang, D., Xia, F., Yang, Z., Yao, L., Zhao, W.: Localization technologies for indoor human tracking. In: 2010 5th International Conference on Future Information Technology (FutureTech), pp. 1–6. IEEE (2010)
11. Farid, Z., Nordin, R., Ismail, M.: Recent advances in wireless indoor localization techniques and system. J. Comput. Netw. Commun. **2013** (2013)
12. Liu, H., Darabi, H., Banerjee, P., Liu, J.: Survey of wireless indoor positioning techniques and systems. IEEE Trans. Syst. Man Cybern. Part C (Applications and Reviews) **37**(6), 1067–1080 (2007)
13. Ruiz-López, T., Garrido, J.L., Benghazi, K., Chung, L.: A survey on indoor positioning systems: foreseeing a quality design. Distrib. Comput. Artif. Intell. 373–380 (2010)

14. Li, H., Chan, G., Wong, J.K.W., Skitmore, M.: Real-time locating systems applications in construction. Autom. Constr. **63**, 37–47 (2016)
15. Zare Mehrjerdi, Y.: Radio frequency identification: the big role player in health care management. J. Health Organ. Manag. **25**(5), 490–505 (2011)
16. Wicks, A.M., Visich, J.K., Li, S.: Radio frequency identification applications in healthcare. Int. J. Healthc. Technol. Manag. **7**(6), 522–540 (2006)
17. Cio, H.: Himss (healthcare information and management systems society), 19th annual 2008 HIMSS leadership survey. Healthcare Information and Management Systems Society
18. Fisher, J.A., Monahan, T.: Tracking the social dimensions of RFID systems in hospitals. Int. J. Med. Inform. **77**(3), 176–183 (2008)
19. Network, R.: RTLS market overview. RFID Netw. (2010)
20. Carrasco, V.N., Jackson, S.S.: Real time location systems and asset tracking: new horizons for hospitals. Biomed. Instrum. Technol. **44**(4), 318–323 (2010)
21. Polycarpou, A.C., Gregoriou, G., Papaloizou, L., Polycarpou, P., Dimitriou, A., Bletsas, A., Sahalos, J.N.: A healthcare application based on passive UHF RFID technology. In: Proceedings of the 5th European Conference on Antennas and Propagation (EUCAP), pp. 2814–2818. IEEE (2011)
22. James, J.T.: A new, evidence-based estimate of patient harms associated with hospital care. J. Patient Saf. **9**(3), 122–128 (2013)
23. Jones, T.L., Schlegel, C.: Can real time location system technology (RTLS) provide useful estimates of time use by nursing personnel? Res. Nurs. Health **37**(1), 75–84 (2014)
24. Lou, J.J., Andrechak, G., Riben, M., Yong, W.H.: A review of radio frequency identification technology for the anatomic pathology or biorepository laboratory: much promise, some progress, and more work needed. J. Pathol. Inform. **2** (2011)
25. Sandberg, W.S., Häkkinen, M., Egan, M., Curran, P.K., Fairbrother, P., Choquette, K., Daily, B., Sarkka, J.P., Rattner, D.: Automatic detection and notification of wrong patientwrong location errors in the operating room. Surg. Innov. **12**(3), 253–260 (2005)
26. Bacheldor, B.: Taiwans chang-gung hospital uses HF RFID to track surgery. RFID J. **1** (2007)
27. Ting, S., Kwok, S.K., Tsang, A.H., Lee, W.: Critical elements and lessons learnt from the implementation of an RFID-enabled healthcare management system in a medical organization. J. Med. Syst. **35**(4), 657–669 (2011)
28. Chen, C.I., Liu, C.Y., Li, Y.C., Chao, C.C., Liu, C.T., Chen, C.F., Kuan, C.F.: Pervasive observation medicine: the application of RFID to improve patient safety in observation unit of hospital emergency department. Stud. Health Technol. Inform. **116**, 311–315 (2005)
29. Stübig, T., Zeckey, C., Min, W., Janzen, L., Citak, M., Krettek, C., Hüfner, T., Gaulke, R.: Effects of a wlan-based real time location system on outpatient contentment in a level I trauma center. Int. J. Med. Inform. **83**(1), 19–26 (2014)
30. Yoo, S., Jung, S.Y., Kim, S., Kim, E., Lee, K.H., Chung, E., Hwang, H.: A personalized mobile patient guide system for a patient-centered smart hospital. Int. J. Med. Inform. **91**, 20–30 (2016)
31. Shirehjini, A.A.N., Yassine, A., Shirmohammadi, S.: Equipment location in hospitals using RFID-based positioning system. IEEE Trans. Inf. Technol. Biomed. **16**(6), 1058–1069 (2012)
32. Fisher, J.A., Monahan, T.: Evaluation of real-time location systems in their hospital contexts. Int. J. Med. Inform. **81**(10), 705–712 (2012)
33. Rezaee, R., Baslyman, M., Amyot, D., Mouttham, A., Chreyh, R., Geiger, G.: Location-based patient-device association and disassociation. Procedia Comput. Sci. **37**, 282–286 (2014)
34. Redondi, A., Tagliasacchi, M., Cesana, M., Borsani, L., Tarrío, P., Salice, F.: Lauralocalization and ubiquitous monitoring of patients for health care support. In: 2010 IEEE 21st International Symposium on Personal, indoor and mobile radio communications workshops (PIMRC Workshops), pp. 218–222. IEEE (2010)
35. Haute, T., Poorter, E.: Performance analysis of multiple indoor positioning systems in a healthcare environment. Int. J. Health Geogr. **15**(1), 7 (2016)

Design of Universal Hardware Node Board for Smart-Home Automation and the IoT

Jan Stepan, Richard Cimler, Jan Matyska and Ondrej Krejcar

Abstract An affordable way of designing and developing an universal node board that can be used with existing Internet gateways or smart home automation frameworks is presented in this paper. Its design allows connecting almost any available sensor, actuator, or control element. The use of the proposed architecture also decreases the time to deploy a system, as well as the time needed to test and analyze various ideas and scenarios. Future improvements and testing are also discussed.

Keywords Microchip · Hardware development · Home automation · IoT node

1 Introduction

The Internet of Things (IoT) promises small devices that will be able to run from a battery for many years and have their data available over a cloud API [8]. Such devices contain various types of sensors and actuators. There are multiple complete solutions or development kits available for rapid prototyping [2]. The vast majority of available solutions now depend on various wireless technologies. The wireless approach is generally optimal in the IoT [6]. However, there are some specific use cases where a wired connection might be more than adequate.

J. Stepan (✉) · J. Matyska (✉) · O. Krejcar (✉)
Faculty of Informatics and Management, University of Hradec Králové,
Hradec Králové, Czech Republic
e-mail: jan.stepan@uhk.cz; jan.stepan.3@uhk.cz

O. Krejcar
e-mail: ondrej.krejcar@remoteworld.net

R. Cimler (✉)
Faculty of Science, University of Hradec Králové, Hradec Králové, Czech Republic
e-mail: richard.cimler@uhk.cz

J. Matyska
e-mail: jan.matyska@uhk.cz

© Springer International Publishing AG, part of Springer Nature 2018   465
A. Sieminski et al. (eds.), *Modern Approaches for Intelligent Information
and Database Systems*, Studies in Computational Intelligence 769,
https://doi.org/10.1007/978-3-319-76081-0_40

In IoT solutions where wired or wireless technology is used, the sensors are connected to the system via nodes. A node is hardware with a micro-controller which handles the sensors or actuators and uses built-in peripherals to communicate with the upper layers of the system. Various wireless and wired solutions are discussed in this paper. The pros and cons of these solutions are pointed out and requirements for a node board described. The design, implementation, and testing of an universal node board for the IoT is the main contribution of this paper.

This paper has the following structure: The following section presents various wireless and wired solutions for the IoT and smart home module development. The next section defines some requirements based on the shortcomings of existing solutions. Software and hardware implementations of the proposed solution are presented in section four. The fifth section summarizes the results of various tests of a HW prototype. The sixth section mentions possibilities of future improvements.

2   Available Solutions

There are a lot of wireless technologies with various levels of power consumption and ranges. Some of them are multi purpose; others are designed especially for IoT applications, e.g. LoRa, Sigfox and IQRF transmitters. The idea of how to rapidly interconnect and prototype various devices is described in [3] and how to approach data acquisition in [5]. Research, as in [1], has also been conducted and shows that even affordable hardware like the Raspberry Pi can be power optimized to serve as a battery powered gateway. The Raspberry Pi can also serve as an end node and handle sensors as described in [13]. Even though the Raspberry Pi is a very universal and versatile development platform, it has its limitations, which are described in [7]. Yun et al. [14] show that IoT and smart home automation solutions tends to be highly fragmented. Therefore some standardization of the M2M protocols needs to be implemented.

Vilajosana et al. [12] describes the OpenMote plattform, which is designed especially for the Internet of Things and rapid prototyping. It implements the upcoming IEEE 802.15.4e commutation standard, working in the 2.4 GHZ band, along with WiFi or Bluetooth Low Energy standards. The hardware works in similarly to the Arduino platform, which means that the OpenMote base board can be attached to a custom PCB with various sensors.

Although each solution and approach is ideal for a different kind of scenario, there are a few limitations to all the solutions, which should be taken in account. Most new buildings are nowadays built by different hybrid construction methods. This means that the ceilings are made from a metal grid layer and the rooms are connected together with a common ground wire. The reason for this is to increase electrical safety, but it also creates a major problem. Grounding the whole building creates a very good Faraday cage and the transmittance of a signal from the outside or even between rooms decreases rapidly. An other shortcoming of wireless technologies is that it is relatively easy to disturb them. This might be caused by accident, but for technologies working on world-wide open frequencies, cheap and effective jammers are available.

In such a situation, a wired solution is more suitable than a wireless one. There is, for example, the very popular Arduino development platform, which includes an opensource compiler and evaluation boards. However, to simplify the development process, the available libraries are optimized for ease of use rather than for performance. Hardware boards also contain an USB driver to simplify the programming, a power circuit, notification LEDs, and some models even have overcurrent protection. When an external bus is required, as in the case of smart home automation, a custom shielded PCB has to be made. The resulting size is going to be significantly bigger and the cost higher than for a custom solution. The goal of this paper is to establish the requirements for an efficient wired node board and describe its implementation.

3 Prototype Requirements

This section specifies the requirements for a node board that can be used for hardware prototyping in smart home systems or IoT projects. The goal of these requirements is to make the prototype as versatile as possible. The main features are the connection type, performance, dimensions, manufacturing, and price.

Connection As mentioned in the Introduction, the use of wireless technologies can increase the difficulty of development, and the signal may not be always available inside new buildings. The prototype should use a wired technology meeting the following requirements. The cabling must be easy to create and the type of connector used should not be very common. This means that the inexperienced user can not accidentally swap cables with something else. The chosen technology should use a bus topology to reduce the costs of the cabling and reduce the need for any kind of active switches. The total connection length of the cabling must be at least 100 meters with a minimum of ten boards chained.

The power adapter has to be realized as an industry standard jack connector. This is going to enable the possibility of powering the board with affordable wall plugs.

Performance The node must certainly contain some kind of microcontroller unit (MCU) to be able to communicate with the upper layers. For common smart home components, such as buttons, lights, relays, and environment sensors, there is no need for especially fast communications. A reaction time of 10 ms (sampling frequency of 100 Hz) is more than sufficient. The same requirement goes for IoT applications. However there are, for example, approaches to determine the heart rate from accelerometers [11], which require a sampling frequency of at least 500 Hz.

Sensors, actuators and controllers The prototype must be able to set digital output pins, read digital input pins, and handle and drive buses used for sensor communication. Those buses are I2C, SPI, and 1-Wire, along with some custom ones.

Dimensions Since the proposed solution is only a development prototype and it is not designated to be sold as a commercial product, the target printed circuit board

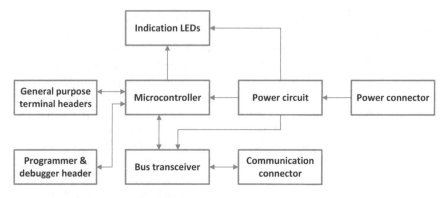

Fig. 1 Block diagram of proposed prototype

(PCB) dimensions are not a limiting factor. Many PCB manufacturing companies offer cheaper prices for PCBs with areas smaller than $100\,\text{cm}^2$.

Manufacturing Surface mount technology (SMT) is almost the only way of manufacturing hardware in commercial products. Surface mount devices (SMD) tends to be much smaller than classical through hole devices (THD). Less drilling is required and machines can assembly the board in a shorter time. However, some surface mounted devices have such a small package that manual soldering is not possible. Because the node board designed must be feasible for use even by unskilled developers and for students as an evaluation kit, THD parts should be used as much as possible. SMD parts would be used only if a commercial solution with all necessary tests is going to be required.

Price The prices of the parts must be taken into an account when choosing technologies. If a board is really affordable, it will open the possibility of adding sensors or automation even to scientific projects with lesser funding.

Summary A solution for the node board has been proposed based on the previously mentioned features. It can be seen in Fig. 1.

4 Node Implementation

The requirements for the prototype were specified and the initial development of an universal node board was carried out. The results and observations of the implementation of the proposed node is presented in this section.

4.1 Microcontroller

The requirements for a new node specified that a wired connection will be used, thus there are lesser performance requirements for the microcontroller. No radio and

Table 1 Comparison of 8bit Midrange PIC MCU suitable for node board

Model	PPS	PWM	DAC	SPI/I2C	UART	TMR	RAM	Flash	MEM	OSC
16F1578	Y	4I	1	1/1	N	2	512	7	128	N
16F15344	Y	4	1	1/1	Y	2	512	7	224	Y
16F15345	Y	4	1	1/1	Y	2	1024	14	224	Y
16F18344	Y	2	1	1/1	N	6	512	7	256	Y
16F18346	Y	2	1	2/2	N	6	2048	28	256	Y
16LF155	N	0	0	1/1	N	2	512	14	128	N
16F1618	Y	2	0	1/1	N	6	512	7	128	N
16F690	N	4	0	1/1	N	2	256	3,5	256	Y
16F1769	Y	4I	4	1/1	N	6	1024	14	128	Y

wireless communication technology needs to be controlled or handled. Therefore, there is less need to select the best fitting MCU.

Although there are plenty of 32bit microcontrollers available, it is more reasonable to pick a simpler architecture for designing an evaluation board with only wired connections. The 32bit ARM Cortex Mx architectures offer a rich instruction set and plenty of peripherals. ARM itself does not manufacture chips, but various companies are licensing Cortex Mx architectures. One of the major shortcomings of the ARM architecture is its complexity. Due to its 32bit nature, there are a lot more configuration registers to set and the most advanced chips are almost impossible to manage without some manufacturer utilities. Another shortcoming is that only NXP offers a two-MCU with THD package. Others do not offer this choice.

For these reasons, an 8bit microcontroller from Microchip was selected. It can be programmed with the really affordable PicKit programmer or with cheap third party programmers. A free compiler is available for general use and third party companies are offering complete IDEs with compilers. Another huge advantage of using them is the PINOUT compatibility between multiple models. This means that, e.g. a 20 PIN 8bit PIC from the Midrange series has the same pin inputs and outputs and that different MCUs may be used for different firmware types. The design of the prototype takes this idea in mind and uses sockets to allow swapping the MCU for another model.

Table 1 shows the variety of MCUs which can be placed on the board. The numbers, except the memory information, are scaled down to show the number of peripherals which are possible to use on the terminal headers of the board. The total number of each peripheral may be higher when an MCU is used as a standalone. The number of ADC channels is not shown because they are available on every terminal header in every selected MCU. Their resolution is 10 bits and when increased resolution is required, external chip over SPI or I2C bus must be used. The acronyms used in Table 1 are the following:

PPS (Peripheral Pin Select)—A feature that allows choosing which periphery is routed on which pin. This is not applicable to all peripherals; the datasheet for the individual MCU must be read before.

PWM (Pulse-width modulator)—Timers on MCUs can be used to generate digital signals with varying periods and duty cycles. Marking the MCU with an 'I' means that its PWM has dedicated timers.

DAC (Digital to analog converter)—A DAC can be used to generate output voltage. It is an alternative to PWM that is more suitable for some use cases, such as driving DC motors.

SPI/I2C —Number of most popular communication buses. Note that the buses are mutually exclusive for some MCUs. Therefore, when an I2C sensor is connected, the SPI bus is unavailable.

UART —The first UART is always used to communicate with RS485 transceivers. This column indicate the presence of a secondary UART. Software UART can be implemented on all MCUs, but HW UART features buffering and is more versatile.

TMR —Number of additional available timers. One 8bit timer is reserved for firmware implementation.

RAM —Size of data memory in bytes.

Flash —Size of program memory in kilobytes.

MEM —Size of non-volatile storage in bytes. The technology used inside the chips varies. It can be traditional EEPROM memory or high endurance Flash sharing address space with the program.

OSC —Every MCU listed has an integrated oscillator inside chip. However there is the possibility of using an external crystal routed on board to have higher clocking precision. Those with this functionality are marked with a 'Y'.

The overview shown in Table 1 does not aim to be complete. There are many more hardware peripherals integrated in an MCU, such as comparators, watch dog circuit, external interrupt triggers, etc. These features are not crucial during the selection of the definitive solution.

4.2 Communication

The communication bus that best satisfies the previously mentioned requirements is RS485. It uses differential voltage driven by operational amps. The bus is used in full duplex mode, where nodes receive commands from the master, which can be any controller equipped with an RS485 transceiver; the nodes send back responses to the master over the second channel. Any existing protocol, such as the industry standard ModBus or ProfiBus, may be used, or a custom protocol can be rapidly developed as described in [10], which is used in our IoT and home automation solution.

4.3 Schematics

The results of the development are shown as schematics in Fig. 2.

The device requires 5 V supplied over a 5 V_IN Jack connector. The input source is filtered through capacitors C4 and C3. Two half duplex MAX485 transceivers are powered from the 5 V input. Two transceivers have been chosen, instead of one full-duplex, to reduce the costs of the components. The node which is connected to a bus as the last element must have short pins TERM1 and TERM2 to attach a termination resistor, as required by the RS485 specification. All remaining nodes must have these pins open. The input voltage is routed either to HT7533 for 3.3 V conversion or directly to the other components. The selection of the route is performed by placing a

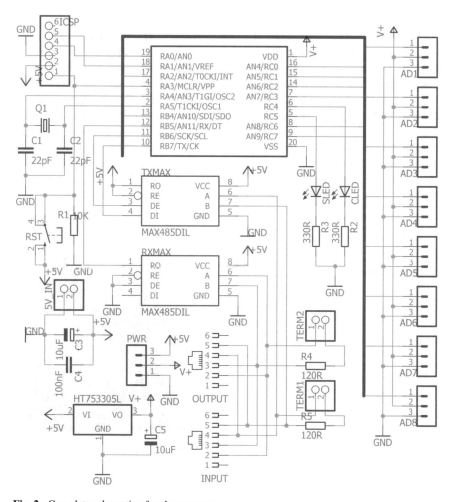

Fig. 2 Complete schematic of node prototype

jumper connector to the PWR header. The PIC MCU works with a selectable voltage, so the use of 5 V or 3.3 V logic for sensors can be changed just by a jumper. With no jumper placed, the MCU will not start. Two notification LEDs are also present, to indicate various states. ICSP header is routed directly to MCU and allows chip programing directly from board. Pins 4 and 5 of the ISCP header can also be used as analog inputs and the header then works as a small antenna. It is possible to use this as a source of entropy to generate real random numbers on board. The price of the presented solution is less than 10 USD, not including printing the PCB, which is very variable, based on the number of printed pieces.

4.4 PCB

Figure 3 shows the result of the PCB routing. Since the schematics are not very difficult, it is possible to create and manufacture the PCB in various shapes. We chose a square design with dimensions 62×62 mm. The spaces between the individual components are large enough to enable comfortable manual soldering. This PCB also complies with the requirement for the prototype size. Terminal header blocks with standard 2.54 mm are routed in triplets. Each triplet has a common ground, a positive voltage, and is routed to the GPIO pin on the MCU. An RJ11 connector is used for data transmission, but in the 6pin version, which is not commonly used. This ensures that the cabling can not be accidentally swapped with Ethernet or phone cable, and allows creating cables very easily. Because the circuit is simple, it is possible to route the PCB in various ways. The only rule that must be followed is placing a crystal with grounding capacitors as close as possible to the MCU input pins.

Fig. 3 Example PCB
designed from schematic

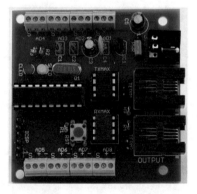

5   Testing

The firmware for the nodes has to comply with the communication protocol described in [10]. Therefore, the node works as a part of the HAuSy framework, designed as a three-layer architecture with fail-over mechanisms, a dynamic rule engine, and almost real-time response times. A Raspberry Pi with a custom shield with an RS485 drive is used as the control unit. The whole architecture is described in more detail in [4].

Even though the power consumption is not critical when the board is powered from a wall plug adapter, measurements showed that the node has a consumption of 11.8 mA in idle state and 14.3 mA when transmitting or receiving data over the bus. Everything was measured without any sensor connected, because power consumption varies from sensor to sensor.

More detailed benchmarks of reaction time were collected in [9], where the testing was performed on the final firmware, but with a breadboard prototype. However, this prototype shared the same connections and components. When using our HAuSY custom protocol with 230400 baudrate, the nodes are able to reply to approximately 950 messages per second when the digital state is being transmitted from eight channels. When a 10 bit analog input state is transmitted, the reply rate is approximately 520 messages per second. The slowdown is caused by the transmission time required to send the additional bytes over an RS485 bus.

6   Future Improvements

Although the developed prototype met all the requirements and has been successfully tested, there are still several possible ways to improve the hardware in the future. Since RS485 requires termination resistors, it would be interesting to try using two free wires on the RJ11 connector in combination with SSR relays and create a mechanism for automatic termination of the last node on the bus. Another approach might be to use those wires for powering the nodes, but this would require adding another voltage stabilizer and using a high voltage, such as 48 V, over the cable to prevent lossed due to the area of the cable.

Other avenue for future improvement might be to conduct durability tests of the designed board in low and high outdoor temperatures.

7   Conclusion

Some of the existing commercial solutions and research approaches to the Internet of Things and smart home automation have been presented in this paper. However, they almost exclusively rely on wireless technologies, which is certainly the right approach in the majority of cases. But there are still good reasons to use wired solutions. Pricing, problematic signal propagation, expenses, or time pressures or reliability could be those reasons.

The requirements for a node board with a wired connection have been specified with the aime of creating the most universal board possible, similar to the Arduino platform but with higher performance, smaller dimensions, and less cost. This requirements also clearly enable the prototype to be used for educational purposes, the IoT, or automation projects, as an added value.

The implemented solution uses the RS485 bus, which might be considered old, but for working with sensors, actuators, and control elements, it offers an appropriate speed while maintaining long cable wiring. The cost of the components for a single board does not exceed 10 USD when using internationally established manufacturers without PCB printing.

The main goal of this paper was to present reasons why sometimes modern wireless IoT or automation solutions might not be the most viable solution. An alternative approach has been presented that is viable for a lot of potential applications.

Acknowledgements The support of the Specific Research Project at FIM UHK is gratefully acknowledged.

References

1. Astudillo-Salinas, F., Barrera-Salamea, D., Vázquez-Rodas, A., Solano-Quinde, L.: Minimizing the power consumption in raspberry pi to use as a remote wsn gateway. In: 2016 8th IEEE Latin-American Conference on Communications (LATINCOM) (2016)
2. Bradle, J., Mesicek, J., Krejcar, O., Selamat, A., Kuca, K.: Arduino as a control unit for the system of laser diodes. In: International Conference on Industrial, Engineering and Other Applications of Applied Intelligent Systems, pp. 569–575. Springer (2017)
3. Hodges, S., Taylor, S., Villar, N., Scott, J., Bial, D., Fischer, P.T.: Prototyping connected devices for the internet of things. Computer **46**(2), 26–34 (2013)
4. Horalek, J., Matyska, J., Stepan, J., Vancl, M., Cimler, R., Sobeslav, V.: Lower layers of a cloud driven smart home system. In: New Trends in Intelligent Information and Database Systems, pp. 219–228 (2015)
5. Kranz, M., Holleis, P., Schmidt, A.: Embedded interaction: interacting with the internet of things. IEEE Internet Comput. **14**(2), 46–53 (2010)
6. Madakam, S., Ramaswamy, R., Tripathi, S.: Internet of things (IoT): a literature review. J. Comput. Commun. **3**(05), 164 (2015)
7. Maksimovic, M., Vujovic, V., Davidovic, N., Milosevic, V., Perisic, B.: Raspberry pi as internet of things hardware: performances and constraints. In: Proceedings of IcETRAN: First International Conference on Electrical, Electronic and Computing Engineering, pp. 1–6 (2014)
8. Pscheidl, P., Cimler, R., Tomášková, H.: Towards device interoperability in an heterogeneous internet of things environment. In: Conference on Computational Collective Intelligence Technologies and Applications, pp. 315–324. Springer (2017)
9. Stepan, J., Cimler, R., Matyska, J., Sec, D., Krejcar, O.: Lightweight protocol for M2M communication. In: Computational Collective Intelligence, pp. 335–344 (2017)
10. Stepan, J., Matyska, J., Cimler, R., Horalek, J.: Low level communication protocol and hardware for wired sensor networks. J. Telecommun. Electron. Comput. Eng. **9**(2–4) (2017)

11. Studnička, F., Šeba, P., Jezbera, D., Kříž, J.: Continuous monitoring of heart rate using accelerometric sensors. In: 2012 35th International Conference on Telecommunications and Signal Processing (TSP), pp. 559–561. IEEE (2012)
12. Vilajosana, X., Tuset, P., Watteyne, T., Pister, K.: Openmote: Open-source prototyping platform for the industrial IoT. In: Ad Hoc Networks, pp. 211–222 (2015)
13. Vujovic, V., Maksimovic, M.: Raspberry pi as a wireless sensor node: Performances and constraints. In: 2014 37th International Convention on Information and Communication Technology, Electronics and Microelectronics (MIPRO) (2014)
14. Yun, J., Ahn, I.Y., Sung, N.M., Kim, J.: A device software platform for consumer electronics based on the internet of things. IEEE Trans. Consum. Electron. **61**(4), 564–571 (2015)

Part VII
Tools and Techniques for Intelligent Information Systems

OpenWebCrypt—Securing Our Data in Public Cloud

Péter Vörös and Attila Kiss

Abstract Privacy is the ability of an individual or group to seclude themselves, or information about themselves, and thereby express themselves selectively. The increasing usage and popularity of web services indicate the likelihood of privacy disclosures. Users generate a high amount of data by using online services, such as social networks, mailing software, even calendars. These data can be sold to marketers, because of their increasing reliance on customer data to create a model of potential customers. While accessing the web services, users unknowingly agree to the privacy policy of the service provider through which they authorize the service providers to collect and share their personally identifiable information. In this paper, we aim to introduce a client-side open-source data securing system.

Keywords Privacy · Web service security · Public cloud · Personal data management · Data breach

1 Introduction

Now when we use smartphones on daily basis, and we're all the time connected to the Internet, it seems to be impossible to avoid companies to collect data from us, however, it seems to be a really big issue to most of us. It's really shocking when one takes a look at his own personal data collected by a service provider. There are plenty of proposals and solutions for keeping our data private, but unfortunately, until now nothing has become widely accepted and used. We believe that the main reason for that is the complexity in the usage or the users' lack of knowledge in the security area. In this paper our goal is to give an easy to use client-side method to make one's data secure in public clouds. In the following of this in Sect. 2 we summarize the work

P. Vörös (✉) · A. Kiss
Eötvös Loránd University, Budapest, Hungary
e-mail: vopraai@inf.elte.hu

A. Kiss
e-mail: kiss@inf.elte.hu

© Springer International Publishing AG, part of Springer Nature 2018
A. Sieminski et al. (eds.), *Modern Approaches for Intelligent Information and Database Systems*, Studies in Computational Intelligence 769,
https://doi.org/10.1007/978-3-319-76081-0_41

479

previously done by others and us in the area of privacy in public services. In Sect. 3 we collected several examples on how our data is exposed and should be protected against different threats. There we also define several terms and state the basics of our model. Then in Sect. 4 we offer methods and solutions for individual users, as well as corporations. Later in Sect. 5 we present the details of our implemented browser add-on designed for end-users. We show our measurements how the extra encryption layer affects the user experience. Lastly in Sect. 6 we conclude the results of the paper, and in Sect. 7 one can see the details of the future plans to continue this project.

2 Related Work

Privacy issues in public clouds have had a huge attention throughout the past years. There are countless surveys and papers about the topic. Most of them state that the main problem is that the providers are not transparent enough, lacking the capabilities for the tracking and auditing the access history of both the physical and virtual servers. Cloud Security Alliance collects the biggest threats of cloud computing from year to year. In the latest (2016) list [1], they collected the 12 highest priority issues. These are the following:

1. Data Breaches
2. Weak Identity Credential and Access Management
3. Insecure APIs
4. System and Application Vulnerabilities
5. Account Hijacking
6. Malicious Insiders
7. Advanced Persistent Threats (APTs)
8. Data Loss
9. Insufficient Due Diligence
10. Abuse and Nefarious Use of Cloud Services
11. Denial of Service (DoS)
12. Shared Technology Issues

In our previous work, we analyzed both the Account Hijacking [2], and DoS/DDoS problem [3, 4]. These fields still require a lot of improvements, but now in this paper we set the focus on the highest threat Data Breaches.

Definition 1 A Data Breach is an incident in which sensitive, protected or confidential information is released, viewed, stolen or used by an individual who is not authorized to do so.

Any data that is stored unencrypted on the servers are potential targets of such incidents, therefore some extra security seems to be a necessary idea. There are numerous papers aiming to summarize the potential cloud threats. In [5–8] they deeply analyze the different security issues, and possible solutions. In [9] they offer a framework for cloud services that satisfies most of the privacy and stability needs. The idea of building our services above a trusted framework is really nice, however, unfortunately it is not widely used. One reason can be the need to rewrite existing applications which is a huge work, so we believe we need some user-side protection instead. Users do not feel safe when they use cloud services, because of the lack of transparency in the background. In a previous survey by Fujitsu Research Institute [10], they measured that a huge number (above 85%) of users were worried about who has access to their data without their permission. In [11] Y. Sun et. al. suggest homomorphic encryption to keep data confidentiality. Homomorphic encryption ensures that the algebraic operation results are consistent both with the clear-text, and the encrypted. This method works really well if we have small chunks of data, and we want to run operations on them. When we have a huge amount of data, this encryption takes too much time, therefore it is not really usable. PaaSword [12] aims to fortify the trust of individuals and corporations in Cloud-enabled services and applications. They offer a Platform-as-a-Service framework with the focus on secure storage of sensitive data. This sounds good, but it may not increase the users' sense of security.

3 Background

As mentioned before, in this paper our aim is data breach, and potential defense against this threat.

Users can rightly fear of unauthorized access both from other users using some exploits, or the service provider itself. In our examples, we will go through two different types of user groups. The first is an individual user and the other is corporation. For our example we will use Google calendar as the target service provider, but others act the same. On Fig. 1 we illustrate the typical server-client HTTPS communication. The client sends encrypted data through the channel, the other part decrypts it and stores/processes the data. In the following we will go through the three parts.

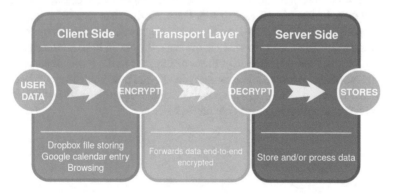

Fig. 1 Client server architecture

3.1 Client Side

The first one is the client side. In our paper, we assume that the client has no vulnerabilities, stores its data and credentials securely so the data cannot be compromised at the client side.

3.2 Transport Layer

We will use network services as the providers meant to, so by default HTTPS is used in browsers.

Definition 2 HTTPS also called HTTP over Transport Layer Security (TLS) is a communications protocol for secure communication over a computer network which is widely used on the Internet. HTTPS provides bidirectional encryption of communications between a client and server, which protects against eavesdropping and tampering with or forging the contents of the communication.

We will not go into HTTPS security issues, we just assume that the transport layer is safe. We rather keep the focus on the stored data in service servers.

3.3 Server Side

After packets arrive at the destination and HTTPS gets decrypted, it processes and stores the data. The servers usually offer some encryption for storage, so for 3rd person, acquiring personal data seems difficult or impossible without knowing the key for the decryption. However there are cases when user sessions are hijacked, and if attackers somehow can pretend to be valid logged in user they can see the decrypted data, just as a legal user would. However this case cannot be ignored, we believe the bigger problem is that nothing stops the provider to access our data. Not just nothing keeps them away to analyze our data, they even have all the rights to do so. Users by using the service accept these that appear in the terms of service: "We collect information about the services that you use and how you use them, like when you watch a video on YouTube, visit a website that uses our advertising services, or view and interact with our ads and content" [13].

3.4 Encryption

As mentioned above, to keep our data confidentiality we need to send/store it pre-encrypted on the server. There are several mentionable encryption mechanisms, like simmetric-, assimetric-, homomorphic encryptions. The choice highly depends on the exact service. In our case we want to keep our solution fast, and simple. So in our prototype, we used a common symmetric encryption AES-256 (Advanced Encryption Standard with 256 bit key length).

Definition 3 Symmetric-key algorithms are algorithms for cryptography that use the same cryptographic keys for both encryptions of plain-text and decryption of ciphertext. The keys, in practice, represent a shared secret between two or more parties that can be used to maintain a private information link.

4 Possible Solutions

Our aim is to give a lightweight client-side method to make our data secure, unanalyzable, unusable by service providers. We differentiate between two different user groups: individuals and corporate users. Usually, the second group deals with more confidential data, but we did not create our groups by different levels of security, instead the place where our extra securing layer fits in.

4.1 Encryption Proxy Server

Corporations usually have very strict rules and try to limit user errors as much as possible. Therefore it is a common characteristic that they offer pre-installed systems fully over-watched by the administrators, revoking admin rights from employees. Corporations shall provide security to their users' data by providing some sort of securing service. It is still important to have an easy-to-use solution, but we expect network administrators to have the required skills to deal with operating an extra proxy server. In our solution encryption proxy server targets these corporations or bigger local networks with the desired amount of trust in their organization.

In this case, we assume that the employees have the trust in their company. This trust is required because in our solution proxy server is able to see unencrypted requests, and this information in the bad hands can be used for various things. In a few words, the proxy server decrypts the sent HTTPS data, encrypts special parts of the payload with a corporate key, and then encrypts it back to HTTPS with its own key, afterwards sends it towards to the service provider via HTTPS.

As one can see on Fig. 2, the internal corporation network has several clients, all of which has its own encryption keys (these are color-coded; Client1: green, Client2: blue, ProxyServer: red). When they communicate with either internal or external servers they use their keys for encryption/decryption. Blank file means the unencrypted request, while color-coded files stand for encrypted/decryptable requests with the specific key.

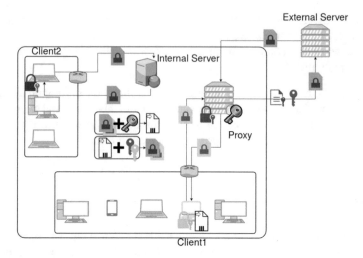

Fig. 2 How data encryption works using a proxy server

Internal servers Internal requests (Client2) are handled without reaching the proxy server because the requests do not go through the boundaries of the protected network so there is no need for extra encryption. A typical request-response looks like this:

1. Client2 sends a request to an internal server
2. Client2 encrypts the whole data with its blue key
3. The request is NOT routed through Proxy server but goes straight to the service provider
4. The service provider processes the request
5. The service provider sends back the encrypted answer to Client2
6. Client2 decrypts the message with its key
7. Client2 has the unencrypted message

External servers We differentiate between two types of external servers: those that require extra encryption, and those that do not. The second group is handled the same way as in the case of the internal servers. Outgoing requests using special services (eg.: Google calendar) are forwarded through an internal proxy server. This server has its own (red) encryption key but also has a master key (gold) that is able to decrypt all the internal requests. How a typical conversation looks like:

1. Client1 sends a request to an extra level of security service
2. Client1 encrypts the whole data with its green key
3. The request is routed through Proxy server
4. Proxy server uses the master key to decrypt the whole request
5. Proxy server uses the corporate encryption key for encrypting the request's payload only (eg.: content of a calendar event)
6. Proxy server encrypts the request with its red key
7. Proxy server sends the request to the service provider

8. The service provider processes the request
9. The service provider sends back the encrypted answer to the proxy server
10. Proxy server decrypts the message with its red key
11. Proxy server decrypts the payload with the corporate key
12. Proxy server encrypts the message with Client1's (green) key
13. Client1 decrypts the message with its key
14. Client1 has the unencrypted message

It can clearly be seen, that clients do not notice the extra layer of security, therefore they do not need any extra configuration, which makes it a perfect solution.

4.2   Browser Extension

Individual users have full control over their machines, they can install software, and browser extensions freely. It is crucial to make our solution easily installable, since we cannot assume any IT pre-knowledge from these users.

As it can be seen on Fig. 3, the encryption and decryption do not necessarily need to be propagated to a dedicated server, but the browser can do it in the background. An example service usage looks like this:

1. Client sends a request to an extra level of security service
2. The client's browser encrypts the payload of the data with a user password or key
3. The client's browser also encrypts the whole data by it's key
4. Client sends the request
5. The service provider processes the request
6. The service provider sends back the encrypted answer to the client
7. The client's browser decrypts the whole message with its key
8. The client's browser decrypts the payload with a user password or key
9. Client has the unencrypted message

Fig. 3 How data encryption works using a browser add-on

5   Prototype

First we would like to state that we plan to make all our solutions, programs available and open-source. As a first step, we implemented a browser extension script which acts like the previously described model for individual users. While browsing performance is crucial, so we also made some measurements about the overhead of running this script. We used Greasemonkey add-on because it is an easy script runner browser add-on that has its own equivalent for all the big browser brands (Tampermonkey for Chrome, Edge for Safari and Violentmonkey for Opera).

After installing the script it starts working invisibly in the background. On Fig. 5 one can see that adding a new entry can be done the same way as without using our script. If we have created events in our calendar previously then those appear as they are. They are not automatically encrypted, but every new event we create will be stored with the proper AES encryption. On Fig. 4a one can see how the encrypted calendar looks like if we choose no to decrypt it. That is what is stored on the servers. And on Fig. 4b this is what the user can see if he uses the correct key for decryption. We don't want to go into details of cracking the AES but state only that it is impossible (or would take way too much time with brute-force methods) to get any data out of the encrypted entries.

We made it easy to export and import the encryption key, so if one wants to use several different machines with multiple browsers it is trivial how to do so.

(a)   **(b)**

Encrypted   Decrypted

Fig. 4   Google calendar before and after decryption

(a)   **(b)**

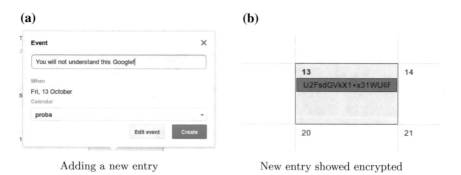

Adding a new entry   New entry showed encrypted

Fig. 5   Adding a new entry that is automatically encrypted on client side, and shown in the calendar without decryption

(a)

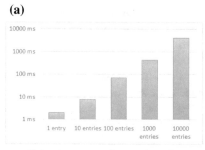

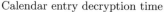

(b)

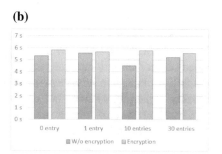

Calendar entry decryption time   Full website loading time

Fig. 6 Encryption measurements

5.1 Measurements

Encryption time does not affect the user experience, because it only happens when one changes an entry or creates a new one. Decryption however is different. We can have several events in a weekly or monthly view so we both calculated the decryption time for one element, hundreds, and thousands of elements, and the overall load time both with and without using our script. For the measurements we used Firefox's development tools: the JS console for encryption time, and the network analyzer for the full page loading.

Test machine For the measurements we used a bussiness laptop Dell Latitude E5530. 4 GB DDR3 RAM, Intel(R) Core(TM) i3-2328 M CPU @ 2.20 GHz 4 cores. We used Mozilla Firefox version 56.0 installed on Ubuntu 16.04.3 LTS.

One can see on Fig. 6a that if we increase the number of entries, the computation time increases linearly, which can result in serious performance loss. Note the logarithmic scales. If we have 10000 entries, the decryption time is just above 4 s, but keep in mind that our target is a calendar which likely does not have more than tens or maybe a hundred entries. However, it can clearly be seen on Fig. 6b that the overall page loading time is only slightly affected by the number of entries. It is clearly visible that the loading time stays around the same no matter how many entries we have, and there is around 0.5 s overhead when we use the extra encryption. This overhead can be explained by the fact that we need to include new libraries to be able to run the encryptions/decryptions, and these scripts also need to be downloaded before.

6 Conclusion

To summarize, we deeply analyzed one of the greatest privacy issues in cloud services, which is the data breach. We offered a client-side encryption mechanism that is able to handle confidential information safely through the network and on the target service providers as well. We differentiated between two user groups: individuals and corporations, that have diverse needs and different IT knowledge. For the corporations, we designed an encryption proxy server solution which is a dedicated server handling outgoing traffic. Individuals who do not have the possibility and skill to maintain a server can install a browser add-on that provides similar functionality as the proxy server, but without the need of extra configuration. As a next step, we showed our results in our prototype browser add-on, we demonstrated how requests are encrypted, and what information the service provider can get from our data. Finally we made some measurements to show that using this browser add-on does not make browsing experience worse, because the extra latency introduced with the encryption can be neglected.

7   Future Work

We need to make some stability improvements in the existing code, then we will publish it to the public open-source. Using Google's provided API also gives us the opportunity to make our Android/iOS app able to handle encrypted calendar entries, which will result in a fully enjoyable encrypted Google Calendar that is ready to use on every device. We are now also working on the proxy server solution, that application also will be released in the near future, also open-source. Our following next big target is going to be Google docs. With its over 10 million users we believe that it can also be a very widely used application. We both plan to support full document encryption, as well as only parts of the document.

Acknowledgment This work was supported by EFOP-3.6.3-VEKOP-16. Authors thank Ericsson Ltd. for support via the ELTE CNL collaboration.

References

1. The treacherous twelve cloud—computing top threats in 2016 (2016). [Online]. https://cloudsecurityalliance.org/group/top-threats
2. Vörös, P., Kiss, A.: Tookie: a new way to secure sessions. In: Recent Developments in Intelligent Information and Database Systems, pp. 195–207. Springer (2016)
3. Csubák, D.., Szücs, K., Vörös, P., Kiss, A.: Big data testbed for network attack detection. Acta Polytechnica Hungarica, vol. 13, no. 2 (2016)
4. Vörös, P., Laki, S., Kiss, A.: Distributed firewall on dataplane against DDOS attack. In: 4th Winter School of Ph.D. Students in Informatics and Mathematics, p. 37
5. Ko, R.K., Lee, B.S., Pearson, S.: Towards achieving accountability, auditability and trust in cloud computing. Adv. Comput. Commun. 432–444 (2011)
6. Ren, K., Wang, C., Wang, Q.: Security challenges for the public cloud. IEEE Internet Comput. **16**(1), 69–73 (2012)
7. Kalloniatis, C., Mouratidis, H., Vassilis, M., Islam, S., Gritzalis, S., Kavakli, E.: Towards the design of secure and privacy-oriented information systems in the cloud: identifying the major concepts. Comput. Stand. Interfaces **36**(4), 759–775 (2014)

8. Sun, D., Chang, G., Sun, L., Wang, X.: Surveying and analyzing security, privacy and trust issues in cloud computing environments. Procedia Eng. **15**, 2852–2856 (2011)
9. Ko, R.K., Jagadpramana, P., Mowbray, M., Pearson, S., Kirchberg, M., Liang, Q., Lee, B.S.: Trustcloud: a framework for accountability and trust in cloud computing. In: 2011 IEEE World Congress on Services (SERVICES), pp. 584–588. IEEE (2011)
10. Fujitsu research institute: personal data in the cloud: a global survey of consumer attitudes (2010). [Online]. http://www.fujitsu.com/downloads/SOL/fai/reports/fujitsu_personal-data-in-the-cloud.pdf
11. Sun, Y., Zhang, J., Xiong, Y., Zhu, G.: Data security and privacy in cloud computing. Int. J. Distrib. Sens. Netw. **10**(7), 190903 (2014)
12. Verginadis, Y., Michalas, A., Gouvas, P., Schiefer, G., Hübsch, G., Paraskakis, I.: Paasword: a holistic data privacy and security by design framework for cloud services. J. Grid Comput. 1–16 (2017)
13. Google privacy policy (2017). [Online]. https://www.google.com/policies/privacy/

A Novel Load Forecasting System Leveraging Database Technology

Chee Keong Wee and Richi Nayak

Abstract Load forecasting has been a key process in electricity utility companies. While there are demands for utilising data mining to meet the requirements of load forecasting, there are substantial challenges in implementing a big data solution. Cost, expertise and new acquisitions are only some of the reasons that hinder this endeavour. The goal of this paper is to propose an interim load forecasting solution to meet the challenge of using big data, data mining, existing hardware and resource expertise while minimizing the cost and overheads.

1 Introduction

All electricity utility companies perform electricity load forecasting to help in their planning and operations, however, they face many challenges. While the utility companies have to manage significant amounts of time series data that have been generated from the monitoring and control systems such as Supervisory Control And Data Acquisition (SCADA) and storing them in network file systems [1, 2], they also need to utilize complex forecasting algorithms including but not limited to ARIMA, ARIMAX, Holt-Winters' exponential smoothing method and Neural network [1].

Statisticians and forecasting engineers usually use Network File System (NFS) for storing and retrieving the load data and use powerful workstations for forecasting processing. Their concern is on load modelling and future prediction. They have a different domain of expertise and may not be skilled to utilize the IT resources available at the corporate infrastructure. Alternatively, for load forecasting, the big data technology with expensive setup, training, and big data

C. K. Wee (✉) · R. Nayak
Science and Engineering Faculty, Queensland University of Technology,
Brisbane, QLD, Australia
e-mail: chee.wee@qut.edu.au

R. Nayak
e-mail: r.nayak@qut.edu.au

© Springer International Publishing AG, part of Springer Nature 2018
A. Sieminski et al. (eds.), *Modern Approaches for Intelligent Information and Database Systems*, Studies in Computational Intelligence 769,
https://doi.org/10.1007/978-3-319-76081-0_42

specialists can be employed. The cost will be huge and turnaround will be substantial to get the proposed new big data set-up to work. Both possible solutions sit on the ends of the spectrum of forecasting functionality and IT investment. Companies may not have the finance or expertise to implement such a large-scale project and the risk of project failure may be high. This paper presents an alternate route for power utility company to venture into data analytics without going down the expensive big data setup. We recommend using existing database infrastructure to enhance load forecasting capabilities. We present a solution integrating the forecasting models that are usually run in R into the database framework and parallelizing it to increase the forecasting throughput. We present the tests and benchmarks in which the forecasting models are run against various substations data and assess their accuracy and the speed of parallelism. Results are promising.

2 Technology Consideration

One of the challenge is to harmonize the domain knowledge of electricity utility industry, data mining forecasting and inhouse IT resources. Each area has their intricate set of theories and procedures that require substantial amount of time and resources to comprehend and develop the systems. Oracle data mining have features in forecasting but requires in-depth understanding of PLSQL command language. It also requires huge investment on licensing, training and consultant engagement. Beside this, Oracle data mining's inbuilt forecasting functions are scoped to its interpretation. In order to change or alter the algorithm to suit more complex requirement, additional coding effort is required [2, 3]. The Oracle R Enterprise (ORE) provides a platform to run R data mining models and parallelise it to scaling up the performance. It requires a file system to store the load data of all the substations. There are several approaches such as use a Network File System (NFS), load the data into the Oracle database or use a Big data solution like Hadoop HDFS [2]. The filesystem option like NFS can meet the need. However, it has to be actively mounted at the failover node all the time, waiting to serve [2]. Oracle data mining suite can support forecasting functionality but it requires an in-depth understanding of PLSQL command language to develop it [2, 3]. It allows storing a large volume of load data in relational databases. Existing metering systems still treat the load data as relational, so tables tend to have over tens of billions of rows in a single table with no partitioning. The existing method is to store the raw load data on the NFS but that occupies too much disk spaces in their raw form. Moreover, storing time-series data in relational tables can be inefficient as it incurs the significant overhead of disk storage and CPU compute to read and compute millions of rows.

R has been proven to be a good and popular statistical language. There is a lot of research made in electricity demand forecasting using R and numerous open sourced libraries exist. However, it has several drawbacks like the support of single core processing, and requirement of loading dataset into memory. It can use multi-core only with add-on packages. Memory availability caps on the volume of data that it can process. So, it is not ideal to handle thousands of substations' load data that are more than hundreds of petabytes each on a desktop [4]. In this paper, we envisage an efficient solution with the existing technology and overcome the shortcomings. We propose to deliver meaningful data-driven decisions while being conscious about cost and limited resources.

3   The Proposed Load Forecasting Solution

We propose a feasible solution that can allow companies to use existing IT resources for electricity load forecasting with minimal investment in hardware and other resources like manpower and licenses. It provides businesses to harness the potential of predictive analytics with minimum risk. It also gives the in-house IT personnel a platform to build upon their knowledge, to transit to the domain of big data and machine learning slowly. The solution leverages existing Oracle database capabilities which are prevalent across existing power utility companies. Oracle R Enterprise (ORE) is a framework of R and database libraries with SQL extension to combine capability of R into Oracle database [3, 5]. Results from the predictive models running with R programs will be stored in the database and, that in turn, can be readily consumed by any reporting applications. Figure 1 illustrates the various stages and process flow of the proposed load forecasting system.

3.1   Proposed System Architecture and Implementation

There are three major sub-systems; the first part is the analytic portion that runs with the Oracle database. The second part is the proposed new Oracle-native file system

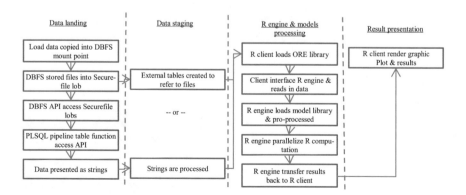

Fig. 1 Proposed system's process flow chart

that will be used to hold the data, and the third part is the high availability setup that comprises of Oracle DataGuard which specify the setup of two servers each with its ownstorage, acting as Primary/hot-standby cluster configuration. For the analytic subsystem, the users' client machines capture the R command/scripts inside the R console and converts them into SQL equivalent before shipping them to the backend database R engine for processing. Upon completion of the code execution, the R server engine will transfer the forecasting results back to the client's R console [6].

For the storage sub-systems, according to the database best practices, Oracle's Automatic Storage Management (ASM) is Oracle disk logical volume manager and is used for this setup as it offers better storage performance than Unix filesystem [2]. Oracle will store the input load data internally as Securefile lob and applies both encryption and compression [7]. For the application side, the R program accesses the data through the database components which in turn extract the data out from the DBFS objects. We propose to use DBFS in conjunction with ORE and Oracle database with physical standby setup. It provides fast performance and a secured file system that runs on top of Linux operating system and stores the data files internally as Securefile lob with compression and encryption [7]. ORE gives a scalable platform to run R programs in the database engine with higher parallelism while DBFS provides the highly available filesystem to store the substation load data as shown in Fig. 2.

In the high availability setup, the primary node that runs the Oracle instance presents the DBFS to the public. The users upload the substation load data to the DBFS with Unix command and store the files internally into its ASM. The primary Oracle instance propagate the updates to the standby node. This is done in near real-time mode through redo transport as shown in Fig. 3. From the client end, Transparent Application Failover (TAF) provides a persistent session to the Oracle instance should there be a switchover/failover so that users can keep their client software connected, although those transactions that have yet to be committed will be rollback [8]. Oracle will store these load data internally as a Securefile lob, applying both encryption and compression for better security and space saving [7].

The forecasting R program runs internally in the database and accesses the substations data stored in the database via two means: either by the external table normally as filesystem access or through DBFS API. The R forecasting program

Fig. 2 Forecasting In ORE with DBFS or DBFS API

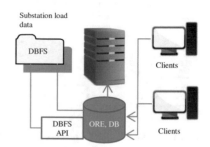

Fig. 3 Architecture of oracle
database, ORE and DBFS

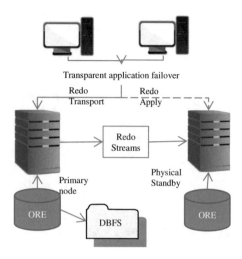

reads in the data, perform pre-processing and run through the forecasting libraries
for generating the forecasting models. Results of these models are consumed by
clients via reporting or visualizing software. One strong advantage of this setup is
the backup/recovery feature. All the substation data that reside in DBFS are stored
into the database's internal Securefile lobs which in term are backed up by enterprise backup system through Oracle's Recovery Manager (RMAN) [8]. This gives
our proposed system very high resilience in business continuity and recoverability.

4 Forecasting Models: Testing and Results

After the ORE has been setup and tested, we run forecasting models to evaluate the
ORE functionality and handling compared to the standalone R client. Models such
as Auto-regressive Integrated Moving average (ARIMA), Holt-winters' exponential
smoothing method and neural network have been popularly used to process complex time series data [9]. Inspired by their success we implemented these three
methods to test the proposed forecasting solution. The dataset used in this paper
belongs to a leading Australian electricity utility operator. It comprises of 237
substations loading data, with each station having around 10 years of power load
data which is captured at 30 min interval. For each station, there are 17520 entries
in a yearly load data. It is a total of 1.49 GB in size. Each data file has 4 unique
columns: date, time, Megawatts (MW) and Megavolts AMP (MVA).

Figure 4 shows a typical 2 years substation data plot with each data point
occurring at 30 min interval. A further decomposition of the substation's data
revealed multi seasonality, from the year down to the individual weeks as shown in
Fig. 5. The load demand is heavily influenced by external factors like weather
temperatures, humidity, work days versus weekends, and seasons including holidays.

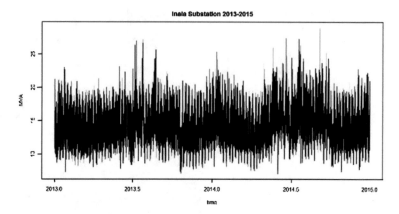

Fig. 4 Plot of the Inala substation's 2 years of data

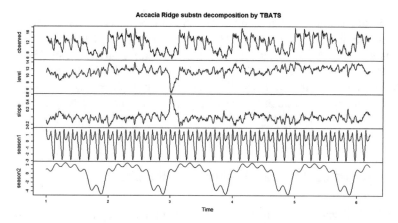

Fig. 5 Decomposition of a typical substation data

Economic condition played a part in determining the trend too. For this paper, we will focus on the consideration of the data usage as a univariate in the forecasting test. We run several models with the data and compare the residuals.

4.1 ARIMA Model Forecasting

The ARIMA model [10] running in ORE has some pre-requisite testing to determine if the load time series is stationary or not and what measure should be used to adjust to make it more stationary. The first test is to subject the data to auto-correlation function (ACF) and partial autocorrelation function (PACF). From Fig. 5, it is visible that the data is highly correlated and non-stationary. We run the

Table 1 ARIMA's parameters combination and results

ARIMA	ME	RMSE	MAE	MPE	MAPE	MASE
c(1, 0, 0)	4.06E-05	0.727	0.562	−0.257	3.951	0.997
c(2, 0, 0)	1.72E-05	0.67	0.498	−0.221	3.49	0.883
c(0, 0, 1)	4.49E-06	1.795	1.451	−2.755	10.69	2.572
c(1, 0, 1)	2.62E-05	0.681	0.513	−0.228	3.597	0.909
c(1, 1, 0)	2.54E-05	0.68	0.504	−0.044	3.522	0.894
c(0, 1, 1)	3.36E-05	0.688	0.516	−0.075	3.611	0.914
c(1, 1, 1)	1.17E-05	0.673	0.498	0.009	3.472	0.883
c(1, 1, 3)	1.23E-05	0.668	0.491	0.011	3.442	0.875
c(2, 1, 3)	1.20E-05	0.668	0.498	0.012	3.442	0.875

model in ORE with different combinations of p, d, q values and results are tabulated in Table 1 where p specifies the time lag of the AR model, d is the integrated part which specifies the amount of differencing, and q set the order of the MA model. From ACF, the estimate of p value for the Auto-regressive (AR) function is > 1 whereas the d value for the differencing should be 1 [9, 11–13].

Models are evaluated using the accuracy measurements such as ME (Mean Error), RMSE (Root Mean Square Error), MAE (Mean Average Error), MPE (Mean Percentage Error), MAPE and MASE. The result showed that ARIMA models with a differencing factor, I, of 1 gives better accuracy and so is the AR's p value. The MA's q value has some positive effect on the ARIMA's residuals too. The combinations of (1, 1, 1), (1, 1, 3) and (2, 1, 3) produced the best results whereas the parameters combination of (1, 0, 0) only activates the AR model which yields 0.99. For the values of (0, 0, 1), it yields a high MASE value. The data is stationary, so having just (1, 0, 1) without differencing, the model's residual isn't too far off as compared to the combo of (1, 1, 1).

4.2 Forecasting with the Holt-Winters' ETS Models

The Holt-Winters' Exponential Trend Smoothing (ETS) model [10] comprises of three smoothing equations: overall smoothing, trend smoothing, and seasonal smoothing. A quick decomposition using the ORE infrastructure showed that the time series have multiple seasons in it.

The data is run with the ETS model and the residual is recorded. Figure 6 showed the forecast of the substation using ETS model. A quick comparison between the results from ETS and ARIMA models showed that the ETS scored better in both MASE and MAPE (Fig. 7). It indicates higher accuracy obtained by ETS as shown in Table 2.

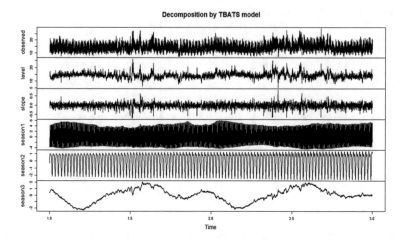

Fig. 6 Decomposition of Inala Substation by TBATS

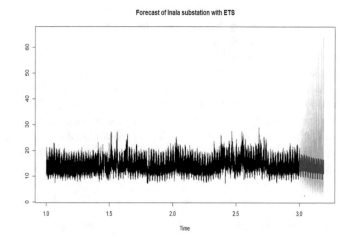

Fig. 7 ETS forecast of substation

Table 2 ETS results compared with ARIMA

	ME	RMSE	MAE	MPE	MAPE	MASE
ETS	0.0103	0.602	0.44	−0.08	3.073	0.78
c(1, 1, 1)	1.17E-05	0.673	0.498	0.009	3.472	0.883
c(1, 1, 3)	1.22E-05	0.668	0.493	0.011	3.442	0.875
c(2, 1, 3)	1.21E-05	0.668	0.493	0.012	3.442	0.875

Table 3 Neural network results with p, k combo

NNAR (p, k)	ME	RMSE	MAE	MPE	MAPE	MASE
(1, 5)	−3.50E-05	0.727	0.563	−0.26	3.956	0.998
(1, 10)	−7.42E-04	0.727	0.563	−0.264	3.956	0.998
(1, 20)	7.21E-06	0.744	0.585	−0.506	4.151	1.037
(1, 30)	−4.82E-04	0.727	0.563	−0.262	3.956	0.998
(2, 5)	−1.51E-05	0.666	0.495	−0.215	3.463	0.877
(2, 10)	−2.13E-04	0.667	0.495	−0.217	3.467	0.878
(2, 20)	3.13E-04	0.667	0.495	−0.211	3.466	0.878
(2, 30)	−2.90E-04	0.663	0.492	−0.215	3.444	0.872

4.3   Forecasting with Neural Network Auto-Regression

Several solutions can be obtained for the feed-forward neural network model by setting variations in the lagged values, p, of the time series data and the number of neurons or nodes, k [9, 12]. The weights are assigned at random before the training starts and it is trained for one-step forecasting. Additional step or forecast period are achieved by running the network iteratively [9, 12]. Neural network modelling treats the data as non-seasonal data which is similar to the Auto-regressive model of ARIMA $(p, 0, 0)$ [10]. We will like to note here that since this is a test of ORE functionality. The focus is not on getting the best model. From the result in Table 3, it can be noted that the higher number of neurons present in the NN, the greater the accuracy of the model will be, and this highlights the complex nature of data. This is not overfitting as the results are reported on unseen test data.

The NN result is compared against the rest of the models' residuals in Table 4. NN yields the best results among the 3 forecasting models.

4.4   Effect of Parallelism (i.e. Multi-core Capability) in Forecasting

The objective of this test is to see the difference in the throughput of model generations between standalone machines versus the multi-core aware ORE server.

Table 4 Results from ARIMA, ETS and NN

	ME	RMSE	MAE	MPE	MAPE	MASE
NN	4.87E-05	0.52	0.375	−0.149	2.649	0.666
ETS	0.01055	0.602	0.44	−0.085	3.073	0.78
c(1, 1, 1)	1.16E-05	0.673	0.498	0.009	3.472	0.883
c(1, 1, 3)	1.22E-05	0.668	0.493	0.011	3.442	0.875
c(2, 1, 3)	1.28E-05	0.668	0.493	0.012	3.442	0.875

Table 5 Performance: desktop with 2 CPU versus multi-core Linux virtual erver

Machine environment	Number of ubstations				
	1 (s)	5 (s)	10 (s)	20 (s)	40 (s)
Desktop	159	1196	2249	4346	9252
No parallelism	145	1092	2053	3967	8446
Parallelism of 4	135	1016	1911	3692	7860
Parallelism of 6	109	820	1542	2980	6345
Parallelism of 8	72	548	1031	1993	4243

Two different hardware platforms are used: desktop with 2 CPU vs multi-core Linux virtual server. All the input files are placed on a shared NFS system. Each test requires the forecast program to run against five sets of test data each with different groups of substations' loads. A forecasting program runs through the set and the time taken for each set is recorded. The test starts with running the models on the desktop to get a baseline performance. The subsequent test is conducted on the ORE server and each test is executed with a different level of parallelism. The R program was set to run with a parallelism of 4, 6 and 8 nodes. As shown in Table 5, results show improvement in the speed with the use of multi-core. The throughput and speed are improved with each set of parallelism set at a higher level.

5   Test Result of DBFS Technology

Followings are the test results and outcomes that have been conducted on the big data setup on a virtual machine that runs on a multi-core desktop. Database filesystem offers the compression functionality that can help to save storage space [7]. For this test, the DBFS feature was configured and mounted on the database. For the trial sample data load, the entire set of the 237 substations' 10 years load data was used. 1.32 GB was obtained from 1.49 GB as training data (with.17 GB of test data) was copied into the DBFS and the objects sizes are queried. DBFS has been set on with compression for all operations, it managed to compress down to 413 Mb. For the DBFS data access test, the substation data is stored in the Oracle DBFS. There are two ways for the R engine inside Oracle database to see the data, one is by the external table and the other by the DBFS API. For external tables, it uses the Oracle directory and external table feature. The advantage of using external table is in its ease of setup. The downside is that it is rather static so if there are more data files that need to be read from the DBFS, then the external tables must be dropped and recreated. Result 1 showed the result retrieved from this option. Note that the data presented are well structured in accordance to create SQL definition and only minor work is required on datatype conversion [2, 7].

Result 1—Retrieved data via external table option				Result 2—Retrieved data via DBFS API option	
VDate	VTime	MW	MVA	Subdata	(Value)
21-Feb-15	2:00:00 PM	6.443479418	6.832417881	19-Nov-14, 9:00:00 AM	17.22305528, 17.9622836
21-Feb-15	2:30:00 PM	6.289191789	6.720857092	19-Nov-14, 9:30:00 AM	16.45388919, 17.39746547
21-Feb-15	3:00:00 PM	6.198203091	6.593361931	19-Nov-14, 10:00:00 AM	18.51022208, 19.538705
21-Feb-15	3:30:00 PM	6.427889225	6.836900636	19-Nov-14, 10:30:00 AM	17.32722239, 18.2138756
21-Feb-15	4:00:00 PM	6.416818441	6.822702785	19-Nov-14, 11:00:00 AM	17.24533334, 18.16291614
21-Feb-15	4:30:00 PM	6.54946575	6.973977097	19-Nov-14, 11:30:00 AM	17.17922235, 18.10874804
21-Feb-15	5:00:00 PM	6.33542853	6.726185153	19-Nov-14, 12:00:00 PM	16.88977766, 17.89510352

For the DBFS API, it is accessed via DBFS Large Binary Object (LOB) function that is called upon by pipelined table function. Result 2 showed the data retrieved via the DBFS API option, note that the data is presented as a long string and the pre-processing effort is required like segmenting, error check and datatype conversion [2, 7].

5.1 Switchover Test

In this section, we tested for key services' availability and functionality; client connection to the Oracle services, R program execution and availability of the DBFS's file system as well as its accessibility from both clients and forecasting application during the failover/switchover. The system has been setup on two servers each with its own storage and in this configuration with one server functioning as the primary server and the other acts as a hot standby. This test is to assess whether the service of the Oracle instance can switchover from the active to the standby node and maintain a persistence database service. We start by enabling the switchover is done via DataGuard manager console, run the dbfs_client service to mount the DBFS manually on the standby node, followed by checking the state of the mounted filesystem. We run several batches of forecasting during the switchover and the outcome was expected; the existing transactions had roll-backed for consistency, so batch re-run is required. The session managed to persist without the need to re-login. We concluded from our test that the Oracle instance managed to switch successfully from one node to another, with all the background processes started on the other node and sustain the database plus forecasting services with minimum disruption. The result showed that the DBFS functionality is a success after the switchover is made [2]. The DBFS was managed to resume its service and present all the substations' load data in it for application consumption.

6 Discussion and Conclusion

This paper presents an alternate system architecture that utilizes current IT investment in the organization so that it can achieve capability in big data while keeping cost low. It reuses existing database technology and open-source data mining language R to provide an effective forecasting solution without venturing into an expensive setup such as Hadoop eco-system. The paper evaluated several forecasting models to establish the choice and flexibility for performing substations' load forecasting. The models and the proposed system have been thoroughly tested to evaluate their effectiveness. For the forecast test, the data has been run through with three models, each with its own set of parameters or combination of parameters and their residuals are then compared. While NN showed that it has a better yield than the rest, the overall computation time took much longer. With the use of ORE which can parallelize the R codes, the turnaround time is significantly reduced. Most of the test outcomes meet our expectation. For the database to access the data, the option of an external table is easier to implement but it lacks the flexibility to accommodate new changes. The other alternative to using the DBFS API package is supposed to be a better way to implement a flexible architecture that can meet this demand. Most of the raw data files have numerous outliers and inconsistent values which need significant pre-processing before they can be used. This is followed by datatype conversion since the data extracted out from the DBFS' Securefile lobs are mainly not in the correct format and unfit for direct consumption. For the storage test, the compression ratio is 3+:1, saving one-third of the storage space as compared to any conventional file system storage. This compression ratio can be further improved by specifying deduplication feature to reduce even further. For the high availability and service resumption test, the result is positive. The database service switched over without any problem, the forecasting program resumes at the standby node, and all the substations' load data are available once DBFS is back online.

References

1. Weron, R.: Modeling and Forecasting Electricity Loads, pp. 67–100. A Statistical Approach, Modeling and Forecasting Electricity Loads and Prices (2006)
2. Kuhn, D.: Pro Oracle Database 12c Administration. 2nd ed. Berkeley, CA: Apress. 1 online resource (xl, p. 714) (2013)
3. Plunkett, T., et al.: Oracle Big Data Handbook. McGraw-Hill Education (2013)
4. Tole, A.A.: Big data challenges. Database Syst J **4**(3), 31–40 (2013)

5. Tierney, B.: Predictive Analytics Using Oracle Data Miner: Develop & Use Data Mining Models in Oracle Data Miner, Sql & Pl/Sql (2014)
6. Hornick, M., Plunkett, T.: Using R to Unlock the Value of Big Data: Big Data Analytics with Oracle R Enterprise and Oracle R Connector for Hadoop (2013)
7. Kunchithapadam, K. et al.: Oracle database filesystem. In: Proceedings of the 2011 ACM SIGMOD International Conference on Management of data. ACM (2011)
8. Kyte, T.: Expert Oracle Database Architecture: Oracle Database 9i, 10g, and 11g Programming Techniques and Solutions. Apress (2010)
9. Hyndman, R.J., Athanasopoulos, G.: Forecasting: principles and practice. OTexts (2014)
10. Makridakis, S., Wheelwright, S.C., Hyndman, R.J.: Forecasting methods and applications. Wiley (2008)
11. Hyndman, R., Khandakar, Y.: Automatic Time Series Forecasting: The Forecast Package for R 7. 2008. https://www.jstatsoftorg/article/view/v027i03Accessed 24 Feb 2016, WebCite Cache (2007)
12. Hyndman, R.J.: Forecasting with Exponential Smoothing the State Space Approach 2008. Available from: http://gateway.library.qut.edu.au/login?http://link.springer.com/openurl?genre=book&isbn=978-3-540-71916-8
13. Alves, A., et al.: Getting Started with Oracle Event Processing 11 g. Birmingham: Packt Publishing. 1 Online Resource v, p. 320 (2013)

A Novel Database Exploitation Detection and Privilege Control System Using Data Mining

Chee Keong Wee and Richi Nayak

Abstract Database objects access by users has always been controlled in individual database or through enterprise authentication system like LDAP. In the event of abnormal usage, there is no way to detect nor remedial the situation. This paper proposes a novel data mining based method that increases the database security through intrusion detection and denial, and centrally controls the object access privilege of users across multitude of databases.

1 Introduction

Security and object privileges are managed in individual databases. As the IT landscape environment grows larger so does the number of databases maintained by an organization which make it more difficult to maintain all the databases' access. While managing privileges across hundreds of databases is already a big challenge, auditing and controlling their privileges is another multi-fold problem [1]. While there are security features built in database software and commercial software solutions that can improve the security, some of the more sensitive information that reside on these databases require a more stringent and proactive security approach [2]. This paper proposes a solution to address this challenge by detecting the anomalies among the usages on databases using data mining and taking remedial action to rectify the situation. We propose a new framework that will extend the security management [3] of the domain of databases especially those that are independent and function in isolation. Databases such as PostgreSQL, Oracle, MySQL or firebird have their own internal security systems and an IT department will incur high overheads in labor and effort when it has more than 100 of these

C. K. Wee (✉) · R. Nayak
Science and Engineering Faculty, Queensland University of Technology,
Brisbane, Australia
e-mail: chee.wee@qut.edu.au; ck.wee@sparq.com.au

R. Nayak
e-mail: r.nayak@qut.edu.au

© Springer International Publishing AG, part of Springer Nature 2018 505
A. Sieminski et al. (eds.), *Modern Approaches for Intelligent Information and Database Systems*, Studies in Computational Intelligence 769,
https://doi.org/10.1007/978-3-319-76081-0_43

databases to manage in the absence of central authentication or access control mechanisms. The novelty of the proposed solution in this paper is not only to mitigate the role of a central access control system but also to prevent the presence of object privileges in the databases with the use of data mining. This will take security measure a step further and harden the databases even further to protect the data assets.

One security concern with presence of user logins in databases is that it is an open area for vulnerability attack. While the databases' internal security system does address that, some organizations feel that this can be even more secured if the users have no privileges granted in the databases to them to begin with. Given that most IT departments manage a big range of heterogeneous systems, the challenge will be even greater [2]. Only with proper authentication will allocate the grants to the users and revoke when they log off. There are commercial software solutions like Identity Access Management (IAM), they mostly work on a centralized user authentication and privileges or roles associations [4, 5]. Users are firstly authenticated before they can access the necessary databases. Users, roles and privileges are present inside the databases as part of the IAM multi-tier architecture deployment. This strategy works well especially when the main authentication and access are managed from a single server. However hackers are getting more sophisticated and they will attempt to bypass the conventional authentication method, attacking and compromising the systems directly [6]. So it pays to be safe, provide as little as possible for the hackers to reach in [7].

Most database security monitoring systems do post-activity reporting and monitoring, but rely on administrators to proactively control the users [8]. It depends on the diligence of the administrators to rectify security breaches. When there are hundreds or thousands of databases in an environment, the turnaround time for the investigation and problem rectification may come too late. Modern databases provide [8] its own security on users and their privileges to the schemas and objects unless central authentication systems like Lightweight Directory Access Protocol (LDAP) or Microsoft's Active Directory are used [9]. Even when they are in use, there is a minute chance that they can be compromised as there are unconventional techniques that have been document to hack through or bypass them [10].

2 Proposed DB Exploitation Detection and Privilege Control System: Overview

To mitigate the new threats against internal databases from within the organization, we propose a novel system framework that will not only detect anomalies but take proactive actions against them. The framework includes two components that perform Database Exploitation Detection (DED) to detect users' activities anomalies and, Privilege Control System (PCS) to control user's accounts and privileges.

The Database Exploitation Detection (DED) module will monitor user activities across the database and will send alert should it detects anomalies in their usage behavior. It will have a predictive model trained with data set of acceptable database user activities and develop a signature repository. It will continuously monitor the databases' user activities and cross reference against it. When it detects anomalies, it will notify the administrator while invoking the PCS to terminate and clean out suspicious user accounts immediately. DED's advantage lies in its ability to analyses user behavior and activities at the database meta-level; examples are the SQL statement that it executes, the objects it tries to retrieve or execute, the logon time that is deemed as acceptable pertaining to a certain business function or role, as well as the application name and desktop that the users use. This is important as hacking activities at the database level need additional information other than IP address, port id or OS command line which IDS use.

The Privilege Control System (PCS) performs user authentication through a set of policies that have been agreed upon by the administrators and system owner on their databases usage. This module authenticates and creates the user accounts in the designated databases followed by granting the privileges dynamically. The user account and privileges will persist within the database while the users are working against them. Once the users log off, the PCS will drop the user accounts and remove all traces of the user privileges from the databases. The advantage of this module is to wipe the database clean out of any user's traces to prevent any possible loopholes of exploitation. The common practice among database administrators is to create both user accounts with privileges to the tables and leave them in the system, then let IAM or database internal authentication mechanism to handle the user logins. This is a significant danger especially when hackers can bypass IAM or database authentication to assume false ID to gain access to the databases.

Figure 1 shows a high-level view of the proposed system. DED uses classification model against past user historical data to find the signature patterns between normal and exploitative pattern. It passes the new user sessions against the model to find signature match. Once it finds a match, an alert will be sent to the administrators for action. He/she review the alerts to be either false alarms or genuine

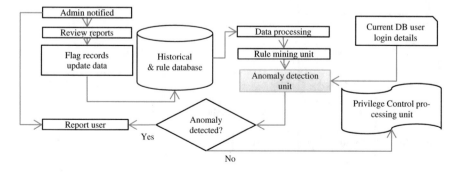

Fig. 1 Proposed DEC and PCS architecture

exploitation and flag them accordingly. This information will be send back into the signature database where the association rule mining module constantly learns and improves its detection accuracy. It will also trigger PCS to take appropriate actions against the suspicious users in the database.

3 Database Exploitation Detection Module

The first stage is to monitor and gather all the current active user sessions that are active against the databases. This forms the initial repository that is used to generate association rules. The session data is comprehensive where the information of interest to this module will be the logon time, machine name, OS user account, Terminal, application used, and SQL ID. SQL ID is a hash value in which it refers to the SQL statement or code that the user is executing. An example of user session is shown in Table 1. Based on this information, the DED module can mine knowledge on the exact schema, objects and DML that the users are performing against the DB.

The data is collected over a period until a certain level is reached. The data will be split into training and testing data sets. The routine of training will be iterative and interactive; iterative training is required to build the list of association rules, and interactive where the administrator will analyze the rules and identify rules as

Table 1 Database user logon information

Logon_Time	Machine	OSUser	Application	Terminal	SQLID
29/03/2017 17:25	Q225491	ja055	toad.exe	Q225491	53c2dfrtgu595
30/03/2017 12:46	CBNEWAS14	SVC-SAPIS	process.exe	CBNEWS14	–
30/03/2017 13:21	SPRQ \Q229065	cw073	sqldeveloper	Q229065	4fhfkpwvf28qt
29/03/2017 4:37	cbns1db02	qeedwhpr	Datastg	cbns1db02	8d8xuxdbrgnju
29/03/2017 14:46	sbnfwua15	svc_SAP..T	websvc	sbnfwua15	bq9bgsgxx6jq
30/03/2017 12:24	cbns1db02	qeedwhpr	sqlplus	cbnf1db02	–
29/03/2017 17:25	Q225491	ja055	sqldeveloper	unknown	4q0js1p8nv2wp
30/03/2017 4:13	cbns1db02	qeedwhpr	websvc	cbnf1db02	–
30/03/2017 7:43	cbns1db02	qeedwhpr	websvc	cbnf1db02	53c2dfrtgu595

normal or abnormal. The DED will first sieve through the load of user session details from all the databases that have been collected and extracted out the features. Association rules will be generated. The user/operator will also input their own set of rules, so the system learns from the input and achieves greater intelligence in detecting anomalies. Those rules that are considered abnormal will form the signatures of abnormal database usage and these will be added into the DED signature databases.

Once the system has reached a satisfactory learning threshold, the testing phase will begin to check the accuracy of the detection. Any errors or gaps will be used in the next phase of rules edit followed by training. The DED will be used against another load of user session details and attempt to match the session details' features with the DED signature databases that it has built up. It then compares them with abnormal/misuse signatures and raises alarms when there is a match indicating a possible intrusion or database exploitation occurrence.

3.1 Association Rules for DED

There is strong correlation between users DB activities and their intention of access. For example, type of applications used, machine name, schema queried, DML performed as well as the time of their activity that took place, do reflect the purposes of access. If the user activities are within the acceptable behavior and parameters, they will not violate the default normal profile. The intent of the association rules mining is to create a correlation of multiple features or attributes from the session information. Figure 2 shows a basic overview of the process flow of using the association rules to define and predict the DB hackers. Each DB

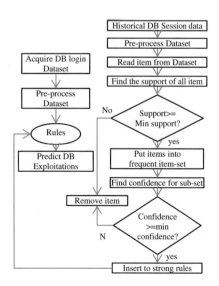

Fig. 2 Using association rules for DB exploration detection process flow

session records comprised of a series of information which we refer to them as items and they are: Username, Logon_time, Machine_name, OS_user, Application_name, Terminal, Logon_time and SQLs.

Let these items be labelled as A, B, etc. A set of these items is called an itemset which we called it X. An association rule between the item sets is expressed as, $A \rightarrow B$ (c, s), where A and B are items from the DB session and they are subset of the itemsets X; $(A, B \subset X)$, c and s are confidence and support of the rule. Support is expressed as a percentage of the item in the session repository, D, which contain Itemsets A and B. this is expressed as; support of the rule, $s = support\ (A \cup B)$.

Confidence is expressed as a percentage of support of Item A, B over the support of A. Confidence, $c = \frac{support(A \cup B)}{support(A)}$.

Table 2 shows a typical example of user session details from a common database. The difference between this and Table 1 is that there is a classification flag which identifies the risk level of the user sessions. Itemsets comprised of items from the user login session; SchemaName is the owner of the database objects that the user is accessing. DBUser is the account in the databases that the user used to log in. OSUser is the account of the operating system in which the user is currently logged in as. Program is the name of the application that the user used to access the database. Hostname is the name of the machine that the user is accessing from. Timing refers to the time that the user is performing his tasks. This will be converted to symbolic representation or discretized into weekday + hours by pre-processing.

We use the top schema level of DB user records to mine the association rules pattern. Some of the logon information forms the essential feature of a session. Other are extension of the information that form the correlation of the feature like schema access, DML performed, resource groups, time of access. This forms the action feature of the session. There are first match to the essential list of pre-approved features. Any items that are not in the database are flagged as "high" like the time of the logon session that occur. The itemsets are mined to form the

Table 2 User logon session and associated risk flag

SchemaName	OSuser	Machine	Program	Time	SQL	Flag
SPLX2230	qenetdpr	erokuv90	sp_ordr@erokuv90	04-09	Select ...	LOW
FDRSTAT	fclkiosk	SOE02197	FCLLite.Net.exe	04-10	Update....	LOW
EIMUPDATE	svc-WL-EIM	ERRDWAS37	JDBC Thin Client	04-10	Insert...	LOW
EIMUPDATE	svc-WL-EIM	ERRDWAS37	JDBC Thin Client	04-09	Exec...	LOW
SYS	root	Unknown	JDBC Connect Client	04-23	Exec..	HIGH
SPLX2105	qenetdpr	erokuv90	sp_ordr@erokuv90	04-12	Select..	LOW
EDW_LDG	SM203	Q39484	JDBC Thin Client	04-13	Select	LOW
OMSMAIN	DP023	ERRDWAS37	JDBC Thin Client	04-02	Delete...	HIGH

Table 3 Example of association rules from the database logon data

Association rule	Meaning
Schema = EDW_LDG → osuser = cw073 → program = sqlplus → hostname = Q223069 → timing = normal → privilege = support_role	The osuser, cw073, is logging in during normal working hour from a valid terminal with a valid program. This is considered a normal database acceptable behaviour
Schema = EDW_VIS_OWNER → osuser = root → program = putty → hostname = Q22396 → timing = offpeak → privilege = dba	The schema is on the list, the osuser is of exceptional high OS privilege, the osuser and program are within the boundary but the timing and privilege is of concern
Schema = SYS → osuser = oracle → program = sqlplus → hostname = unknown → timing = offpeak → privilege = sysdba	All the items here are of exceptionally high privilege that is not normal

association rules. An example of the association rules created on the login data is detailed in Table 3 using data from the database's logon as in Table 2.

Itemset = {$SchemaName, $Osuser, $Machine, $Program, $Time, $SQL where $<item_class> is the instances that belong to the item_class.
Feature set in this example is expressed as:
Itemset = {EDW_LDG,cw073, sqlplus, q223069, sp_ord@erouv90, 04-01, select employees from HR.Payroll'}

The user session data can be represented as a transaction data as shown in Table 4. Frequent patterns can now be obtained. From the example above, we find the frequent patterns by calculating the support as follows:

Itemset (OMSMAIN, sqlplus.exe, 04-09) = 3
Itemset (OMSMAIN, 'select * from pole', 04-09) = 3

Table 4 User login details in transaction database TDB

TID	User sessions details
100	EDW_LDG, cw073, sqlplus, q229069, 04-09, 'select * from edw_ldg_owner. Msf620'
200	Fdrstat, cw073, sqlplus, q22609, 05-09, 'update MSF920 where ...'
300	SYS, root, putty.exe, unknown, 'update payroll where ...'
400	OMSMAIN, sm203, toad.exe, q339506, 04-09, 'select * from pole'
500	OMSMAIN, dp023, sqlplus.exe, q495873, 04-09, 'select * from pole'
600	OMSMAIN, fg056, sqldeveloper.exe, q40596, 03-08, 'update pole where..'

4 Proposed Privilege Control System

The proposed PCS follows the Least Privilege Principle [11]. It allocates users with the least level of privilege access to do the job. No additional privileges will be allowed unless they are necessary. It is based on the concept of the denial of privilege presence in the databases until a level of authentication has been met before permission can be granted to the user. For all users across this proposed database landscape, they don't have any assigned roles or objects access privileges in the databases. The proposed Privilege Control System (PCS) is a multi-tier architecture with agents running on the databases' OS servers that maintain a persistent connecting session to the databases while remain connected to the main central server. The agent will function as a gateway to monitor the user connecting session and upload their logon details back to the central. Access will be granted when the user fulfils the pre-set criteria; hour of logon, approved machines, approved list of software/application to access. If he fails the authentication, his account will be timed locked, and an alert will be sent to the administrator. When the user meets the requirements, the PCS will check the user against the list of applications' access charts. These charts contain user/role and object access privileges in a matrix-relationship. The PCS will gather the user-to-objects privileges to generate out two sets of SQL commands and send to the agent; one set will grant the required permission to the user and allow him to progress on with his works, the second set is to revoke all the permissions that were previously granted to him after he logs off. This step cleans out all the users' privileges from the databases after his activity leaving nothing behind except his DB accounts.

4.1 PCS Architecture

Figure 3 shows the overview of the multi-tiered PCS. The PCS agents are daemons running on the servers that host the databases and they maintain persistent connection, monitoring the sessions 24 × 7. The centralized logon and logoff trigger control all users in the local databases. When a user logs in, the SYS' trigger will activate a procedure that gather his user session details; logon time, client machine used, name of client application. These details are captured from the database's system views (for oracle it is v$session). The procedure will send the information back to the central PCS via links that are present between server and clients [2, 12].

The PCS server holds a user logon approval repository, and this is checked by the server against the user's details. Should the user's information fulfil the criteria, the user is flagged as approved and this will proceed to the next step. But if the user fails the check, the server will signal the agent to disconnect the user and disable his account. The next step is to produce the list of grants versus the objects that the user is entitled to. In the PCS, the repository holds a series of matrix-relationship templates that maps the privileges that the users supposed to get versus the schema

Fig. 3 Proposed privilege
control system

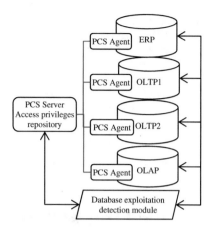

objects on that database which the users is to have. The PCS then generates two set of database commands; one to grant and the other to revoke. The grant set will be executed by the agent to allow the user to proceed ahead and use the database for his work. The other set is kept by the agent and wait for the user to finish his work. Once the user logs off, the logout trigger captures his logout and run the revoke privilege set of commands to clean them out. There are a few user accounts that will be exempted from this control: Schema owners that own objects, accounts that need to perform background activities like replication, ETL, scheduled events or jobs. All the resource groups are ring fenced so there will be no overlapping of privileges [13]. Figure 4 illustrates the sequences of activities that happen between databases and PCS. Note that all the procedures are handled by individual databases' internal packages and functions; logon triggers, PLSQL packages to database specific features like spatial, XML, etc.

Fig. 4 Privilege control
process flow

Fig. 5 Complex
inter-schemas/objects
relationship

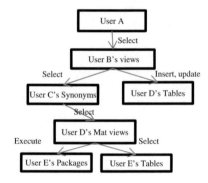

4.2   User's Privilege Representation

User are not allowed to access the database objects that belong to the schema owners, so specific grants will need to be present before the users can access the objects. However, not all databases are created equal and there are cases where permission to certain object like a view will entail a series of access privilege that need to be present down the stream.

Consider Fig. 5 for example, which depicts the inter-dependencies between different users' objects and privileges. Such complex hierarchies are fairly common among systems and this poses a challenge to the PCS concept; that is to have a proper representation of the hierarchical series of privileges in proper SQL format which can be created or granted as well as revoke or remove as required.

5   Experiments and Results

Both the DED and PCS modules are tested individually and separately before combining them together for the final integrated test. The DED is tested for its ability to build the frequent patterns and classification models from a set of user session historical data followed by the accuracy test on a separate set of user session data. The PCS will be tested for its ability to create and revoke user's related objects with trial login to evaluate the logic of privilege allocation/de-allocation. The integration test is conducted against the two modules combined to evaluate the functionality of both modules together against a set of user login scenario. The final test brings both modules together to assess their integration into the proposed system that is shown in Fig. 1, and test their functionality. This prototype testing is conducted through a central R program which has the capability of using the association rules libraries and make remote procedure call to the database via the database client library. New user session logs are ingested into the program and the observations are made based on the results that are produced by both DED and PCS module, both of which have different output between R studio and database backend. The dataset comprised of 500

Table 5 Final DED and PCS test results

		DED test	
		Passed	Failed
PCS test	Passed	1, 2, 3, 4, 5, 6, 7, 8	9, 10
	Failed	–	–

database login sessions, with 376 normal and 124 anomalous types. Over 1800 rules were generated by the DED with accuracy at 84%. Several test runs are conducted to detect DB exploitation with a list of test transactions that comprise of users accessing several schemas through multiple programs on both weekends and weekdays. The result is tabulated in Table 5 after processing with Privilege Control System. Based on the result, the basic integration is complete with no errors and the modules perform without any software failures. The result showed that there is still room for improvement on the prediction of the DED which fired off the PCS to deny the users.

The failures encountered by DED was due to misclassification of two activities, misclassifying them as false positive and false negative, with one normal session misclassified as anomalous whereas the anomalous session was not picked up. For future improvement, the dataset will need to be larger and more inclusive. The algorithm will have to be tuned to reduce misclassification so as to increase its sensitivity and specificity. Techniques like direct adjustment approach, the permutation-based approach or holdout approach prescribe methods such as shuffling the class labels of the records in the dataset as well as adjusting the p-values of the association rules [14].

6 Conclusion

This paper proposes a novel intrusion detection technique based on data mining for databases. The goal is to improve the security of databases especially when there are increasing reports of mission critical production databases that suffered information leakage due to hacking, disgruntled employees, abuses of privileges that occur from both internal and external entities. Details like customers' credit cards, passwords, or corporate R&D blueprints, financial records, classified documents and others, all had leaked out when hackers manage to breach through the barriers of security systems and firewall. So any new steps in the security that help to minimize data loss while meeting to the CC's database security standard [15] will be highly valued.

With this proposed DED/PCS, the database's security landscape is consolidated into a central location without the need of spending large amount for LDAP or Identity Access Management solution. It monitors the databases' activity continuously for abnormal user activities and takes appropriate actions as programmed to drive the accessibility of the malicious user out of the systems cleanly and effectively. That minimizes potential risks that hackers could exploit. Users will have more difficulty to bypass the control and they will be caught in the act much faster as compared to passive alerting, the amount of damages can be minimized drastically with PCS doing ring-fencing the database more rigidly, preventing cross schema confusion and implicit privileges that potentially have been missed. However, no system is totally secured but getting rid of all the pre-granted privileges on any databases is a significant step to secure the databases asset as well as attempting to monitor any user exploitation within the databases. This paper sets the foundation of the concept of DED/PCS.

References

1. Connolly, T.M., Begg, C.E.: A constructivist-based approach to teaching database analysis and design. J. Informat. Syst. Educat. **17**(1), 43 (2006)
2. Bertino, E., Sandhu, R.: Database security-concepts, approaches, and challenges. IEEE Trans. Depend. Secure Comput. **2**(1), 2–19 (2005)
3. Dubois, É., et al.: A systematic approach to define the domain of information system security risk management. In: Intentional Perspectives on Information Systems Engineering, pp. 289–306. Springer (2010)
4. Kruk, S.R., et al.: D-FOAF: distributed identity management with access rights delegation. In: Asian Semantic Web Conference, Springer (2006)
5. Gaedke, M., Meinecke, J., Nussbaumer, M.: A modeling approach to federated identity and access management. In: Special Interest Tracks and Posters of the 14th International Conference on World Wide Web, ACM (2005)
6. Liu, Y., Chen, T.-y.: Security strategies of database microsoft SQL server and its realization. Comput. Eng. Design **1**, 019 (2003)
7. Ramaswamy, C., Sandhu, R.: Role-based access control features in commercial database management systems. In: Proceedings of 21st National Information Systems Security Conference. Citeseer (1998)
8. Khanuja, H.K., Adane, D.: Database security threats and challenges in database forensic: a survey. In: Proceedings of 2011 International Conference on Advancements in Information Technology (AIT 2011). http://www.ipcsit.com/vol20/33-ICAIT2011-A4072 (2011)
9. Lampson, B.W.: Computer security in the real world. Computer **37**(6), 37–46 (2004)
10. Pritchett, W.L., De Smet, D.: Kali Linux Cookbook. Packt Publishing Ltd (2013)
11. Schneider, F.B.: Least privilege and more [computer security]. IEEE Secur. Priv. **99**(5), 55–59 (2003)
12. Kuhn, D.: Pro Oracle Database 12c Administration. 2nd ed., Berkeley, CA: Apress. 1 online resource xl, p. 714 (2013)
13. Sandhu, R.S., Samarati, P.: Access control: principle and practice. IEEE Commun. Mag. **32**(9), 40–48 (1994)
14. Liu, G., Zhang, H., Wong, L.: Controlling false positives in association rule mining. Proc. VLDB Endow. **5**(2), 145–156 (2011)
15. (CC), T.C.C.f.I.T.S.E.: Base protection profile for database management systems. In: Common Criteria Portal (2016)

Agent Programming Languages and Logics in Agent-Based Simulation

John Bruntse Larsen

Abstract Research in multi-agent systems has resulted in agent programming languages and logics that are used as a foundation for engineering multi-agent systems. Research includes reusable agent programming platforms for engineering agent systems with environments, agent behavior, communication protocols and social behavior, and work on verification. Agent-based simulation is an approach for simulation that also uses the notion of agents. Although agent programming languages and logics are much less used in agent-based simulation, there are successful examples with agents designed according to the BDI paradigm, and work that combines agent-based simulation platforms with agent programming platforms. This paper analyzes and evaluates benefits of using agent programming languages and logics for agent-based simulation. In particular, the paper considers the use of agent programming languages and logics in a case study of simulating emergency care units.

Keywords Multi-agent systems · Logic · Simulation

1 Introduction

Agent-Oriented Programming (AOP) is a programming paradigm where programs are composed of agents. Similar to objects in Object-Oriented Programming (OOP), agents maintain a mental state and react to input by performing actions and changing their mental state. Some agents are also assumed to be intelligent agents, meaning that they pursue goals and exhibit social behavior by communicating with other agents. Agent programming languages are programming languages that are designed for development of multi-agent systems with AOP. Examples of platforms that use agent programming languages include Agent-0 [1], 3APL [2], 2APL [3], Jason [4], JACK [5, 6] and GOAL [7]. The notions of belief, desire and intention (BDI) are

J.B. Larsen (✉)
DTU Compute, Technical University of Denmark, 2800 Kongens Lyngby, Denmark
e-mail: jobla@dtu.dk

© Springer International Publishing AG, part of Springer Nature 2018   517
A. Sieminski et al. (eds.), *Modern Approaches for Intelligent Information and Database Systems*, Studies in Computational Intelligence 769,
https://doi.org/10.1007/978-3-319-76081-0_44

key components in these languages, as they respectively denote what the agent believes, what the agent would like to achieve, and what the agent is currently working towards achieving. Formalizations of a BDI model in modal logics provide syntax and semantics for the model. Thus logic provides a theoretic framework for specification and verification of agent programs.

The BDI paradigm has also been used in agent-based simulation (ABS). The purpose of ABS compared to multi-agent systems is to gain insight into how global properties emerge from a system of local interacting processes. Examples of ABS platforms include Mason [8], Repast [9] and GAMA [10]. ABS platforms generally do not use above mentioned agent programming languages but some of them provides a framework for making models with the BDI paradigm [11]. A BDI model allows agents to exhibit more complex behavior than purely reactive models but without the computational overhead of cognitive architectures. It is generally also easier for domain experts to specify their knowledge in terms of a BDI model compared to an equations-based model, and a BDI model supports explainable behavior. Adam and Gaudou [12] present an extensive analysis and evaluation of approaches to integrating BDI models in ABS. They highlight the previously mentioned benefits of BDI models as a way to implement *descriptive agents* which use richer and more complex models than reactive agents.

This paper presents an analysis and evaluation of using recent advances in agent programming languages and logics, in particular frameworks for implementing social behavior, in ABS. The paper first presents a summary of AOP, ABS platforms and work on integrating BDI models in simulation platforms based on the work of Adam and Gaudou [12]. It then describes research in frameworks and meta-models for implementing virtual environments and social behavior in agent programming languages. It evaluates potential benefits of using a framework for implementing agent organizations in ABS, and finally discusses further use of MAOP in ABS. The criteria used in the evaluation are in terms of:

1. How the framework supports descriptive agents.
2. How reusable the framework or meta-model is.
3. How useful the framework or meta-model is for analysis.

The evaluation is based on previous work on using the agent organization framework AORTA [13] to create a simulation model for an emergency care unit [14].

2   AOP, Logic and Agent-Based Simulation

AOP was originally proposed by Shoham [1] as a specialization of OOP. Shoham motivated AOP with cases in which multiple entities interacted with each other in order to manufacture cars and reserve plane tickets. In AOP, each entity (now called an agent) maintains a mental state of beliefs, capabilities and decisions that have dedicated terms with a formal syntax. Communication with other agents occurs through speech-act inspired messages. Some of the approaches to programming languages designed for AOP include:

- AgentSpeak(L) [15] and Agent-0 [15] in which an agent has a database of plans or rules for choosing actions that match its current mental state. The agent programming platform Jason [4] implements AgentSpeak(L).
- Languages based on logic programming such as 3APL [2], 2APL [3], and GOAL [7].
- Jack [5, 6] which extends Java with agent programming keywords.
- A combination of XML and Java. This approach is used in the agent programming platform Jadex [16].

These programming languages use BDI as a common paradigm for a mental model but as it can be seen, they have very different approaches to implementing it. The BDI paradigm comes from philosophy and the mental model can be given formal syntax and semantics with epistemic logics. Other logics such as first-order logic and temporal logics can be used to specify world models of concepts and dynamics. Given a specification it is then possible to use logic reasoning to verify properties of the specification. Thus logic provides a theoretical framework for specifying and verifying properties of agent programs. In the programming languages AgentSpeak(L), 3APL, 2APL and GOAL, the agents also use logic to do reasoning in their decision making.

ABS is an approach to simulation that takes the perspective of the individuals that inhabit the simulated system. ABS is useful in cases where it is easier to describe a system in terms of interacting agents rather than as a global process [17]. A critical part of ABS is a scheduling mechanism which ensures that all agents are synchronized in a finite sequence of time steps. ABS platforms provide frameworks for ABS and typically features tools for visualizing the simulation, data extraction tools and analysis tools. Commonly used ABS platforms include:

- Mason [8] which is a Java based discrete-event simulation platform that has been extended with ABS.
- Repast [9] which is a suite of tools in multiple programming languages for implementing ABS.
- GAMA [10] which features an XML based language GAML for implementing agents. GAMA also features tools for using GIS data in the simulation.

The ABS platforms typically have tools for implementing reactive agents but little support for implementing proactive behavior. This works well for many cases but as argued by Adam and Gaudou [12], there are also cases, often those involving human agents, where more descriptive agent models are useful for gaining insight into the decision making. The BDI paradigm provides a framework for implementing descriptive agents that are still fairly efficient. There have been three general approaches to implementing BDI in ABS:

- Extending agent programming platforms with ABS features. Bordini and Hübner does this with Jason [18].

- Extending ABS platforms with BDI modeling features. Caballero et al. does this with Mason [19].
- Combining ABS platforms with agent programming platforms. Padgham et al. [20] does this with Repast and JACK, and Singh et al. [21] designs a framework for integrating any two platforms with each other.

The benefit of the last approach is that it leverages features from both platforms but with a cost of computational power in keeping agents synchronized between the platforms. Besides mental models for the individual agents, there is also work on implementing meta-models for the environment and social behavior such as the MASQ meta-model by Dignum et al. [22]. Meta-models for environments and social behavior are covered further in the following section.

3 From AOP to MAOP

Much of the early research in agent programming languages has been focused on the internal agent architectures with different approaches to programming languages based on the BDI paradigm and speech-act communication. Both the environment that the agents inhabit and the social skills of the agents have been designed and programmed for specific domains. Recent research has gone into making more reusable frameworks and meta-models for creating environments and agent societies [23, 24]. Notable examples include:

- CArtAgO (Common Artifact Infrastructure for Agent Open environment) [25] which is a Java-based framework for developing and running virtual environments based on the Agents and Artifacts meta-model. In this meta-model, the agents use artifacts to communicate with other rather than only by speech-acts. The artifacts provide an interface for the communication that allows for also non-BDI agents to communicate with BDI agents. The framework has been integrated with Jason, 2APL and JADEX [26, 27].
- EIS (Environment Interface Standard) which is a Java-based framework for connecting agent programming platforms with environments. It is not a meta-model for environments but it acts as an interface for agent programming platforms to environment platforms such as CArtAgO-based platforms.
- OperA [28] which is a meta-model for agent organizations. In agent organizations, the agents are assigned roles that puts a structure on how the agents can use their abilities to communicate and carry out actions. The Eclipse plugin Operetta [29] is a tool for design, verification and simulation of OperA models.
- $\mathcal{M}$OISE$^+$ [30] which is also a meta-model for implementing agent organizations. $\mathcal{M}$OISE$^+$ is integrated with CArtAgO and Jason in the JaCaMo platform [31].
- AORTA [13, 32] which is a meta-model that enables individual agents to reason about organizations described in OperA. It is designed for adding organizational reasoning capabilities to BDI agents and has been integrated with Jason [33].

Table 1 Summary of main characteristics of MAOP frameworks and meta-models

Framework/meta-model	Main characteristic
CArtAgO	Virtual environments with Agents and Artifacts meta-model
EIS	Interface between AOP platform and environment
OperA	Agent organization meta-model
$\mathcal{M}$OISE$^+$	Agent organization meta-model
AORTA	Meta-model for enabling reasoning with OperA in agents

Table 1 summarizes the main characteristics of these frameworks and meta-models. A common feature of the examples is that they support more open heterogeneous systems of agents: agents can enter and exit the system freely even though they use different internal mechanisms for decision making. The frameworks and meta-models put an emphasis on Multi-Agent Oriented Programming (MAOP) with system level frameworks rather than traditional AOP with agent-level mental models and speech-act communication. Use of MAOP is not common in ABS literature. A possible reason for this might be that openness is less important in simulation where the purpose is to gain insight in a given system. A potential benefit of MAOP though is that it can offer reusable tools for implementing environments and social behavior. Using MAOP with a foundation in logic would also allow for specification and validation of simulation models similar to the work presented by Jensen [34] on verification of organization-aware agents in AORTA.

4 Case Study: Emergency Care Units

In this section, the use of agent programming languages and logic in simulation is analyzed and evaluated in a case study with emergency care units based on previous work [14]. Emergency care units are responsible of providing care for acute patients. Hospitals often have an entire department dedicated to emergency care and the number of incoming patients has been increasing in recent years. Due to limited funding, it is necessary for the department to work efficiently by establishing clear guidelines including work procedures, roles and responsibilities. Such guidelines are not complete though and the doctors and nurses are flexible in taking actions that comply with the guidelines. In some cases they might even go against the guidelines. For example if there are no nurses available and a secretary has no other currently urgent tasks, the secretary might help a patient with taking a urine sample even though that is the job of a nurse. The dynamic nature of the decision making makes it difficult to keep track of what consequences the decisions can have and what can be done to avoid potential issues such as long waiting lists or staff overwork hours. Simulation

522 J. B. Larsen

could provide insight into these things given that the simulation model describes the interaction in the emergency department with sufficient accuracy.

The first thing to notice in this case study is that it involves quite different kinds of actors. There are individual human actors such as doctors, nurses and patients that show goal achieving behavior. The doctors and nurses are responsible for providing emergency care and the patients arrive at the department with the goal of receiving emergency care. The human actors also make use of various tools and IT-systems. Finally secretaries, nurses and doctors communicate with institutions such as other departments in the hospital or rescue teams. A BDI model offers flexibility to implement these different actors. The human actors can be implemented as agents that show proactive behavior, the tools and IT-systems can be implemented as reactive agents, and the other departments and institutions can be implemented as agents that have goals and can send requests to the emergency department. Such a model can be implemented in an agent programming platform with simulation features such as Jason or an ABS platform with a BDI model extension such as Mason.

The second thing to notice in this case study is the significance of established work procedures, roles and responsibilities, which serve as guidelines rather than complete plans. The staff are expected to follow the guidelines but also to fill out with details not covered by the guidelines. The staff are assumed to be intelligent and be able to act independently in a manner that fits the situation, deviating from the guidelines in extraordinary situations. In other words, the guidelines describe norms and goals that the agents are obliged but not enforced to adhere to. The BDI paradigm on its own does not offer a formalism for modeling such behavior. While BDI agents are flexible and act toward solving goals, they are enforced to follow the plans they are given. Instead it makes sense to apply meta-models for agent organizations that have frameworks for encoding organizational knowledge explicitly in the model. As mentioned earlier, some of these meta-models are already integrated with agent programming platforms. Previous work presented an AORTA model of the acute patient treatment process [14]. In the AORTA meta-model, we can encode organizational knowledge in terms of roles, objectives and sub-objectives, role dependencies and conditions. Each agent then maintains two knowledge bases: one with personal knowledge and one with organizational knowledge. The organizational knowledge base describes the stages that the patient goes through, which staff members are involved in each stage and a selection of conventions that the agents are expected to follow. When deliberating which action to perform, an agent can then reason about if an action complies or violates any obligations of the agent. Updating the knowledge base is done accordingly to general rules of the meta-model when the agents perform actions. The agents can also perform organizational actions, such as enacting roles, which will update their knowledge base accordingly. In addition, the explicit representation of organizational knowledge supports specification and verification of the organizational agent model. To summarize the evaluation in accordance to the criteria proposed in the introduction, using an AORTA model in ABS would give:

1. Descriptive agents that have a mechanism to include organizational reasoning in their decision making. They are descriptive in the sense that they implement

complex social behavior and support explainable behavior. An agent can use AORTA to reason about what other agents expect of the agent, and what it can expect of the other agents.

2. A reusable meta-model that can be integrated in any agent programming platform.

3. Formal syntax and semantics in logic that can be used for specification and verification of the organizational agent model. Logic reasoning can provide insight into social relations which are otherwise hard to identify or reason about.

5 Discussion

The BDI paradigm on its own only provides generalized methods for implementing internal agent reasoning. It does not provide generalized methods for implementing important aspects of multi-agent systems such as organizations and environments. The previous section analyzed potential benefits that the AORTA meta-models can provide for ABS in the emergency care unit scenario. In this section we recap that analysis and discuss potential benefits of applying the other frameworks and meta-models listed in Table 1 for ABS.

CArtAgO provides a framework for implementing agent environments in Java, which is commonly used in ABS platforms, using the Agents and Artifacts meta-model. In domains where people interact through physical objects such as whiteboards or telephones, CArtAgO would provide a generalized framework for encoding these objects. In the case study with emergency care units, the physical location and availability of information communication technologies can have a major influence on the workflow. CArtAgO has been implemented in Jason and has been used to an increasing extent in MAS. As it is Java based, it could potentially also be implemented for dedicated ABS platforms that support BDI models. The Agents and Artifacts meta-model also provides theoretical foundation for specification and verification of agent environments.

EIS provides a Java framework for integrating agent programming platforms with environments. This is useful for implementing systems where the internal agent reasoning logic and the environment logic are separated from each other. The separation allows for more openness, as agents can then be integrated in the environment no matter how their internal reasoning works like. As mentioned earlier, openness is less of a concern in ABS than MAS so, although the framework is reusable, we do not see an immediate benefit of using EIS in ABS.

OperA provides a meta-model for designing and analyzing agent organizations. As the evaluation in the previous sections shows, there are clear benefits of applying organization meta-models to domains with human organizations. Making a model of the organization in OperA would provide a basis for implementing ABS with AORTA agents that perform organizational reasoning. $\mathcal{M}$OISE$^+$ provides an alternative meta-model for agent organizations. Its integration with CArtAgO and Jason in JaCaMo could provide a framework for implementing ABS with both environment and organization models.

The AORTA meta-model, which was evaluated in the previous section, provides a basis for implementing organizationally aware agents in ABS platforms. Doing so would give ABS that supports descriptive agents that replicate organizational behavior in terms of roles and norms. In domains with human organizations, such as in the hospital case, simulation with organizationally aware agents should provide more accurate outcomes than with only the BDI paradigm. There are already implementations of AORTA in Jason, which to some degree supports ABS, and since AORTA has well defined semantics and operational rules, it can be implemented in dedicated ABS platforms that support BDI models. The formal syntax and semantics in logic also supports specification and verification of the organizational agent model.

In ABS of social systems, there is also a growing interest in frameworks and meta-models for social values. A social value represents a concept that an agent cares about and it will generally perform actions that promotes its social values. Simulation with social value models have gained interest as a way to implement social behavior that agents do exhibit without explicitly reasoning about them. Although there is work on meta-models for social values, there still remains much to be done in terms of formalization and implementation in ABS platforms.

6   Conclusion and Future Work

There is active research into providing better frameworks for implementing BDI models in ABS. They generally use one of the methods:

- Implementing simulation features in agent programming platforms [18].
- Implementing BDI models in ABS platforms [19].
- Combining ABS platforms with agent programming platforms [20, 21].

The third method has the advantage that it can make use of advances in tools for both ABS and AOP platforms. As argued by Adam and Gaudou [12], the cost of high computational power might also become negligible as computers get more powerful. Research in agent programming languages and logics has given frameworks and meta-models for implementing environments and social behavior. These are designed to be reusable and their logical foundation can be used for specification and verification of ABS models. The paper has given an analysis and evaluation of using agent programming languages and logics in a case study based on emergency care with the AORTA meta-model. The meta-model gives descriptive agents that can include organizational reasoning in their decision making, is reusable, and has a formal syntax and semantics in logic that can be used for specification and verification. We also discussed potential benefits of using some of the other MAOP frameworks shown in Table 1 for ABS.

To the author's knowledge, there are still few reusable frameworks and meta-models for implementing social behavior in ABS. Future work include implementing an ABS of the emergency care unit scenario with agents that exhibit social behavior, and using logic for specifying and verifying properties of meta-models for social behavior.

Acknowledgements This work is part of the Industrial PhD project *Hospital Staff Planning with Multi-Agent Goals* between PDC A/S and Technical University of Denmark. I am grateful to Innovation Fund Denmark for funding and the governmental institute Region H, which manages the hospitals in the Danish capital region, for being a collaborator on the project. I would also like to thank Jørgen Villadsen, Rijk Mercuur and Virginia Dignum for comments on the ideas described in this paper and comments on a draft.

References

1. Shoham, Y.: Agent-oriented programming. Artif. Intell. **60**(1), 51–92 (1993)
2. Hindriks, K.V., De Boer, F.S., Van Der Hoek, W., Meyer, J.J.C.: Agent programming in 3APL. Auton. Agent. Multi-Agent Syst. **2**(4), 357–401 (1999)
3. Dastani, M.: 2APL: a practical agent programming language. Auton. Agent. Multi-Agent Syst. **16**(3), 214–248 (2008)
4. Bordini, R.H., Hübner, J.F., Wooldridge, M.: Programming Multi-agent Systems in AgentSpeak Using Jason. pp. 1–273. (2007)
5. Winikoff, M.: Jack intelligent agents: an industrial strength platform. In: Bordini, R.H., Dastani, M., Dix, J., El allah Seghrouchni, A. (eds.) Multi-Agent Programming: Languages, Platforms and Applications, pp. 175–193. Springer (2005)
6. Busetta, P., Ronnquist, R., Hodgson, A., Lucas, A.: JACK intelligent agents—components for intelligent agents in Java. AgentLink News Lett. **2**, 2–5 (1999)
7. Hindriks, K.V.: Programming rational agents in goal. In: El Fallah Seghrouchni, A., Dix, J., Dastani, M., Bordini, R.H. (eds.) Multi-Agent Programming: Languages, Tools and Applications, pp. 119–157. Springer (2009)
8. Luke, S., Cioffi-Revilla, C., Panait, L., Sullivan, K., Balan, G.: MASON: a multiagent simulation environment. Simul. Trans. Soc. Model. Simul. Int. **81**(7), 517–527 (2005)
9. North, M.J., Collier, N.T., Ozik, J., Tatara, E.R., Macal, C.M., Bragen, M., Sydelko, P.: Complex adaptive systems modeling with Repast Simphony. Complex Adapt. Syst. Model. **1**(1), 3 (2013)
10. Amouroux, E., Chu, T.Q., Boucher, A., Drogoul, A.: GAMA: an environment for implementing and running spatially explicit multi-agent simulations. Lect. Notes Comput. Sci. **5044**, 359–371 (2009)
11. Kravari, K., Bassiliades, N.: A survey of agent platforms. Jasss J. Artif. Soc. Soc. Simul. **18**(1), 11 (2015)
12. Adam, C., Gaudou, B.: BDI agents in social simulations: a survey. Knowl. Eng. Rev. **31**(3), 207–238 (2016)
13. Jensen, A.S., Dignum, V.: AORTA: adding organizational reasoning to agents. Proc. 13th Int. Conf. Auton. Agent. Multiagent Syst. (AAMAS 2014) **2**(3), 1493–1494 (2014)
14. Larsen, J.B., Villadsen, J.: An approach for hospital planning with multi-agent organizations. In: Rough Sets: International Joint Conference, IJCRS 2017, Part II, pp. 454–465. Springer (2017)
15. Rao, A.S.: AgentSpeak(L): BDI agents speak out in a logical computable language. In: Van de Velde, W., Perram, J.W. (eds.) Agents Breaking Away: Proceedings of the 7th European Workshop on Modelling Autonomous Agents in a Multi-Agent World, MAAMAW '96 Eindhoven, The Netherlands, 22–25 January 1996, pp. 42–55. Springer (1996)

16. Pokahr, A., Braubach, L., Lamersdorf, W.: Jadex: A BDI reasoning engine. In: Bordini, R.H., Dastani, M., Dix, J., El Fallah Seghrouchni, A. (eds.) Multi-Agent Programming: Languages, Platforms and Applications, pp. 149–174. Springer (2005)
17. Siebers, P.O., Macal, C.M., Garnett, J., Buxton, D., Pidd, M.: Discrete-event simulation is dead, long live agent-based simulation!. J. Simul. **4**(3), 204–210 (2010)
18. Bordini, R.H., Hübner, J.F.: Agent-based simulation using BDI programming in Jason. In: Multi-Agent Systems: Simulation and Applications, pp. 451–476. CRC Press (2009)
19. Caballero, A., Botia, J., Gomez-Skarmeta, A.: Using cognitive agents in social simulations. Eng. Appl. Artif. Intell. **24**(7), 1098–1109 (2011)
20. Padgham, L., Scerri, D., Jayatilleke, G., Hickmott, S.: Integrating BDI reasoning into agent based modeling and simulation. In: Proceedings of the Winter Simulation Conference, pp. 345–356 (2011)
21. Singh, D., Padgham, L., Logan, B.: Integrating BDI agents with agent-based simulation platforms. Auton. Agent. Multi-Agent Syst. **30**(6), 1050–1071 (2016)
22. Dignum, V., Tranier, J., Dignum, F.: Simulation of intermediation using rich cognitive agents. Simul. Modell. Pract. Theory **18**(10), 1526–1536 (2010)
23. Birna Van Riemsdijk, M.: 20 years of agent-oriented programming in distributed AI: history and outlook. Splash 2012: Agere 2012—Proceedings of the 2012 Acm Workshop on Programming Systems, Languages and Applications Based on Actors, Agents, and Decentralized Control Abstractions, pp. 7–10. (2012)
24. Weiss, G.: Multiagent Systems—2nd Edition. MIT Press (2013)
25. Ricci, A., Viroli, M., Omicini, A.: CArtAgO: a framework for prototyping artifact-based environments in MAS. In: Third International Workshop on Environments for Multi-agent Systems III, E4mas 2006. Selected Revised and Invited Papers (Lecture Notes in Artificial Intelligence, vol. 4389), pp. 67–86 (2006)
26. Piunti, M., Ricci, A., Braubach, L., Pokahr, A.: Goal-directed interactions in artifact-based MAS: Jadex agents playing in CARTAGO environments. Int. Conf. Intell. Agent Technol. **2**(2008), 207–213 (2008)
27. Ricci, A., Bordini, R.H., Piunti, M., Hbner, J.F., Acay, L.D., Dastani, M.: Integrating heterogeneous agent programming platforms within artifact-based environments. Proc. Int. Joint Conf. Auton. Agent. Multiagent Syst. **1**, 222–229 (2008)
28. Dignum, V.: A Model for Organizational Interaction: based on Agents, founded in Logic. SIKS Dissertation Series 2004-1. Ph.D. Thesis, Utrecht University (2004)
29. Aldewereld, H., Dignum, V.: OperettA: organization-oriented development environment. Lect. Notes. Comput. Sci. **6822**, 1–18 (2011)
30. Hübner, J.F., Sichman, J.S., Boissier, O.: Developing organised multiagent systems using the MOISE+ model: programming issues at the system and agent levels. Int. J. Agent-Oriented Softw. Eng. **1**(3/4), 370–395 (2007)
31. Hübner, J.F., Boissier, O., Kitio, R., Ricci, A.: Instrumenting multi-agent organisations with organisational artifacts and agents. Auton. Agent. Multi-Agent Syst. **20**(3), 369–400 (2010)
32. Jensen, A.S., Dignum, V., Villadsen, J.: A framework for organization-aware agents. Auton. Agent. Multi-Agent Syst. **31**(3), 387–422 (2017)
33. Jensen, A.S., Dignum, V., Villadsen, J.: The AORTA architecture: Integrating organizational reasoning in Jason. Lect. Notes Comput. Sci. **8758**(3), 127–145 (2014)
34. Jensen, A.S.: Model checking AORTA: verification of organization-aware agents. arXiv:1503.05317 (2015)

A Tool to Compute the Leakage of Multi-threaded Programs

Tri Minh Ngo and Quang Tuan Duong

Abstract This paper studies the security of multi-threaded programs and presents a tool for analyzing quantitative information flow (QIF) for multi-threaded programs written in a core imperative language. The aim of the tool is to measure the leakage of secret data in case a program leaks secret information. The tool is based on a method of the quantitative analysis where an attacker is able to select a scheduling policy to attack the program. The scheduling policy is used to construct the execution model of the program. We outline the workings of the tool and summarize results derived from running the tool on a range of case studies.

1 Introduction

Information flow analysis is a technique to assess the security of software, and has recently become an active research topic. In general, the approach of information flow security is based on the notion of *interference* [3]. Informally, the *interference* exists inside a program when the private data affect the value of public data. Therefore, an attacker might guess the value of private data from observing the public data. Non-interference, i.e., the absence of interference, is often used to prove that a system is secured.

In information flow analysis, variables in a program are assigned security levels. The basic model comprises two distinct levels, i.e., *publicly observable* data are indicated as *low* while *private* data are indicated as *high*. Non-interference prohibits any information flow from a high security level to a low security level.[1] For example, the following program,

---

[1] This model can be generalized in an obvious way, i.e., security levels can be viewed as a *lattice* with information flowing only *upwards* in the lattice.

T. M. Ngo (✉) · Q. T. Duong
The University of Danang - University of Science and Technology, Da Nang, Vietnam
e-mail: tringominh@gmail.com

Q. T. Duong
e-mail: quang.duong1910@gmail.com

© Springer International Publishing AG, part of Springer Nature 2018
A. Sieminski et al. (eds.), *Modern Approaches for Intelligent Information and Database Systems*, Studies in Computational Intelligence 769,
https://doi.org/10.1007/978-3-319-76081-0_45

if $(S > 0)$ then $O := 1$ else $O := 2,$

where S stores private information and O stores public information, is rejected by qualitative security analysis, since an attacker is able to learn information about S based on the value of O.

Non-interference is required for applications where the users need their private data strictly protected. However, many practical applications might *intentionally* violate non-interference by leaking *minor* information. Such systems include password checkers (*PWC*), cryptographic operations etc. For instance, when an attacker tries to guess the password: even when the attacker makes a wrong guess, secret information has been leaked, i.e., it reveals information about what the real password is *not*. Similarly, there is a flow of information from the plain-text to the cipher-text, since the cipher-text depends on the plain-text. These applications are rejected by non-interference.

However, the insecure property will happen only when it exceeds a specific threshold, or amount of interference. If the interference in the system is small enough, e.g., below a threshold given by specific security policy, the system is considered to be secure. The security analysis that requires to determine how much information flows from high level, i.e., secret data, to low level, i.e., public output, is known as quantitative information flow. It concerned with measure the leakage of information in order to decide if the leakage is tolerable.

Qualitative information flow analysis, i.e., non-interference, aims to determine whether a program leaks private information or not. Thus, these absolute security properties always reject a program if it leaks any information. *Quantitative* information flow analysis offers a more general security policy, since it gives a method to tolerate a *minor* leakage, i.e., by computing how much information has been leaked and comparing this with a threshold. By adjusting the threshold, the security policy can be applied for different applications, and in particular, if the threshold is 0, the quantitative policy is seen as a qualitative one.

This paper presents a computational tool of the quantitative information flow analysis for multi-threaded programs discussed in [6]. Given the source code of a program and the initial distribution on variables, the tool builds up a model of its execution, which is the state-space incorporating probability distribution transformations on variables inside the program, and then the tool computes the leakage if it is exists in the program. A brief introduction of quantitative information flow analysis will be given in Sect. 2, followed by Sect. 3 which presents the overall structure of the simulation tool. In the next section, we produced some results based on our own case studies. Finally, we draw conclusions.

2   Background of Quantitative Information Flow

Quantitative information flow applied information theory to the analysis, in which it considered the program as a communication channel [4, 5, 7, 8]. To aim for simplicity and clarity, rather than full generality, following the traditional approaches [1, 7], our models of analysis are based on the following basic settings. First, we assume that programs always terminate, and the attacker knows the source code of the program. We restrict to programs with a single *high security input* S. Since the high security output is irrelevant, programs only give a *low security outcome* O. Our goal is to quantify how much information about S is deduced by the attacker who can observe the traces of O. We also assume that the sets of possible values of data are finite, as in the traditional approaches.

Secondly, we assume that there is a *priori, publicly-known uniform* probability distribution on the high values. Our approach focuses on the situation that the attacker tries to guess the value of S based on the observation of O in the *one-try guessing* model. In other words, by seeing O, the attacker might be able to deduce something about S [6]. In the following examples, we argue that it is necessary to look at the trace of O, i.e., the sequence of states obtained during the execution of the program, in stead of only looking at the finale state of O.

Consider the following examples where S is a 3-bit binary number.

Example 1 (Program C1) Given that $(100)_b$ and $(011)_b$ are the binary forms of 4 and 3, respectively.

$$O := 0;$$
$$O := S \, \& \, (100)_b;$$
$$O := S \, \& \, (011)_b;$$

Let $(s_3 s_2 s_1)_b$ denote the binary form of S. The analysis based only on the final-state observation of O judges that C1 leaks 2 bits of private data, i.e., $O = (s_3 s_2 s_1)_b \, \& \, (011)_b = (0 s_2 s_1)_b$. However, it is clear that this program leaks the secret *completely*, i.e., due to the leakage in the intermediate state, i.e., $O = (s_3 s_2 s_1)_b \, \& \, (100)_b = (s_3 00)_b$.

Thus, an appropriate model of quantitative security analysis needs to consider the leakage given by a sequence of publicly observable data obtained during the program execution. However, notice that the leakage in intermediate states does not always contribute to the overall leakage, as in the following example,

Example 2 (Program C2)

$$O := 0;$$
$$O := S \, \& \, (001)_b;$$
$$O := S \, \& \, (011)_b;$$

The overall leakage of this program trace is only 2 bits, since the leakage in the intermediate state, i.e., the last bit s_1, is also leaked in the final state. This example shows that leakage of a program trace is not simply the sum of the leakage of each step, as in the approach of Chen and Malacaria [2]. The following part addresses how we can compute a program-trace leakage.

Let $(s_3s_2s_1)_b$ denote the binary form of S. We assume that the priori distribution of S is uniform. The execution of this program results in just one trace, i.e., $\langle(000)_b\rangle \longrightarrow \langle(00s_1)_b\rangle \longrightarrow \langle(0s_2s_1)_b\rangle$, where a state $\langle\rangle$ is represented by the public value O.

Assume that $s_2 = 1$ and $s_1 = 1$, the obtained trace is

$$\langle(000)_b\rangle \longrightarrow \langle(001)_b\rangle \longrightarrow \langle(011)_b\rangle.$$

At the initial state $\langle(000)_b\rangle$, the attacker's initial uncertainty is represented by the priori uniform distribution of S, i.e., $\{0 \mapsto \frac{1}{8}, 1 \mapsto \frac{1}{8}, 2 \mapsto \frac{1}{8}, 3 \mapsto \frac{1}{8}, 4 \mapsto \frac{1}{8}, 5 \mapsto \frac{1}{8}, 6 \mapsto \frac{1}{8}, 7 \mapsto \frac{1}{8}\}$. At state $\langle(001)_b\rangle$, the attacker learns that the last bit of S is 1. Thus, the distribution of S changes, for example, S cannot be 0. Hence, the updated distribution at state $\langle(001)_b\rangle$ is $\{0 \mapsto 0, 1 \mapsto \frac{1}{4}, 2 \mapsto 0, 3 \mapsto \frac{1}{4}, 4 \mapsto 0, 5 \mapsto \frac{1}{4}, 6 \mapsto 0, 7 \mapsto \frac{1}{4}\}$. Similarly, at the final state $\langle(011)_b\rangle$, the updated distribution is $\{3 \mapsto \frac{1}{2}, 7 \mapsto \frac{1}{2}\}$.[2]

Based on the final distribution of S, the attacker derives that the value of S is either 3 or 7. His uncertainty on secret information is reduced by the knowledge gained from observing the trace.

Since the program is a distribution transformer, the distribution of private data at the *initial state* of a trace can present the *initial uncertainty* of the attacker about the secret, and the distribution of private data at the *final state* can present his *final uncertainty*, after the trace has been observed. Thus, we can define the leakage of a program trace as,

Leakage of a program trace = Initial uncertainty − Final uncertainty.

2.1 Measure of Uncertainty

Given a distribution of private data, the *best* strategy of the one-try guessing model is to choose the value with the maximum probability. 'Best' means that this strategy induces the smallest probability of guessing wrongly. Let **S** be the set of possible values of S, the value that affects the notion of uncertainty is $\max_{s\in \mathbf{S}} p(s)$. If $\max_{s\in \mathbf{S}} p(s) = 1$, the uncertainty must be 0, i.e., the attacker already knows the value of S. Thus, the *notion of uncertainty* is computed as the *negation* of the logarithm of $\max_{s\in \mathbf{S}} p(s)$, i.e., uncertainty $= -\log \max_{s\in \mathbf{S}} p(s)$, where the negation is used to ensure the *non-negative uncertainty*.[3]

This measure coincides with the notion of Rényi's min-entropy.

---

[2] We leave out the elements that have probability 0.

[3] The quantity of uncertainty is always non-negative, which is different from the quantity of information flow.

Definition 1 The Rényi's min-entropy of a random variable X is defined as [7]: $\mathcal{H}_{Rényi}(X) = -\log \max_{x \in X} p(X = x)$.

As indicated above, only the element with the highest possibility is given to credit in calculating min-entropy. That is, instead of treating every element equally, we focus on the one which has the biggest effect to the result.

Thus, given a distribution of S, the attacker's uncertainty about the secret in this analysis is: Uncertainty $= \mathcal{H}_{Rényi}(S)$.

Therefore, the leakage of a program trace T is,

$$\mathcal{L}(T) = \mathcal{H}_{Rényi}(S_T^i) - \mathcal{H}_{Rényi}(S_T^f),$$

where $\mathcal{H}_{Rényi}(S_T^i)$ is Rényi's min-entropy of S with the initial distribution, and $\mathcal{H}_{Rényi}(S_T^f)$ is Rényi's min-entropy of S with the final distribution, i.e., the distribution of the secret at the final state in T.

Thus, following our measure, in $C2$, $\mathcal{L}(T) = -\log \frac{1}{8} - (-\log \frac{1}{2}) = 2$. This value matches the intuitive understanding that $C2$ leaks 2 bits of private information.

Consider $C1$. Assume that $s_1 = s_2 = s_3 = 1$, the execution of $C1$ results in the trace $\langle (000)_b \rangle \longrightarrow \langle (100)_b \rangle \longrightarrow \langle (011)_b \rangle$.

At state $\langle (100)_b \rangle$, the attacker learns that the first bit of S is 1. Thus, the distribution of S is $\{4 \mapsto \frac{1}{4}, 5 \mapsto \frac{1}{4}, 6 \mapsto \frac{1}{4}, 7 \mapsto \frac{1}{4}\}$. At the final state $\langle (011)_b \rangle$, the distribution is $\{7 \mapsto 1\}$, which is different from the final distribution given by $C2$. Hence, $\mathcal{L}(T) = -\log \frac{1}{8} - (-\log 1) = 3$. This result also matches the real leakage of the program, i.e., the attacker is able to derive S precisely from the execution trace of $C1$.

2.2 Leakage of a Multi-threaded Program

In order to apply this approach to analyze the information leakage of a multi-threaded program, the target program is described as a state-transition model under the control of a probabilistic scheduler. The execution of multi-threaded program is modeled by a state-transition model, which can be represented as a tree, where each possible state of the execution is a node in the tree. The result of the programs execution yields into a set of possible traces, from the initial state to various possible final states, due to many possible choices of the scheduler. The computation is based on the probability transition between states along a particular trace, taking into account the intermediate states along each trace.

In order to compute the leakage of a program C with a scheduler δ, Ngo and Huisman [6] was proposed a mathematical expression using Rényi's min-entropy. Intuitively, the execution of a multi-threaded program C under the control of a scheduler δ results in a set of traces $Trace(A_\delta)$. Therefore, the leakage of C is computed as the *expected* value of its trace-leakages, i.e.,

$$\mathcal{L}(C,\pi) = \sum_{T \in Trace(\mathcal{A}_\delta)} p(T) \cdot \mathcal{L}(T)$$

$$= \sum_{T \in Trace(\mathcal{A}_\delta)} p(T)(\mathcal{H}_{R\acute{e}nyi}(S_T^i) - \mathcal{H}_{R\acute{e}nyi}(S_T^f)),$$

where π is the initial distribution of S. Since $\mathcal{H}_{R\acute{e}nyi}(S_T^i)$ is the same for all $T \in Trace(\mathcal{A}_\delta)$, for notational convenience, we simply write it as $\mathcal{H}_{R\acute{e}nyi}(S^i)$. Thus, we can rephrase the above expression as follows,

$$\mathcal{L}(C,\pi) = \mathcal{H}_{R\acute{e}nyi}(S^i) - \sum_{T \in Trace(\mathcal{A}_\delta)} p(T) \cdot \mathcal{H}_{R\acute{e}nyi}(S_T^f).$$

3 The Structure of the Tool

The program is written on MATLAB. Figure 1 describes the basic structure of the system. The inputs of the tool are the source code of the program and the prior probability distribution on the possible values of private data. The tool simulates the scheduler by selecting randomly a thread of the program to execute. The tool then calculates the distribution transformations of secret variables, collects the cumulative probability of each trace, and then measures the secret information leakage.

The main function of the tool is the state space exploration. Based on values of public data, the program updates the probability distribution of private data. Each state is stored as a *struct*, which can be described as {*state, observe, secret, probDist, probTrace*}, where the variable *state* identifies the state itself in the execution tree,

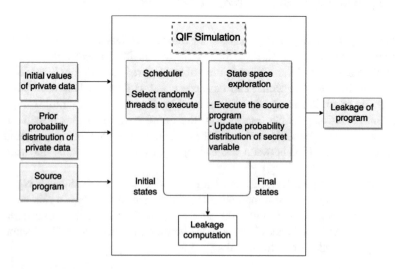

Fig. 1 Overall structure of the tool

Fig. 2 Basic execution of the tool

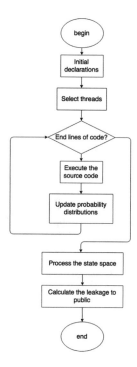

observe stores the values of public data, *secret* stores the values of private data, *probDist* stores the update probability distribution of private data, *probTrace* is the cumulative probability of a trace.

Figure 2 gives the basic flow of the QIF simulation. The tool reads the users input, creates the state space, executes the commands, and then updates the probability distributions. After all code of sources program are completely executed, the tool creates a set of all final states. Based on the information of these states, the computational leakage is computed.

4 Result

We tested our tool with four different examples. The table below demonstrates these examples and results produced by the tool.

Case study 1:

$$O := 0;$$
$$\{\text{if } (O = 1) \quad \text{then} \quad O := S/4 \quad \text{else} \quad O := S \bmod 2\} \parallel O := 1;$$
$$O := S \bmod 4;$$

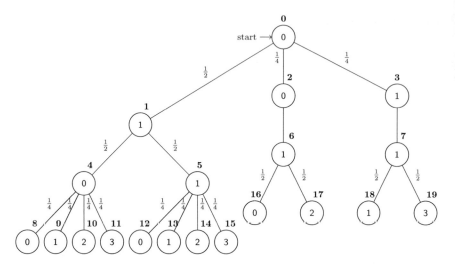

Fig. 3 Model of the program execution

where S is a 3-bit unsigned uniform integer, and $\|$ is the parallel composition operator of the two threads.

The execution of this program with a uniform scheduler is illustrated by a model $\mathcal{A}$ in Fig. 3. The model consists of 20 states that are numbered from **0** (the initial state) to **19**. The contents of each state is the value of O in that state, e.g., in the initial state, the value of O is 0, which corresponds to the first command of the program: $O := 0$.

Let C_1 and C_2 denote the left and right threads. Since we consider the uniform scheduler, either thread C_1 or C_2 can be picked next with the same probability $\frac{1}{2}$. If the scheduler picks C_2 before C_1, $\mathcal{A}$ evolves from state **0** to state **1**, where $O = 1$. If C_1 is picked first, $\mathcal{A}$ might evolve from state **0** to either state **2** or state **3** with the same probability $\frac{1}{4}$. Since the value of O in state **0** is 0, the command $O := S \mod 2$ is executed. Since the possible values of S are $\{0, \dots, 7\}$, the outcome O might be 0 (state **2**) if $S \in \{0, 2, 4, 6\}$, or 1 (state **3**) if $S \in \{1, 3, 5, 7\}$.

At state **1**, $\mathcal{A}$ might evolve to either state **4** or state **5** with the same probability. Since currently, O is 1, the command $O := S/4$ is executed. Thus, O might be 0 if $S \in \{0, 1, 2, 3\}$, or 1 if $S \in \{4, 5, 6, 7\}$.

The PKS $\mathcal{A}$ evolves from one state to another until the execution terminates, i.e., when the last command $O := S \mod 4$ is executed.

The attacker's initial uncertainty about S is denoted by the uniform distribution, i.e.,

$$\pi = \{0 \mapsto \frac{1}{8}, 1 \mapsto \frac{1}{8}, 2 \mapsto \frac{1}{8}, 3 \mapsto \frac{1}{8}, 4 \mapsto \frac{1}{8}, 5 \mapsto \frac{1}{8}, 6 \mapsto \frac{1}{8}, 7 \mapsto \frac{1}{8}\}.$$

At state **1**, the distribution of S is still uniform, since the attacker learns *nothing* from the command $O := 1$. At state **4**, since the execution of $O := S/4$ results in 0, the attacker learns that the true value of S must be in the set $\{0, 1, 2, 3\}$, with the same probability. Thus, the updated distribution of S at this state is: $\{0 \mapsto \frac{1}{4}, 1 \mapsto \frac{1}{4}, 2 \mapsto \frac{1}{4}, 3 \mapsto \frac{1}{4}\}$.

In the next step, the outcome of $O := S \mod 4$ helps the attacker to derive S precisely, e.g., at state **8**, since $O = 0$, the distribution of S is: $\{0 \mapsto 1\}$. Similarly, the attacker is also able to derive S precisely, basing on the final distributions at states **9**, ..., **15**.

At state **2**, the execution of $O := S \mod 2$ results in 0. Thus, the distribution of S at this state is: $\{0 \mapsto \frac{1}{4}, 2 \mapsto \frac{1}{4}, 4 \mapsto \frac{1}{4}, 6 \mapsto \frac{1}{4}\}$. This distribution remains unchanged at state **6**, since no information is gained from the execution of $O := 1$. At state **16**, the update distribution of S is: $\{0 \mapsto \frac{1}{2}, 4 \mapsto \frac{1}{2}\}$, since the execution of $O := S \mod 4$ results in 0. The same form of distributions is also obtained at states **17, 18, 19**.

Among the 12 possible traces, 8 traces have the final uncertainty 0, i.e., $\mathcal{H}_{R\acute{e}nyi}$ $(S_T^f) = -\log 1 = 0$, and the other 4 traces have the final uncertainty 1, i.e., $\mathcal{H}_{R\acute{e}nyi}$ $(S_T^f) = -\log \frac{1}{2} = 1$. The probability of traces with the final uncertainty 0, i.e., traces end in state **8**, ..., **15**, is equal to the probability of traces with the final uncertainty 1. Thus, according to this analysis,

$$\mathcal{L}(C, \pi) = 3 - (\frac{1}{2} \cdot 0 + \frac{1}{2} \cdot 1) = 2.5 \text{ (bits)}.$$

This value coincides with the real leakage of the program. It is clear that the last command $O := S \mod 4$ always reveals the last 2 bits of S. The first bit might be leaked with probability $\frac{1}{2}$, depending on whether the scheduler picks thread $O := 1$ first or not. Thus, with the uniform scheduler, theoretically, the real leakage of this program is 2.5 bits.

Leakage given by the tool is also 2.5 bit.

Case study 2:

```
O := 0;
{if (O = 1)  then  O := S/2  else  O := S mod 2} || O := 1;
```

where S is a 2-bit unsigned uniform integer. Leakage given by the tool $= 1.0$ bit, the same as the theoretical analysis.

Case study 3:

```
S := S  mod 2 ;
O := S || {O := 1 || O := 0};
```

where S is a 2-bit unsigned uniform integer. Leakage given by the tool $= 1.0$ bit, the same as the theoretical analysis.

Case study 4:

$$O := 0;$$
$$\{\texttt{if}\ (O = 0)\quad \texttt{then}\quad O := S/8\ \ \texttt{else}\quad O := S \bmod 2\} \parallel O := 1;$$
$$O := S \bmod 4;$$
$$O := S \parallel O := 0;$$

where S is a 4-bit unsigned uniform integer. Leakage given by the tool $= 1.5$ bit, the same as the theoretical analysis.

Based on these case studies, we summarize some main points as follows: (1) The tool gives a same result of leakage as computed by the theoretical analysis, and (2) For a particular program, the execution time of the tool is affected by the size of the variables.

5 Conclusions

In this paper, we presented a simulation tool to compute automatically the leakage of multi-threaded programs. The tool models the execution of a multi-threaded program under the control of a scheduler by a probabilistic state transition system. States of the transition system are labeled with probability distributions of private data. This distribution reflects the attacker's knowledge about the secret value. The distribution of private data changes from state to state along a trace, depending on the relation between private data and public data, and also on the command executions resulting in such public data. We demonstrated the structure of the tool as well as its basic working flow. We created some examples to verify the tool, compared their results with the values from the theoretical analysis.

This simulation tool can be considered as a first step of the goal to analyze the security of programs automatically, i.e., to check whether a program is secure or not, and in the case it is insecure, to compute the value of leakage. The input code to the tool is C-language. However, based on this work, a tool which can verify the security of programs of different programming languages is also feasible.

Acknowledgements This research is funded by Funds for Science and Technology Development of the University of Danang under grant number B2016-DN02-13.

References

1. Alvim, M.S., Andrés, M.E., Chatzikokolakis, K., Palamidessi, C.: Foundations of security analysis and design vi. In: Quantitative Information Flow and Applications to Differential Privacy, pp. 211–230. Springer (2011)

2. Chen, H., Malacaria, P.: The optimum leakage principle for analyzing multi-threaded programs. In: Proceedings of the 4th International Conference on Information Theoretic Security, ICITS'09, pp. 177–193. Springer (2010)
3. Goguen, J.A., Meseguer, J.: Security policies and security models. In: IEEE Symposium on Security and Privacy, pp. 11–20 (1982)
4. Malacaria, P.: Risk assessment of security threats for looping constructs. J. Comput. Secur. **18**, 191–228 (2010)
5. Malacaria, P., Chen, H.: Lagrange multipliers and maximum information leakage in different observational models. In: Proceedings of the Third ACM SIGPLAN Workshop on Programming Languages and Analysis for Security, PLAS '08, pp. 135–146. ACM (2008)
6. Ngo, T.M., Huisman, M.: Complexity and information flow analysis for multi-threaded programs. Eur. Phys. J. Spec. Top. **226**(10), 2375–2392 (2017)
7. Smith, G.: On the foundations of quantitative information flow. In: Proceedings of the 12th International Conference on Foundations of Software Science and Computational Structures, FOSSACS'09, pp. 288–302. Springer (2009)
8. Zhu, Y., Bettati, R.: Anonymity vs. information leakage in anonymity systems. In: Proceedings of the 25th IEEE International Conference on Distributed Computing Systems, ICDCS'05, pp. 514–524. IEEE Computer Society (2005)

Erratum to: Fuzzy Ontology Modeling by Utilizing Fuzzy Set and Fuzzy Description Logic

Xuan Hung Quach and Thi Lan Giao Hoang

Erratum to:
Chapter "Fuzzy Ontology Modeling by Utilizing Fuzzy Set and Fuzzy Description Logic" in: A. Sieminski et al. (eds.), *Modern Approaches for Intelligent Information and Database Systems*, **Studies in Computational Intelligence 769, https://doi.org/10.1007/978-3-319-76081-0_2**

In the original version of the book, Reference [15] has been removed from the reference list in chapter "Fuzzy Ontology Modeling by Utilizing Fuzzy Set and Fuzzy Description Logic". The erratum chapter and the book have been updated with the change.

The updated online version of this chapter can be found at
https://doi.org/10.1007/978-3-319-76081-0_2

© Springer International Publishing AG, part of Springer Nature 2018   E1
A. Sieminski et al. (eds.), *Modern Approaches for Intelligent Information and Database Systems*, Studies in Computational Intelligence 769,
https://doi.org/10.1007/978-3-319-76081-0_46

Author Index

Printed in the United States
By Bookmasters